BASIC CARPENTRY

Basic Carpentry

John Capotosto

RESTON PUBLISHING COMPANY, INC.

A Prentice-Hall Company

Reston, Virginia 22090

Library of Congress Cataloging in Publication Data

Capotosto, John.
  Basic carpentry.
  xiii, 525 p., illus.
  Bibliography: p. 513-514
  Includes index.
  1. Carpentry. I. Title.
TH5604.C34   694   74-34063
ISBN 0-87909-064-2

10 9 8 7 6 5 4 3 2   Printed in the United States of America

TABLE OF CONTENTS

Table of Contents

Table of Contents

Table of Contents

ACKNOWLEDGEMENTS

The following companies and associations were very cooperative in supplying photos, illustrations, and pertinent material to the author. Their assistance is greatly appreciated.

* * *

Acoustical & Insulating Materials Assoc., Park Ridge, Ill.
American Ladder Institute, Chicago, Ill.
American Plywood Assoc., Tacoma, Wash.
Andersen Corp., Bayport, Minn.
Armstrong Cork Company, Lancaster, Pa.
Asphalt Roofing Manufacturers Assoc., New York, N.Y.
Atkins, Barrow & Graham, Inc., Champaign, Ill.
Automated Building Components, Inc., Miami, Fla.
Berger Instruments, Boston, Mass.
The Black & Decker Mfg. Co., Towson, Md.
Bostitch Division of Textron, Inc., East Greenwich, R.I.
Bruce, E.L., Co., Memphis, Tenn.
California Redwood Assoc., San Francisco, Calif.
The Ceco Corp., Chicago, Ill.
The Celotex Corp., Tampa, Fla.
Certain-Teed Products Corp., Valley Forge, Pa.
Dacor Mfg. Co., Worcester, Mass.
Deer Park Lumber Co., Deer Park, N.Y.
Dexter Lock, Grand Rapids, Mich.
Estwing Mfg. Co., Rockford, Ill.
Evans Products, Co., Portland, Oreg.
Fimbel Door Corp., Hillside, N.J.
Flint, A.W., Co., New Haven, Conn.
Forest Products Laboratory, U.S. Department of Agriculture, Madison, Wisc.
Formica Corp., Cincinnati, Ohio
GAF Corp., New York, N.Y.

Acknowledgements

Georgia-Pacific Corp., Portland, Oregon
Gold Bond Building Products, Buffalo, N.Y.
Gypsum Assoc., Chicago, Ill.
Hand Tools Institute, New York, N.Y.
Ideal Co., Waco, Texas
International Paper Co., Long Bell Div., Little Rock, Ark.
Keystone Steel & Wire, Peoria, Ill.
Koppers Co. Inc., Pittsburgh, Pa.
The Majestic Co. Inc., Huntington, Ind.
Marsh Wall Products, Chicago, Ill.
Masonite Corp., Chicago, Ill.
Modern Materials Corp., Detroit, Mich.
Morgan Co., Oshkosh, Wisc.
National Particleboard Assoc., Silver Spring, Md.
National Assoc. of Home Builders, Washington, D.C.
National Forest Products Assoc., Washington, D.C.
National Oak Flooring Mfgrs. Assoc., Memphis, Tenn.
National Woodwork Mfgrs. Assoc., Chicago, Ill.
Patent Scaffolding Co., Fort Lee, N.J.
Pemko Mfg. Co., Emeryville, Calif.
Pfister, Herbert, New York, N.Y.
Pittsburgh Corning Corp., Pittsburgh, Pa.
Porter, H.K., Co. Inc., Pittsburgh, Pa.
Red Cedar Shingle & Handsplit Shake Bureau, Seattle, Wash.
Reuten, Fred, Inc., Closter, N.J.
Rockwell International, Pittsburgh, Pa.
Rodman Industries, Marinette, Wis.
Shakertown Corp., Winlock, Wash.
Simpson Timber Co., Seattle, Wash.
Southern Forest Products Assoc., New Orleans, La.
The Stanley Works, New Britain, Conn.
Steel Scaffolding & Shoring Institute, Cleveland, Ohio
Sumner Rider & Assoc., New York, N.Y.
TECO, Washington, D.C.
Thomasville Furniture Industries, Inc., Thomasville, N.C.
U.S. Forest Service, Pacific Northwest Forest & Range Experimental
    Station, Seattle, Wash.
United States Gypsum Co., Chicago, Ill.

Acknowledgements

U.S. Plywood, New York, N.Y.
Ventarama, Port Washington, N.Y.
Wausau Homes, Inc., Wausau, Wis.
Western Red Cedar Lumber Assoc., Portland, Oreg.
Western Wood Moulding & Millwork, Portland, Oreg.
Western Wood Products Assoc., Portland, Oreg.
Weston Instruments, Newark, N.J.
Weyerhaeuser Co., Tacoma, Wash.
Wild Heerbrugg, Inc., Farmingdale, N.Y.
Wilton Corp., Des Plaines, Ill.
Wisconsin Knife Works, Inc., Beloit, Wis.

Part One

BUILDING MATERIALS, TOOLS & EQUIPMENT

Hand Tools

Although power tools are important in carpentry work, hand tools are equally important. The carpenter's tool box must have a good selection of both. Functionally most hand tools fit into these categories: measuring, cutting, boring, fastening, and finishing.

Measuring Tools

The common measuring tools used most by carpenters are shown in Fig. 1-1. These include *rules* (folding or zigzag, steel rules and tape rules), *squares,* (try square, combination square, T bevel and rafter square), *levels* (line level, carpenter's level and mason's level), *plumb bobs, dividers* and *marking gauges*. The proper use and care of these tools is important because good work cannot be done well or safely with dull, worn or unreliable instruments.

The folding rule, sometimes called the zigzag rule, is either 6- or 8-feet long, is usually made of wood, and is graduated in inches and common fractions. Some also have stud markings every 16 inches which make them useful for framing work. Others have brass extensions which make them useful for taking inside measurements; Fig. 1-2 illustrates the various graduations on rules. They are useful for taking or for laying off long or short measurements, Fig. 1-3.

The flexible rule is usually made of light steel and is available in 6-, 8-, 10-, and 12-foot lengths. Long tapes are available in 50- and 100-foot lengths and are made of metal or cloth. Ordinarily they are made to retract into a metal case.

The try square is a basic tool used in carpentry for checking the squareness of boards and for laying out straight lines perpendicular to an edge. Some have a 45-degree bevel in the handle for use in laying out miters. Their blades may vary in length from 4 to 12 inches.

The combination square is similar to the try square, has a sliding head and can be used for making square and miter layouts. The sliding head also makes it useful as a gauging tool.

Hand Tools

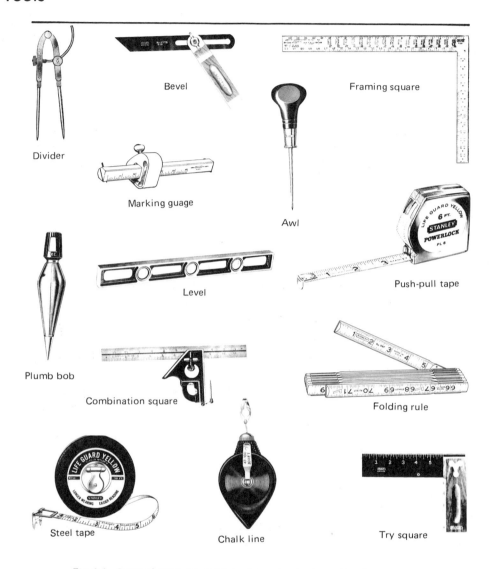

Fig. 1-1 Some of the most common measuring tools used by the carpenter.

The T bevel is adjustable so that any angle may be measured or drawn. The carpenter uses it mostly in hip and valley rafter layouts.

The rafter square is also known as a builder's square, framing square, or carpenter's square. It is one of the most important measuring tools for the carpenter and its use is fully covered in Chapter 4.

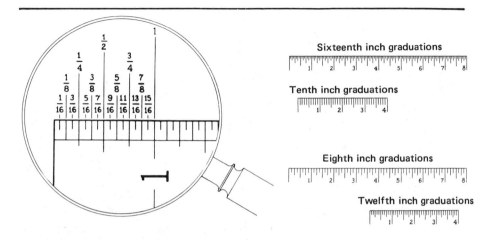

Sixteenth inch graduations

Tenth inch graduations

Eighth inch graduations

Twelfth inch graduations

Fig. 1-2 The various graduations used on carpenters rules are shown here.

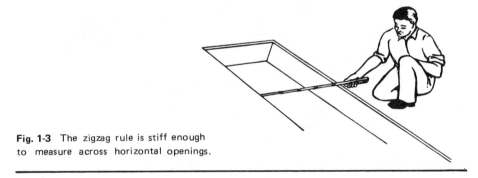

Fig. 1-3 The zigzag rule is stiff enough to measure across horizontal openings.

You will employ several levels in building construction, including the line level, the carpenter's level, and the mason's level. The line level is a small instrument with a bubble centering in a liquid-filled glass vial. The pair of hooks on its top enable the carpenter to suspend the level on a taut string or line between two points to check for horizontal trueness. Widely employed in construction work, you will find it useful for laying out foundations and for excavating and grading. But to make sure that a surface is even or plumb, use the carpenter's level. Constructed of wood or lightweight metal, it has one or more sets of bubble tubes or vials. The bubbles running in vials lengthwise on the carpenter's level are used to check horizontal surfaces, while the crosswise bubble is used to check vertical surfaces. The mason's level is

Fig. 1-4 The longer the level, the more accurate the measurement.

similar but much longer. Some are over 6-feet long; Fig. 1-4 shows a long level being used to check the plumb of a wall under construction.

The plumb bob is made of metal, is fairly heavy, and has a pointed end. A cord is attached to the opposite end so that you can suspend it to hang true. It thus accurately indicates a point in relation to a vertical plane.

Fig. 1-5 The chalk line is used to lay out long straight lines.

Use a chalk line to mark long straight lines. First rub the line thoroughly with chalk, then stretch it between two points and snap it smartly so that a straight chalk mark offsets on the surface being laid out, Fig. 1-5.

Use dividers, also called wing dividers, to draw arcs, circles and dividing lines, and to transfer irregular outlines. Most dividers have two steel points; on some models, one point can be replaced with a pencil.

The awl has two functions: it can be used for marking lines on the surface of materials and for piercing wood surfaces to make starter holes for small screws.

Use the marking gauge to make lines parallel to the edge of a board. Its sliding head can be moved along the length of the tool and locked with a setscrew.

Boring Tools

The two basic hand tools for making holes in wood are the **hand drill** and the **bit brace**. For holes up to 11/64 of an inch in diameter, use the hand drill normally. When holes 3/16 of an inch in diameter or larger are required, you need the brace. Use a ratchet brace when confined quarters prevent the normal swing of the regular brace handle.

The chuck or moveable jaw of the brace is designed to hold bits with square tapered shanks. Auger bits are suitable for boring holes up to one inch in diameter, Fig. 1-6. For larger holes, use expansive bits. Expansive bits have adjustable cutters and are suitable for boring holes from 7/8 to 3 inches in diameter. You can use screwdriver bits and countersinks with the brace also.

For holes under 3/16 inches in diameter, you must use the hand drill. This tool looks somewhat like an egg beater and will take regular twist drills.

Hand Tools

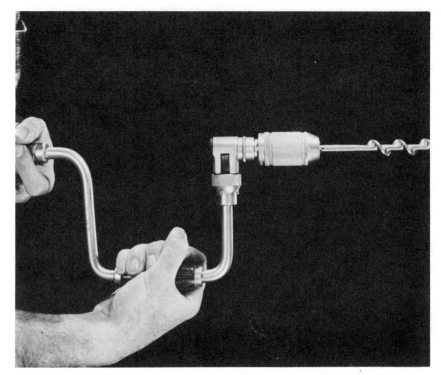

Fig. 1-6 The bit brace is useful for boring large holes in wood.

Another small drill popular with carpenters is the automatic or **push drill**. When you push the handle its mechanism rotates the drills. These drill handles are usually hollow and are used for storing the small drill points. Figure 1-7 shows drill sizes for various screws; Fig. 1-8 shows various drilling and boring tools.

Drill Points to use in Automatic Drill for Wood Screws									
Number of Screw	0	1	2	3	4	5	6	7	8
Body Diameter of Screw	$\frac{1''}{16}$	$\frac{5''}{64}$	$\frac{3''}{32}$	$\frac{3''}{32}$	$\frac{7''}{64}$	$\frac{1''}{8}$	$\frac{9''}{64}$	$\frac{5''}{32}$	$\frac{11''}{64}$
Drill to use for first hole for the smooth shank of screw	$\frac{1''}{16}$	$\frac{5''}{64}$	$\frac{3''}{32}$	$\frac{7''}{64}$	$\frac{1''}{8}$	$\frac{1''}{8}$	$\frac{9''}{64}$	$\frac{5''}{32}$	$\frac{11''}{64}$
Drill to use for pilot hole for threaded end of screw	X	X	$\frac{1''}{16}$	$\frac{5''}{64}$	$\frac{5''}{64}$	$\frac{3''}{32}$	$\frac{7''}{64}$	$\frac{1''}{8}$	$\frac{1''}{8}$

Fig. 1-7 Chart shows the drill sizes needed for wood screws.

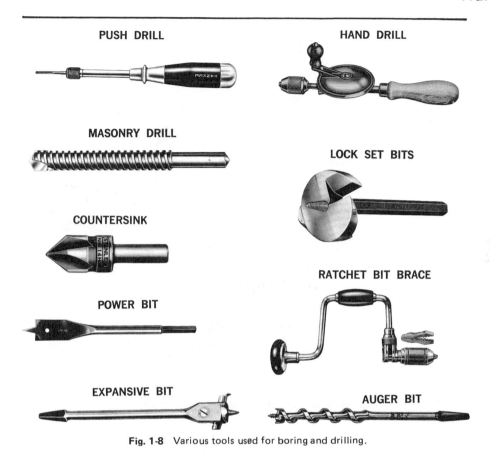

PUSH DRILL

HAND DRILL

MASONRY DRILL

LOCK SET BITS

COUNTERSINK

POWER BIT

RATCHET BIT BRACE

EXPANSIVE BIT

AUGER BIT

Fig. 1-8 Various tools used for boring and drilling.

Handsaws

There are five basic hand-operated saws for carpenters: handsaw, backsaw, compass saw, coping saw and hacksaw, Fig. 1-9. They are available in various lengths and tooth sizes. A saw blade is designated by the number of teeth (points) per inch; thus, a 10-point saw will have 10 teeth per inch, Fig. 1-10. Generally, the more points to a saw, the finer the cut.

Handsaws are either for crosscutting or for ripping. The shape, angle and set of the teeth differ considerably between the two types. Use the *crosscut saw* for cutting across the grain of the wood, and the *ripsaw* to

Hand Tools

HAND SAW

COPING SAW

BACK SAW

HACK SAW

COMPASS SAW

Fig. 1-9 Saws most frequently used by carpenters.

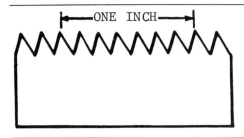

|←——ONE INCH——→|

Fig. 1-10 To determine the tooth size, count the number of teeth per inch.

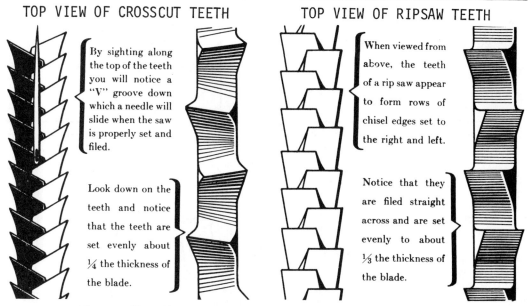

TOP VIEW OF CROSSCUT TEETH

By sighting along the top of the teeth you will notice a "V" groove down which a needle will slide when the saw is properly set and filed.

Look down on the teeth and notice that the teeth are set evenly about ¼ the thickness of the blade.

TOP VIEW OF RIPSAW TEETH

When viewed from above, the teeth of a rip saw appear to form rows of chisel edges set to the right and left.

Notice that they are filed straight across and are set evenly to about ⅓ the thickness of the blade.

Fig. 1-11 The difference between crosscut and ripsaw teeth. *(H.K. Porter Co.)*

cut along the grain. Crosscut teeth are sharp and pointed and slice the wood fibers while the ripsaw teeth are like little chisels which cut or chop the wood into minute chips; Fig. 1-11 shows the difference between crosscut and ripsaw teeth.

The ripsaw cuts with a chisel-like action, Fig. 1-12. On the push stroke the teeth cut the wood into particles, first on one side then on the other side of the blade. On the return stroke (toward the user), the ripsaw pulls this sawdust up and out of the kerf. The kerf is the groove or width of the saw cut.

The teeth of the crosscut saw score the wood as the saw is drawn across the grain (toward the user). On the push stroke the edges of the teeth begin paring the groove deeper and clear the sawdust from the kerf, Fig. 1-13.

The proper methods for holding the crosscut and the ripsaw are shown in Fig. 1-14. Never force a saw. Instead, operate them with long easy strokes and never over nails or screws.

Backsaws are easily recognizable since they have a reinforced metal back for rigidity. They are especially suited for making precise cuts in molding. They vary in length between 10 and 20 inches.

The *keyhole saw* has a narrow tapered blade which can be used for

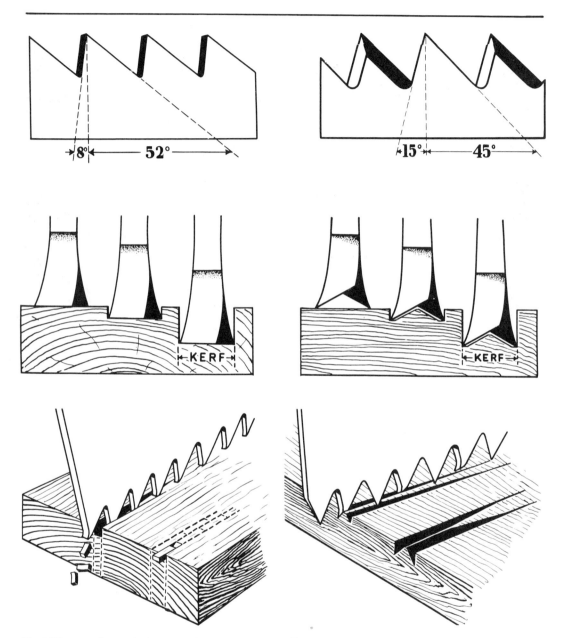

Fig. 1-12 **top,** shape of ripsaw teeth; **center,** cross section of rip teeth; **bottom,** how ripsaw cuts. *(H.K. Porter Co.)*

Fig. 1-13 **top,** shape of crosscut teeth; **center,** cross section of crosscut teeth; **bottom,** how cross-cut saw cuts. *(H.K. Porter Co.)*

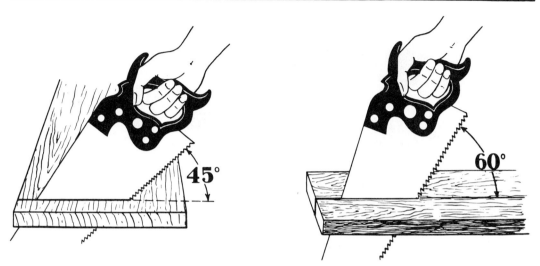

Fig. 1-14 The proper method of cutting with a handsaw: **left,** cross-cutting; **right,** ripping. *(H.K. Porter Co.)*

making irregular cuts or inside holes after a starting hole is drilled. Special blades for cutting metal and wood are also available. *Compass saws* are usually slightly wider and not as sharply pointed. They are also used to cut holes and curves.

For fancy cuts in thin wood, use the *coping saw*. It is ideal, takes thin blades and can be used for making internal cuts in materials as varied as wood, metal and plastic.

Use the *hacksaw* for cutting metals. It has an adjustable frame or bow which accomodates various blade lengths. You can cut nails, bolts, and screws easily with this saw.

Employ the *miter box* to hold work that has to be cut or mitered to match at joints, such as the corners of frames and moldings. You can set the guides on the box so that any desired angle from 45 to 90 degrees can be easily sawn, Fig. 1-15.

Finishing Tools

Carpentry work involves many finishing operations in which the wood or other construction materials must be trimmed, beveled, smoothed and shaped precisely. Power tools are available for these operations, but in many cases you will find it more practical to employ

Hand Tools

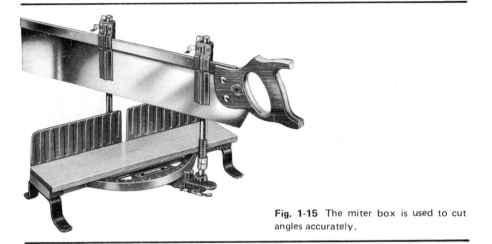

Fig. 1-15 The miter box is used to cut angles accurately.

hand tools such as chisels, planes, scrapers, files, and knives, Fig. 1-16. Your carpenter's tool box should include several planes, a spokeshave, and a set of chisels.

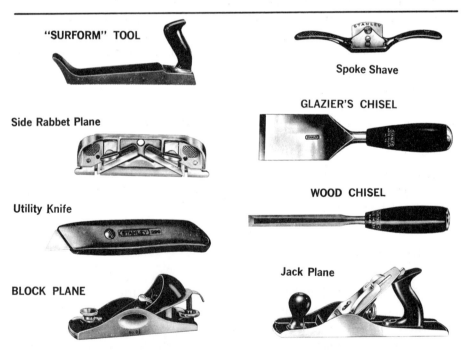

"SURFORM" TOOL

Spoke Shave

GLAZIER'S CHISEL

Side Rabbet Plane

WOOD CHISEL

Utility Knife

BLOCK PLANE

Jack Plane

Fig. 1-16 Edge-cutting tools used by the carpenter.

Finishing Tools

Planes cut and smooth rough surfaces and bring work down to size. They are made in various sizes, each for specific work; Fig. 1-17 illustrates the parts of a *jack plane*. The *jointer plane* is the longest of

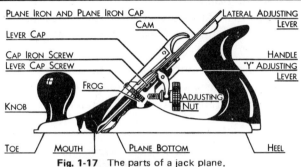

Fig. 1-17 The parts of a jack plane.

all planes (20 to 24 inches), and is especially suited for truing up long boards and doors. The *fore plane* and jack plane are next in size (14 to 18 inches). The *smoothing plane* is 4- to 9-inches long and is generally used for finishing after the longer planes have done their preliminary work. The *block plane* is designed for cutting end grain. Its blade is set at a lower, more acute angle (12 degrees) than the 20-degree angle of conventional planes in order to better cut across the tougher end grain.

A set of *wood chisels* for cutting wood and a *cold chisel* for metal cutting are useful tools for carpentry work. Chisels are made in widths from 1/8- to 2-inches. For safety's sake always keep both hands well back of the cutting edge when using a chisel, Fig. 1-18.

The *spokeshave* is a two-handled plane used for rounding spokes and similar curved surfaces. The tool's handles enable the carpenter to control the fine shaping of pieces of small diameter.

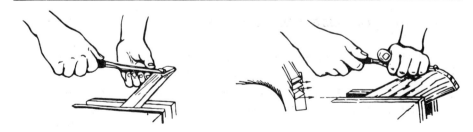

Fig. 1-18 Safety rule: Use sharp chisels and always keep both hands well back of their cutting edges.

Surform tools are also used for shaving, shaping and smoothing operations. Made in many shapes and sizes, they can be used on wood, plastics, and soft metals. Some of these tools have the appearance of a cheese grater. They cut rapidly and are versatile, Fig. 1-19.

Files and *wood rasps* are also used in carpentry for shaping and smoothing operations. Rasps are coarser than files and are used mostly for rough shaping. Use angular files for sharpening saws and other metal cutting tools.

Scrapers are used for smoothing wood surfaces after planing since they can remove any ridges left by the plane. Use them also when either cross-grained or wavy-grained wood makes planing too difficult.

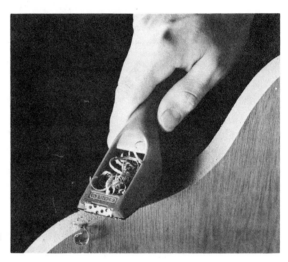

Fig. 1-19 Shaving tool can be used for many types of cutting and trimming.

Fastening or Assembling Tools

Hammers, screwdrivers, wrenches, clamps, staplers and vises are all fastening or assembling tools used in expert carpentry, Fig. 1-20.

Hammers are made in various sizes and shapes, each having been developed for specific work. Two of the most important you will use are the curved claw and the straight claw. The curved claw hammer is used for nailing and removing nails. The straight claw or ripping

Fastening or Assembling Tools

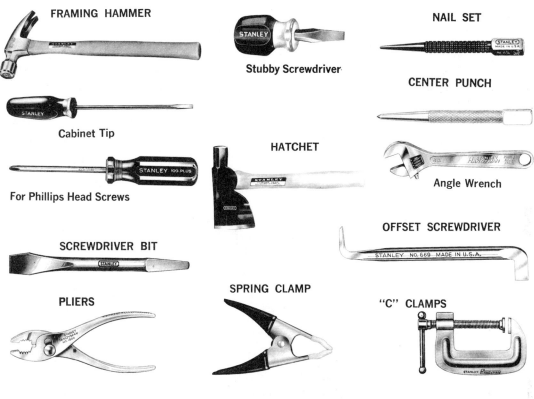

FRAMING HAMMER

Stubby Screwdriver

NAIL SET

CENTER PUNCH

Cabinet Tip

HATCHET

For Phillips Head Screws

STANLEY 100 PLUS

Angle Wrench

SCREWDRIVER BIT

OFFSET SCREWDRIVER

STANLEY NO. 669 MADE IN U.S.A.

PLIERS

SPRING CLAMP

"C" CLAMPS

Fig. 1-20 Assembling tools used in carpentry.

hammer can also be used for removing nails, but it is best suited for ripping apart fastened pieces. In this case the claw functions as a wedge; Fig. 1-21 shows the parts of a claw hammer.

Hammers are sized according to the weight of their heads. Claw hammers range from 7 to 20 ounces. The 16- and 20-ounce sizes are widely used by carpenters. The hammer face can be either flat or bell shaped. The bell-shaped hammer face is slightly convex and less likely to show hammer marks on the work, but it requires somewhat more precise handling than the flat face. It is generally used on finish work where appearance is important.

Quality hammers have heads made of drop forged steel. Cheap hammers have cast heads, are dangerous, and should never be used since they tend to loosen or even break unexpectedly. Wear safety glasses or

Hand Tools

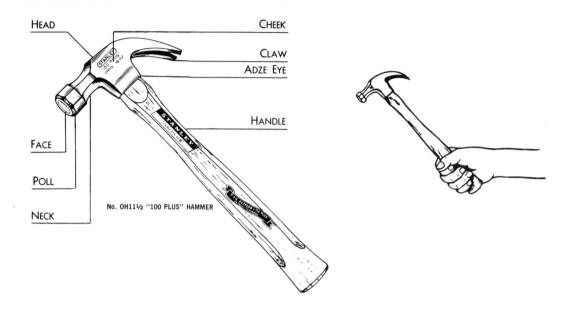

HEAD

CHEEK

CLAW

ADZE EYE

HANDLE

FACE

POLL

NECK

No. OH11½ "100 PLUS" HAMMER

Fig. 1-21 left, the parts of a claw hammer; **right,** the proper way to hold a hammer.

goggles when using hammers and other striking tools, Fig. 1-22.

Hatchets are generally used for rough work such as trimming stakes, rough fencing, and hand-hewn beams and mantels. Often used inexpertly, they are the cause of many accidents. Use them carefully.

Screwdrivers vary in length as well as tip size and shape. The two most common shapes are the slotted and cross-slotted or Phillips. To function properly, the tip should fit the screw slot snugly, and the end must be square and no wider than the slot. The size of a screwdriver is generally determined by its length as measured from tip to ferrule. A good selection of screwdrivers for the carpenter would include 4-, 6-, and 8-inch sizes with both slotted and Phillips tips. Use the longest screwdriver convenient for the job. Never hold the work in one hand while using the screwdriver on it with the other hand. Also, never use a screwdriver as a chisel.

Wrenches and *pliers* are commonly used by the carpenter. They include open-end and adjustable wrenches, and slip-joint as well as grip-jaw pliers. A special clamp is available to convert grip-jaw pliers into a portable vise.

Clamps are versatile tools utilized in many kinds of carpentry. C clamps, bar clamps, parallel and spring clamps, are just a few of the

Fig. 1-22 Eye protection is important when using striking tools.

many types available. C clamps and parallel clamps (also called hand screws) are the most widely used by carpenters. They can be employed to hold glued work while it sets, Fig. 1-23. Such clamps are also used to hold work pieces together temporarily while some other function is performed.

Nailers and *staplers* are fastening tools widely used to increase productivity. Some are hand operated while others are powered by compressed air or electricity. Special nailers and tackers are also available for installing flooring, underlayment, tiles, roofing, and other building products.

Care of Hand Tools

The proper care and maintenance of carpentry tools is important to top-level performance. Never place your sharp-edged cutting tools in a tool box without some sort of protection for the cutting edges. Plane

Fig. 1-23 Handscrews (clamps) are used to hold work while glue sets.

blades should be retracted fully when not in use; chisels ought to be stored in a plastic or leather pouch; and saws' edges protected with a grooved stick held in place with rubber bands. If possible, keep tools in a portable compartmented tool chest that permits easy access to the most-used items. Give all raw metal tools a light coat of oil to prevent rust.

SAFETY RULES

* Never use files without handles.
* Before use, always be sure hammer heads are securely fastened to their handles.
* Always keep cutting tools sharp; dull ones may slip.
* Discard striking or struck tools if they are excessively worn.
* Do not allow chisel and punch heads to "mushroom."

* Do not hold nails or tacks in your mouth while working.
* Always wear safety goggles when using hammers and other striking tools.
* When using sharp-edged tools, always strike, chop, chisel or carve with the cutting action away from yourself.

GLOSSARY

chuck: a hollow, moveable device to grip and hold augers and bits; usually found on drills, but also used on other tools to which blades or auxiliary equipment is attached.

crosscutting: sawing wood across its grain.

kerf: a cut or groove made by a saw blade.

mitering: any angular cutting done to join materials at other than a 90-degree angle.

plumb line: a weighted cord used for locating a point on a surface from another point above; also used for marking vertical lines.

ripping: sawing wood with or along its grain direction.

2

Power Tools

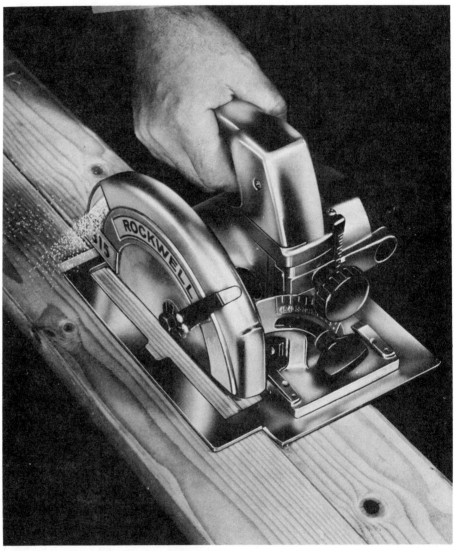

Fig. 2-1 The portable saw is an efficient tool.

Although carpenters still use many hand tools in their trade, they also do much of their work with power tools, Fig. 2-1. Electrically-powered boring, cutting and shaping tools enable the modern carpenter to produce better work more efficiently. Ripping the length of a 3-inch thick 14-foot board takes considerable time and effort with a handsaw. An electric saw sails through the same job in seconds.

Nonelectric power tools are also available to the carpenter. These are generally driven by compressed air and are not as bulky as electrically-driven tools. However, they are seldom used on small construction work.

It must be pointed out that while safe work habits are essential when working with any tool, special care is required with electric power tools. Shock hazard is ever present, especially in carpentry work. A workman standing on damp ground can easily receive a severe electric shock if his tools are improperly grounded. Many carpenters even foolishly remove the grounding leg from the power cords of their tools. Never do this. A grounded tool is a safe tool; an ungrounded one can be a killer.

Power tools are either portable or stationary. If the tool can be carried to the work, it is classified as portable. Stationary tools are usually floor- or bench-mounted and, in most cases, the work is carried to the tool.

Safety must be first and foremost on the carpenter's mind whenever he uses power tools. Tools must be sharp and fit for the work at hand, and the material must be supported properly. Large boards or panels should rest on strong, stable supports. Never force the working of a tool and be sure that the piece being worked on is free of nails, pebbles, and other foreign matter. Always, before plugging in a power tool, check that the switch is off.

Portable Power Tools

The *portable saw* is also known as a circular saw, electric saw, and builder's saw, Fig. 2-2. It is one of the most useful tools employed by the carpenter. In addition to handling blades for cutting wood, it takes others specially made for cutting metals, plastics, and even masonry. The special blades are embedded with bits of tungsten carbide so that they can cut tough materials. The size of a portable saw is determined

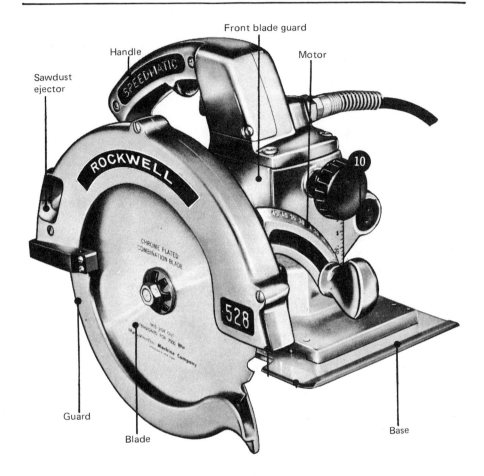

Front blade guard

Handle

Motor

Sawdust ejector

Guard

Blade

Base

Fig. 2-2 Parts of a portable saw.

by the largest blade it will take. An 8" saw will take blades up to 8 inches in diameter, and so on.

The depth and bevel of cut is adjustable by regulating the saw base. Normally the blade should protrude about 1/4 of an inch below the work. All adjustment screws must be tight before using the saw. When the blade is changed, be sure to unplug the machine first, of course.

While most portable saws have adjustable guides for ripping, few carpenters actually use them. They prefer to cut "free-hand", guiding the saw by eye. Ripping, crosscutting, miter, plywood, plastic, and combination blades are commonly available for the portable saw. The

Fig. 2-3 When using a portable saw on prefinished panelling, cut with face-side down. *(U.S. Plywood)*

Fig. 2-4 This saw has an improved sawdust ejection system; safety glasses offer added protection.

combination blade is designed to rip and crosscut. A variety of carbide-tipped blades are available. Many carpenters prefer them since they remain sharper up to twenty times longer than regular blades. They are more fragile than regular blades, however, and you must handle them more carefully.

Because their blades cut from the bottom up as they rotate, portable saws leave splinters on the top side of a cut. This rarely matters when wood is being sawn for rough framing, but it is important when paneling or other delicate work is being done. So, if splintering matters, keep the "good" side of the work down as illustrated in Fig. 2-3. Note that the left hand is kept well away from the blade. Bear in mind also that portable saws have trigger switches which can easily be tripped accidentally if the saw is not picked up correctly.

NOTE: Many texts describe how to make plunge cuts (internal cutouts in a panel) with a portable saw. This is very dangerous and should not be done. Saber saws are ideal and fairly safe when used for plunge cuts, but the portable saw is not. To use a portable saw for plunge cuts, the left thumb and index finger must be held dangerously close to the revolving blade when starting the cut. Do *not* attempt it!

The careful worker uses safety glasses to protect his eyes, Fig. 2-4. When the left hand is not holding work, it should be placed on the saw knob—if there is one.

Power Tools

SAFETY RULES FOR THE PORTABLE SAW

* Rest the saw base on the work firmly and be sure that the blade is not touching the work when starting a cut.
* Do not allow the work to bind against the blade when cutting since this can "throw" the saw. Support the work properly.
* Do not force the saw. A tool is "forced" when it is fed through the work too fast. This causes the blade to slow down and could damage the bearings and motor.
* Be sure the switch is off when the power plug is inserted.
* Do not cut short pieces on the portable saw. The piece together with your thumb can be turned into the blade by uneven thrust.
* Do not stand in the "line of cut." If the saw binds, it can kick back and severely cut the user's legs.
* Use caution when cutting loose knots; they can cause the saw to kick.
* Keep both hands on the saw. If that is not practical, keep the free hand well away from the blade.
* Do not talk to anyone or look away when operating the saw.
* Adjust the saw blade so that it projects about 1/4 of an inch through the bottom of the work.
* Unplug the cord when changing blades.

The portable *electric drill* is used primarily for boring holes. You can also use it for grinding, sanding and cutting when it is fitted with special accessories. The parts of an electric drill are shown in Fig. 2-5. A beading cutter fitted to an electric drill converts it to a shaper. The drills most widely used by carpenters are the 1/4" and 3/8" sizes. Drill size is determined by the largest drill shank the chuck will accept.

Some drills have a variable speed control allowing speeds from zero to 2,000 rpm. Some are equipped with a reversing switch. These can be used for driving and removing screws. Some drills are made with right-angle drives for drilling in tight places, Fig. 2-6.

SAFETY RULES FOR THE ELECTRIC DRILL

* Never leave the key in the chuck.
* Never force a drill.
* Use safety goggles when drilling metal or masonry.

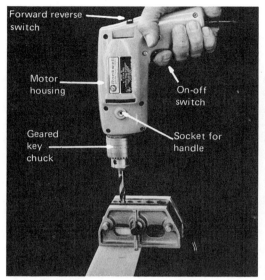

Forward reverse switch

Motor housing

On-off switch

Geared key chuck

Socket for handle

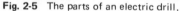

Fig. 2-5 The parts of an electric drill.

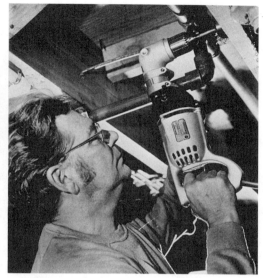

Fig. 2-6 A heavy-duty right angle drill for getting into tight spaces. *(Black & Decker)*

* Never use a square-tang bit in an electric drill.
* Do not stop a coasting drill by hand.
* Wait for a coasting drill to stop before laying it down.
* Do not hold small work in the hand when drilling.
* When drilling very thin metal, the drill bit will have a tendency to "grab." Use a backing block and be careful, otherwise the drill can be thrown from your hands.

The **saber saw** is used for a variety of cutting operations. It has a reciprocating, up-and-down action and cuts on the upstroke. It can be used to cut curves, no matter how intricate. Many types and sizes of blades are available for cutting wood, metal, and other materials. Special blades are also available for flush cutting up to the edge of a wall.

Most saber saw bases can be tilted for bevel cutting. The principal parts of the saw are shown in Fig. 2-7. Just like the circular saw, the saber saw splinters material on the saw side; therefore you must face the finished side of a panel away from the saw during cutting. You can also use this saw for metal cutting. Support the work close to the cutting line and be sure to wear safety goggles.

Although the saber saw is one of the safest power tools, certain safety precautions should be observed.

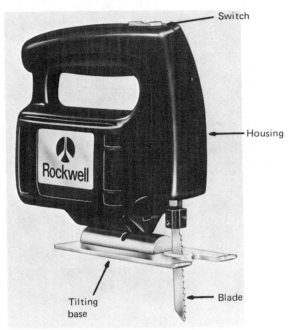

Fig. 2-7 The parts of a saber saw.

SAFETY RULES FOR THE SABER SAW

* Never place the free hand under the work.
* Start the motor before the blade contacts the work.
* Keep the hand not holding the work well away from the cutting line, especially when bevel cutting.
* Never change blades without disconnecting the power cord.

The *reciprocating saw* is similar to the saber saw in principle. A heavy-duty saw, it is suitable both for rough carpentry and construction work. Fitted with a long blade, the reciprocating saw can easily cut through finished walls. It is good for notching timbers and joists and general roughing-in work. Blades are available up to 12-inches long.

To make pocket cuts without drilling a blade entry hole, tilt the saw forward so that the base rests against the work. Then turn on the power, and slowly tilt the saw backward until the blade enters the work. It is best to use a coarse blade for this operation, Fig. 2-8.

Safety rules for the saber saw also apply to the use of the reciprocating saw.

Fig. 2-8 Making a pocket cut with a saber saw.

The **electric plane** will do everything the hand plane does, but faster and with much less effort. It is ideal for planing window and door casings, doors, and shelves. The high cutter speeds (20,000 to 25,000 rpm) produce a smooth glass-like cut. Most electric planes have adjustable fences for bevel cutting. The unit shown in Fig. 2-9 is being used to surface the edge of a door.

Fig. 2-9 A portable electric plane being used on a door edge.

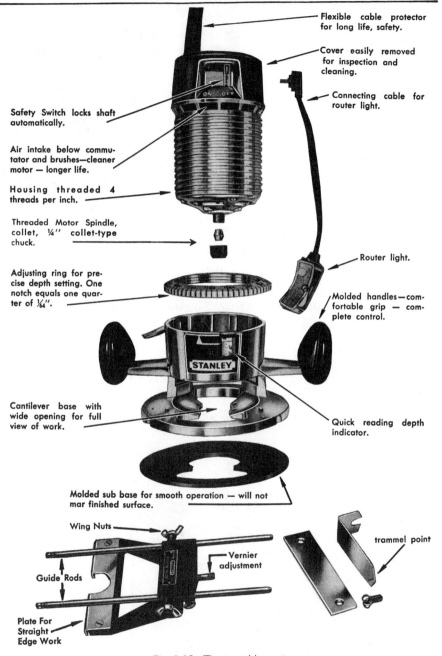

Flexible cable protector for long life, safety.

Cover easily removed for inspection and cleaning.

Connecting cable for router light.

Safety Switch locks shaft automatically.

Air intake below commutator and brushes—cleaner motor — longer life.

Housing threaded 4 threads per inch.

Threaded Motor Spindle, collet, ¼" collet-type chuck.

Router light.

Adjusting ring for precise depth setting. One notch equals one quarter of ¹⁄₆₄".

Molded handles—comfortable grip — complete control.

Cantilever base with wide opening for full view of work.

Quick reading depth indicator.

Molded sub base for smooth operation — will not mar finished surface.

Wing Nuts

trammel point

Vernier adjustment

Guide Rods

Plate For Straight Edge Work

Fig. 2-10 The portable router.

The ***router*** is used for cutting and shaping wood and is also useful for trimming plastic laminates. It takes shaped cutters for decorative work and straight cutters for rabbeting and for cutting dadoes.

Basically, the router consists of a high speed motor (25,000 rpm), a chuck for holding the cutting tool and an adjustable base, Fig. 2-10. Adjust the depth of the cut by raising or lowering the base. Bits for the router come in many shapes and sizes. Some have pilots for guiding the cutter around the edge of the material, Fig. 2-11.

On straight work the direction of feed for routers is from left to right, and on circular work the feed should be counterclockwise.

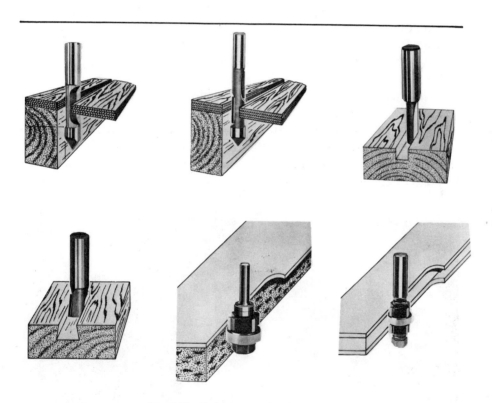

Fig. 2-11 Various router bits illustrated.

Top: Left—pilot panel bit
Center—pilot panel bit, stagger tooth
Right—straight bit

Bottom: Left—stair routing bit
Center—carbide tipped laminate trimmer
Right—double flush trimmer

SAFETY RULES FOR THE ROUTER

* Keep both hands on the tool.
* Keep cutters sharp.
* Disconnect the motor when changing cutters.
* Keep fingers well away from a moving cutter.
* Wear safety goggles.
* Be sure the bit is chucked tightly.
* Tighten the base lock.
* Hold the tool so that the bit is clear of work when power is turned on.
* Do not force the cut.
* Do not cut into nails or screws.
* Wait for the cutter to stop before lifting it from work.
* Unplug the machine when it is not in use.

Belt and *disk sanders* are portable sanders used for stock removal. *Finish sanders* are for smoothing the work surface. They remove very little stock and are easy to use. Generally, finish sanders are driven either in an orbital or an oscillating motion. Orbital sanders can be moved in any direction, but oscillating sanders should be moved with the grain of the wood.

Fig. 2-12 The belt sander in use.

Belt sanders require somewhat more skill to operate than finish sanders. Since they are fast-cutting, they must be kept moving continuously and not held in one spot, Fig. 2-12. Each stroke should overlap the previous stroke, by one-half the belt width. The weight of the machine provides the proper pressure on the work. Do not press down on the sander.

SAFETY RULES FOR SANDERS

* Wear snug-fitting clothes when using belt and disk sanders; if practical, roll up shirt sleeves.
* Be sure to secure the end of one's trouser belt.
* Keep both hands on the handles.
* Do not set the machine down until it stops coasting.

Stationary Power Tools

The *radial arm saw* is a table saw that serves basically as a cut-off tool. It can also be used for ripping and with various accessories for shaping, dadoing, rabbeting, and other operations. Fig. 2-13 shows a typical radial arm saw.

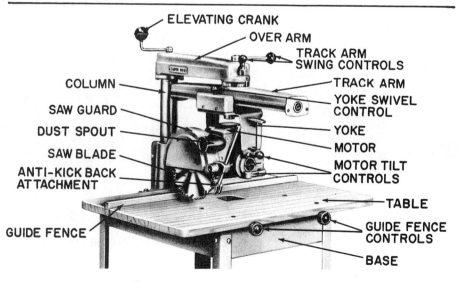

Fig. 2-13 The parts of a radial arm saw.

Power Tools

This tool consists of a base, a vertical column, and a horizontal overhead arm. Unlike an ordinary table saw, its blade cuts from on top of the work rather than from below. A yoke supporting the cutting assembly (motor and circular saw blade) travels along the overarm. This assembly can be tilted to make angular cuts. Also, the arm can be positioned to make various miter cuts. The depth of cut is adjusted by raising and lowering the overarm.

For crosscutting operations, hold the work stationary while the saw blade moves over the work. For ripping operations, hold the saw blade in a fixed position and push the work through the saw. A groove or plow cut is made with the grain. When the cut is across the grain, it is called a dado; Fig. 2-14 shows a plowing operation with a dado blade. Ripping is done in the same manner.

A fence serves as a backstop for crosscutting and mitering operations. The fence is locked in place but it can be moved to accomodate various size boards.

Cross Cutting: When the radial saw is at rest, the normal position of its blade is at the rear of the arm behind the fence, ready for crosscutting. Place the material to be cut against the fence. Hold it firmly, keeping fingers well away from the blade, then turn on the power and pull the blade across the work, Fig. 2-15. Return the blade to its normal position, then turn off the power. For duplicate cuts, clamp a stop to the fence.

Fig. 2-14 A plowing or grooving operation on the radial arm saw. Note safety gripper fingers.

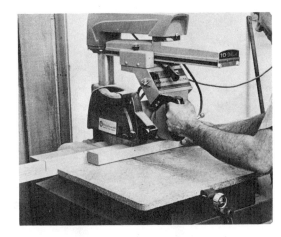

Fig. 2-15 Crosscutting on the radial arm saw.

Mitering: This operation is similar to crosscutting except that the arm of the radial saw is positioned to the desired angle. During miter-cutting, the work has a tendency to creep or slide. To prevent creeping, install anchor points along the fence. Such anchor points are simply screw ends pointed with a file. The screws themselves are driven from the rear of the fence so that their points protrude slightly along the fence.

Ripping: The operation involves lengthwise cuts through a board in the direction of the grain. Position the yoke so that the blade is parallel to the fence or at right angles to its normal position. Then lock the yoke in position and push the work through the saw blade. Adjust the anti-kickback fingers to ride the surface of the board. When ripping narrow stock, be sure to use a push stick to keep the fingers well away from the blade.

Beveling: The radial saw arm can also be used to bevel as it rips and crosscuts. The crosscut bevel is made by tilting the yoke downward to the desired angle. In order to tilt the yoke sufficiently, elevate the cutting head so that the blade clears the table. Then, when the angle is set, lock the yoke and lower the blade so that it cuts into the table about 1/16 of an inch. Follow the normal crosscutting procedure. Do the ripping operation in the same manner, except that you must rotate the yoke so the blade is parallel to the fence.

Compound Angle Cutting: In carpentry work, there are times when you will have to cut either a double miter or a compound angle. In such a case, tilt the radial saw blade and swing the arm to the desired angle. Otherwise follow normal beveling and mitering procedures.

Miscellaneous Sawing Operations: Dadoes are cut by replacing the radial saw blade with a dado head. Cross-dadoing, angle-dadoing, and plowing (grooving) are then performed in the same manner as when sawing. Set the depth of cut as desired.

SAFETY RULES FOR THE RADIAL ARM SAW

* Never keep tools or scrap lumber on saw table.
* Be sure the saw blade is back against the stop before turning on power.
* Always feed work into the cut when ripping. (This is the end opposite the anti-kickback fingers.)
* Keep fingers well away from rotating blade.
* Hold work securely.

Power Tools

* Use holding stick when cutting small pieces.
* Never cut pieces "piggy-back" (one on top of the other).
* Be sure ends of long boards are supported.
* Do not remove guard and anti-kickback devices.
* Try to avoid ripping large sheets of plywood on the radial arm saw. Use a table saw or portable saw instead.
* Do not attempt to rip or crosscut warped boards.
* Do not attempt to cut while the saw is coasting.

The *table saw* is also a type of circular saw and is sometimes called that. It and the radial arm saw are the two most important stationary tools used by the carpenter in house construction. Used on site, both machines save the carpenter much time and energy, and both are available in portable versions which can be easily transported in a station wagon or small pickup.

The table saw, like the radial arm saw, is suitable for ripping and crosscutting. You can also do mitering, plowing, and shaping with it. A tilting arbor permits bevel cuts up to 45 degrees. Fig. 2-16 shows a typical table saw with the blade projecting up through the center slot.

Saw size is rated by blade diameter and may vary from 8 to 14 inches. The 10″ saw is generally used by carpenters and contractors. Some table saws are available in combination with a jointer, further increasing their usefulness.

The table saw is comparatively easy to use. Make rip cuts by setting the fence to the desired width. Make miter and crosscuts with the miter gauge. In normal use, the blade projects about 1/8 of an inch above the work. Long pieces are best cut with the aid of an outfeed table such as a sawhorse or other stable support. An auxiliary fence with points similar to those described for the radial arm saw is useful in preventing creep when cutting miters, Fig. 2-17. Dado and grooving cuts are made with a dado head. Rabbet cuts can be done by making two rip cuts. Taper cuts are made with simple step jigs.

SAFETY RULES FOR THE TABLE SAW

* Roll up shirt sleeves.
* Use a push stick when ripping narrow stock.
* Do not remove guard or splitter plate unless absolutely necessary.
* Use clamps or jigs to hold small work.

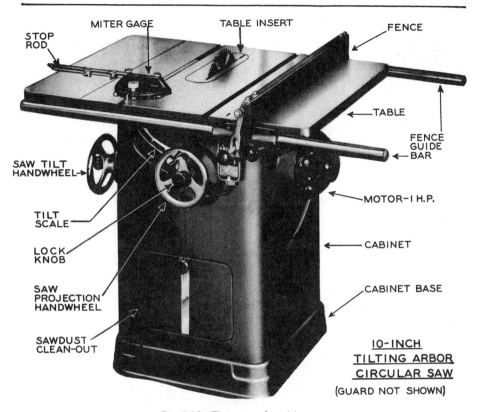

STOP ROD

MITER GAGE

TABLE INSERT

FENCE

SAW TILT HANDWHEEL→

TILT SCALE

LOCK KNOB

SAW PROJECTION HANDWHEEL

SAWDUST CLEAN-OUT

TABLE

FENCE GUIDE BAR

MOTOR-I H.P.

CABINET

CABINET BASE

10-INCH TILTING ARBOR CIRCULAR SAW

(GUARD NOT SHOWN)

Fig. 2-16   The parts of a table saw.

* Never hold on to free end of crosscut work.
* Stand to one side of blade while working.
* Never cut work freehand. Use the gauge or fence.
* Never clear scraps near blade while machine is running.
* Never cut round stock on the table saw.
* Never reach across the blade.
* Set blade 1/4″ above the top of board being cut.
* Do not cut warped stock on the table saw.
* When ripping, push work past the blade at end of cut.
* Unplug machine when changing blades.

The *motorized miter saw* is a powered version of the miter box. It is ideal for builders and carpenters. It makes fast, accurate, square, and

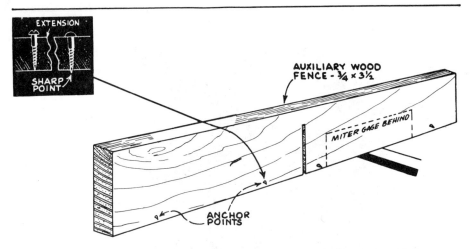

Fig. 2-17 A simple device for preventing "creep" when mitering on the table saw.

miter cuts in wood, composition and lightweight aluminum. A compact machine, you can easily transport it.

The *jointer* is a planing tool you will use mostly to smoothen the edges and surface of boards cut on the circular saw. The size of a jointer is determined by the maximum width of a board it will surface.

It has an adjustable fence that permits bevel cuts. Control the depth of cut by raising or lowering the front table of the tool.

The jointer can be used for making rabbet cuts as well as for surfacing work. With special jigs you can also use it to cut tapers and tenons.

SAFETY RULES FOR THE JOINTER

* Never remove the guard unless absolutely necessary.
* Never joint thin stock without using a backup board.
* Never joint pieces shorter than 10-inches.
* Always cut with the grain of the wood.
* Be sure work is free of nails or other foreign matter.
* Use push stick on narrow pieces.
* Never hold stock directly over cutter head.
* Never run end-grain pieces less than 12-inches on the jointer.

Tool Sharpening Techniques

Properly sharpened tools are important to carpentry. No matter how good or expensive a tool may be, it is worthless if you do not keep it sharp. Not only do sharp tools cut faster and more accurately, but they are safer and easier to use.

Sharpening methods vary, depending on the tool being sharpened. Sharpen edged-tools, such as chisels, plane blades, knives and drill bits with an abrasive stone. Auger bits and saws when sharpened by hand are best done with a file. Often, all that is required is a simple honing to restore a tool's edge; but grinding is often necessary to renew a bevel or to remove nicks.

When you use a grinding wheel, take care not to burn or soften the metal. Dip the tool edge into water frequently to keep it cool. Grind plane blades and chisels to an angle of 25 degrees ordinarily. After grinding whet or hone the tool with an oilstone. To do this, place the blade with the existing bevel in full contact with the stone. Then raise it slightly (about 5 degrees) and rub it back and forth over the stone at the same angle. Lubricate your stone with light oil or kerosene to float off the particles and keep it clean. After the beveled edge is honed, place the flat side of the blade on the stone and stroke it sideways a few times to remove any fine feathering on the edge, Fig. 2-18.

Normally, saw blades can be sharpened by touching up the points with a file. However, after repeated filings, the blade will require jointing. This means evening-up the teeth so they are all the same height

WHET THE PLANE IRON ON THE OIL STONE TO PRODUCE THE REAL SHARP CUTTING EDGE
HOLD THE PLANE IRON IN THE RIGHT HAND WITH THE LEFT HAND HELPING
PLACE THE BEVEL ON THE STONE WITH THE BACK EDGE SLIGHTLY RAISED
MOVE THE PLANE IRON BACK AND FORTH

REMOVE THE WIRE OR FEATHER EDGE BY TAKING A FEW STROKES WITH THE FLAT SIDE OF THE PLANE IRON HELD FLAT ON THE STONE AVOID THE SLIGHTEST BEVEL ON THIS SIDE
IF A NICK OR A SHINY EDGE OF BLUNTNESS CAN BE SEEN, REPEAT BOTH PROCESSES OF WHETTING

FINISH WITH A FEW STROKES ON A LEATHER STROP TO PRODUCE A KEENER EDGE

Fig. 2-18 The method of sharpening a plane iron.

again. Because it is exacting work and special equipment is needed, most carpenters are satisfied to touch up their saws themselves and then send them out when jointing and setting are required. The points are "set" by bending them alternately and sideways of the blade's center at a slight but precise angle.

When sharpening a saw blade, hold it securely to prevent vibration. Special vises are available for this purpose.

To file a crosscut saw, use a slim tapered triangular file and follow the original tooth shape. The face of the tooth has a slight bevel which you must keep. File rip saw blades with a mill file straight across the face.

Pitch or resin on saw blades lessens their cutting efficiency and causes rapid dulling of the blades because of the resulting overheating. Keep saws clean at all times. Special cleaners are available, or you can use kerosene or mineral spirits.

GLOSSARY

bevel: an angular cut; or degree of inclination on an edge.

carbide: in woodworking, usually refers to tungsten carbide, a very hard metal alloy.

compound angle: two mitered angles.

creep: in woodworking, the tendency of the work to move while being cut, especially when mitering.

dado: a rectangular groove across the grain of a board.

fence: an adjustable guide on a tool.

honing: removing the burr left by sharpening with a grinding wheel.

jointing: evening up the teeth (points) of a saw blade.

kickback: a violent reaction of a machine tool when work is fed to or withdrawn incorrectly.

oscillating: a back-and-forth motion which on tools like sanders permits the tool to be worked easily in all directions.

plow (plough): a lengthwise groove in a piece of wood.

portable tool: a tool which can be carried about easily.

push stick: a piece of wood used to push work through a saw; it is used to keep the fingers away from the blade.

rabbet: a rectangular groove cut in the edge of a board and intended to receive another board.

rpm: revolutions per minute of a motor or other mechanism.

stationary tool: a non-portable tool; usually mounted on a table or the like.
taper: two converging lines or edges.
variable speed: adjustable speed.
whetting: sharpening a tool on an oilstone or other substance.

3

Transits & Levels

Special optical instruments are widely used in house and building construction generally. They provide an accurate and efficient method of laying out building lines and determining grade elevations. Features of the various surveying instruments may differ, but basically they all operate on the same principle.

The Builder's Level

The builder's level, also called a dumpy level, has a telescope which rotates sideways to measure horizontal angles. It contains a spirit level, adjusting screws, an index pointer, and other parts shown in Fig. 3-1. A self-leveling or automatic level is shown in Fig. 3-2, and Fig. 3-3 shows a level in use at a building site.

Fig. 3-1 Builder's level has its telescope fixed in the horizontal position; this telescope turns sideways for measuring horizontal angles. *(Berger Instruments)*

Fig. 3-2 The automatic level requires the centering of one bubble to set the line of sight. *(Wild Heerbrugg, Inc.)*

Fig. 3-3 The dumpy level (builder's level) in use at a construction job.

The Transit Level

This instrument is similar to the builder's level but its telescope, in addition to turning sideways, can also move up and down to measure vertical angles, Fig. 3-4. It is used to lay out level and plumb lines. Fig. 3-5 shows a transit level in use.

Sturdy tripods support both the transit and the builder's level. Some are made with adjustable legs, for use on sloping ground.

Instrument Setup: After setting the tripod firmly on the ground, carefully mount the surveying instrument and fasten it to the tripod's head or turntable. Be sure to tighten the mounting screw securely.

In order to function properly, the instrument must provide a true line of sight. This means that a line passing through the telescope objective remains perfectly horizontal—no matter in what direction the telescope is pointed. If it is "off" level, all measurements will be inaccurate. Swing the telescope so that it is directly over the first two leveling screws. Then turn the screws either toward or away from each

Transits & Levels

Fig. 3-4 The transit-level; the telescope turns sideways and up-and-down to measure vertical and horizontal angles. *(Berger Instruments)*

Fig. 3-5 Carpenter using a transit-level; note that he keeps his hands off the instrument while he sights.

other, as in Fig. 3-6. The bubble will start to move; in most cases, in the same direction as the left thumb. Adjust the screws until the bubble is perfectly centered between the graduations on the vial.

After leveling in one direction, swing the telescope 90 degrees over the second set of leveling screws and level again. Recheck both positions several times. When it is perfectly level, the bubble will remain centered, regardless of where the telescope points.

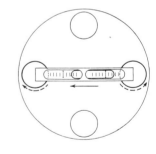

Fig. 3-6 In most instruments, the bubble moves in the same direction as the left thumb.

Leveling Rods

Leveling rods are used to measure vertical differences in elevation. These instruments are especially useful when sighting over long distances. Basically they consist of two narrow sliding sticks or sections which are held upright and are graduated in feet and inches.

The levelman can operate a self-reading rod and read the figures through his telescope. But a target rod with crosslines may also be used. It is read at the rod by the rodman, Fig. 3-7. The rodman moves the target up or down on signal from the levelman and then calls off the readings thus found on the rod's vernier scale. The rodman usually carries a small spirit level which he uses to make sure that he is holding the rod plumb (vertical).

For small home construction where no great distances are involved, a folding rule may be used instead of the leveling rod. To make it stiff enough to remain upright, hold it against a piece of 1 × 2 furring.

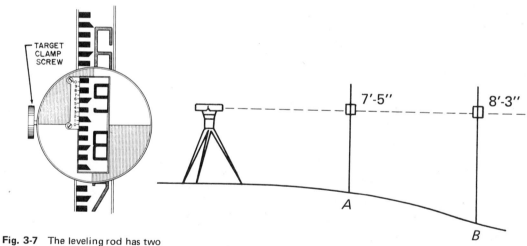

Fig. 3-7 The leveling rod has two sliding sections and a target.

Fig. 3-8 Using the level to measure the difference in elevation demonstrates that Point *B* is 10-inches lower than Point *A*.

Measuring Elevation Differences: To find the difference in elevation between two points (leveling), set up the level as shown in Fig. 3-8. The readings are then used to calculate the difference in elevation. If the ground slopes considerably, set up the level midway between the two points, as shown in Fig. 3-9. Place the rod at point *A* and take a reading. Then swing the telescope 180 degrees and with the rod at point *B* take another reading. The difference between the two readings will be the difference in elevation.

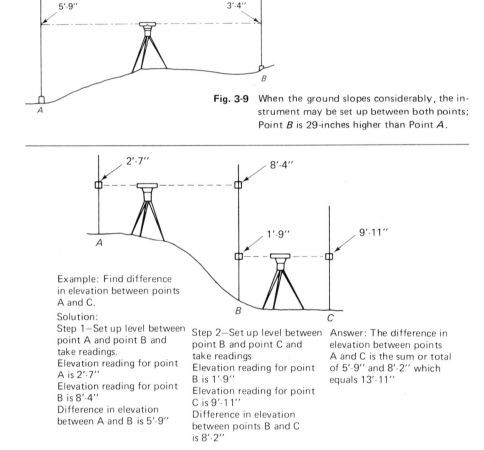

Fig. 3-9 When the ground slopes considerably, the instrument may be set up between both points; Point *B* is 29-inches higher than Point *A*.

Example: Find difference in elevation between points A and C.

Solution:

Step 1—Set up level between point A and point B and take readings.
Elevation reading for point A is 2'-7"
Elevation reading for point B is 8'-4"
Difference in elevation between A and B is 5'-9"

Step 2—Set up level between point B and point C and take readings
Elevation reading for point B is 1'-9"
Elevation reading for point C is 9'-11"
Difference in elevation between points B and C is 8'-2"

Answer: The difference in elevation between points A and C is the sum or total of 5'-9" and 8'-2" which equals 13'-11"

Fig. 3-10 Method of finding difference in elevation between Point *A* and Point *C*.

If the distance between two points is too far apart for accurate readings, or the difference in elevation is too great for sightings, set up the level as shown in Fig. 3-10. Then take readings between the various points. Add the difference in height between each point or station to find the total difference in elevation.

To use the level for locating batter boards or grade stakes, set it up centrally within the building lines, Fig. 3-11. Then put elevation marks on the stakes as indicated by the crosslines of the telescope. Batter boards may then be installed, using the elevation marks as a reference.

If a level mark or stake cannot be placed at the required level, mark the stakes as shown in Fig. 3-12.

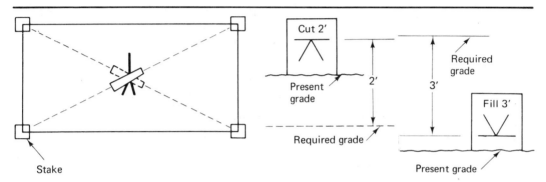

Fig. 3-11 Level is centrally located to set grade stakes.

Fig. 3-12 Method of marking stakes when it is not practical to set them at the required height.

Setting straight lines: Employ the transit to set straight lines for boundaries, fences, and roadways. It is especially useful on irregular or sloping terrain. Set the instrument over a reference point, and swing it to the desired direction. Then by raising or lowering its telescope, you can align stakes perfectly, as shown in Fig. 3-13.

Vertical angles: The vertical arc of the surveying telescope is also used for measuring vertical angles above or below the horizontal.

Plumbing: You can use the transit for plumbing walls, columns, and other vertical objects. To check the plumb of the corner post of a frame building as in Fig. 3-14, proceed as follows:

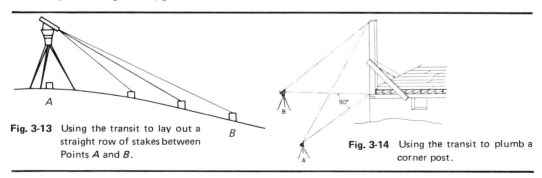

Fig. 3-13 Using the transit to lay out a straight row of stakes between Points *A* and *B*.

Fig. 3-14 Using the transit to plumb a corner post.

Set the transit at a point about equal to the height of the corner post. (This distance is not critical.) Then level the instrument and sight a point at the base of the post, placing the center of the cross wires on it. Elevate the telescope, sighting along the length of the post. The vertical wire or hair should stay "on target." If your sighting on the post moves off center and away from the wire, the post is off plumb. Adjust the post accordingly. Select another position 90 degrees away from the first position; repeat the procedure. If necessary, adjust the post again. When the reading holds, an exactly plumb post results.

Horizontal angles: Use either the transit or the builder's level for measuring or laying out horizontal angles.

To measure a horizontal angle (Fig. 3-15), set the instrument over a reference point *A*. Rotate the instrument to align the vertical cross hair with point *B*. Set the horizontal circle to zero; next swing the telescope so the vertical cross hair is in line with point *C*. The index pointer indicates the angle *BAC*.

To lay out the building lines, the procedure is similar to the above, except that the angles are already known. In the case of square or rectangular buildings, the corners are 90-degree angles. Line *A-B* in Fig. 3-16 represents a known building line. Set the instrument up over point *A*, then swing along line *A-B*. Set the horizontal circle index to zero. Next swing the instrument 90 degrees and establish line *A-C*. Tape off the required distance and drive a stake. Place the instrument over point *C* and, sighting back to point *A*, set the horizontal circle to zero again. Swing the instrument 90 degrees to establish line *C-D*. Tape the required distance along *C-D* to complete the layout.

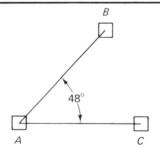

Fig. 3-15 Method for measuring horizontal angle *BAC* with a surveying instrument.

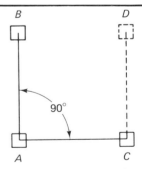

Fig. 3-16 Laying out building lines: Sight line *AB*, then swing scope 90 degrees, and mark off point *C*; set instrument over Point *C*; align with *A* then swing 90 degrees to locate stake *D*.

Vernier Scale: Transit levels are equipped with vernier scales which permit accurate readings in fractions of a degree. A typical vernier scale is shown in Fig. 3-17. All measurements in the vertical and horizontal plane are based on the divisions of a circle as follows:

$$\begin{aligned} \text{circle} &= 360 \text{ degrees} \\ 1 \text{ degree} &= 60 \text{ minutes} \\ 1 \text{ minute} &= 60 \text{ seconds} \end{aligned}$$

To read a horizontal angle, set the upper scale at zero by matching the index marks. Then swing the scope to the line to be measured. Lock the scope and note the reading. As an example: the index might be past the 44 mark but not reach to the 45 mark. This means that the measurement lies between 44 and 45 degrees. Note that some lines in the vernier come close to each other and one line is closer to another than any of the others. That fourth line is the one to read. (Each division or line on the vernier is equal to 5 minutes.) Since the fourth line is the closest, it is read as 20 minutes. Therefore, the reading is 44 degrees, 20 minutes. The vertical is read in a similar manner.

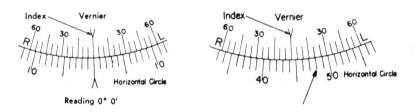

Fig. 3-17 The vernier scale: **Left,** scale zeroed; **Right,** reading of 44 degrees, 20 minutes. *(Berger Instruments)*

GLOSSARY

dumpy level: a builder's level
elevation: the height above a reference line.
levelman: the man operating the transit level or scope.
line of sight: an imaginary line from the observer's eye toward a distant point.
rodman: the man operating the leveling rod.
transit level: an instrument for measuring and laying out vertical and horizontal lines.
vernier: a scale used to measure subdivisions of a larger scale.

4

Framing Square

The framing square is one of the most useful tools in the carpenter's possession. It helps to solve quickly many of the difficult problems found in layout work. When you use it properly, this important tool will replace much complicated mathematical work. It is especially useful in roof framing where it is used to determine lengths of common, hip, valley and jack rafters. It is also used to determine top, bottom and side cuts for rafters. A table for board measure enables the carpenter to determine at a glance the contents in board measure of any size board or timber. The framing square is also useful for finding the lengths of common braces, for laying out stairs, for laying out octagons, and more.

The square is made in the form of a right angle, Fig. 4-1. The two legs are called the body and tongue. If the ends of the legs are connected, they form a right triangle as shown in Fig. 4-2. The square is therefore based on the principles of a right triangle. The framing square consists of these parts: body, tongue, heel, face and back.

The *body* is the longer and wider leg. Generally it is 24 inches long and 2 inches wide. It is sometimes called the blade.

The *tongue* is the shorter and narrower leg. It is usually 16 inches long and 1-1/2 inches wide.

The *heel* is the outside corner where the body and tongue meet. At times the inside corner may also be called the heel.

The *face* of the square is the side with the manufacturer's name or trademark stamped on. If you hold the square with the body in your

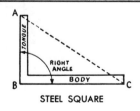

STEEL SQUARE

Fig. 4-1 The framing square consists of a body and tongue which form a right angle (90 degrees).

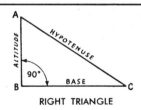

RIGHT TRIANGLE

Fig. 4-2 When the ends of the body and tongue are connected, they form a right triangle.

Framing Square

left hand and the tongue in your right, the face will be the uppermost side, Fig. 4-3.

The **back** of the framing square is the side opposite the face. Fig. 4-4 illustrates the face and back of a square with markings.

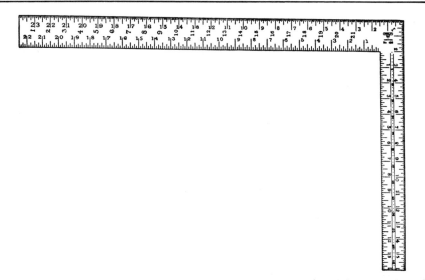

Fig. 4-3 The face side of the square is up when the body is to the left and the tongue is to the right of the holder.

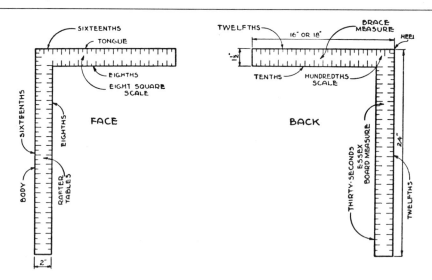

Fig. 4-4 The markings of a typical framing square.

Scales and Tables

Most squares have two types of markings: scales and tables. The scales are on all outer and inner edges and are marked off in inches and fractions of an inch. Their markings are listed in the table that follows:

PART	EDGE LOCATION	SCALE MARKINGS
Face of body	Outside edge	Inches & sixteenths
Face of body	Inside edge	Inches & eighths
Back of body	Outside edge	Inches & twelfths
Back of body	Inside edge	Inches & sixteenths
Face of tongue	Outside edge	Inches & sixteenths
Face of tongue	Inside edge	Inches & eighths
Back of tongue	Outside edge	Inches & twelfths
Back of tongue	Inside edge	Inches & tenths

Hundredth Scale: It is located on the back of the tongue near the heel and is a line one inch long, divided into 100 parts. See Fig. 4-5. The longer lines indicate 25 hundredths and the shorter ones are 5 hundredths. Directly below the divisions is a sixteenth scale. This allows the conversion of hundredths to sixteenths at a glance and without using a divider. This is very useful when determining rafter lengths since the figures in the rafter table on the square's face are given in hundredths.

Octagon Scale: Also called the eight scale, it is located on the face of the tongue along the center. This scale is used to obtain the measurements to shape a square timber into an eight-sided one.

To shape an octagon from a square timber, 8″ × 8″, draw two center lines as *AB* and *CD* in Fig. 4-6. Using the scale on the face of the tongue, set a divider to as many spaces on the scale as there are inches in the width of the timber. That would be 8 spaces on the scale in this case. Lay this space on each side of the centerlines as shown, then

Fig. 4-5 The hundredth scale is found on the back of the square near the heel.

Framing Square

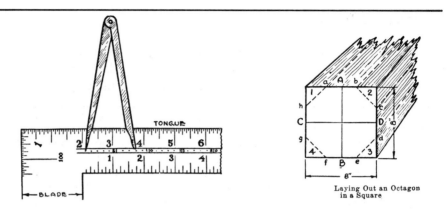

Laying Out an Octagon in a Square

Fig. 4-6 The steel framing square can be used to lay out an octagon on square timbers.

connect the points with lines. The resulting figure will be an eight-sided eight-cornered octagon.

Essex Board Measure: A table on the back of the square's body gives the number of board feet contained in pieces of wood one inch thick and of varying standard sizes. The figure 12 on the outer edge represents a one inch board by 12 inches wide and is the starting point for all calculations. Directly under the figure 12 are figures from 8 to 15: these represent board lengths in feet. The graduations on both sides of the 12 on the inch scale indicate the width of the piece. The figures under these inch graduations are the board measure. The numbers to the left of the line give the full board measure and the numbers to the right are in twelfths. The inch marks along the outer edge of the square are used in combination with the seven parallel lines, Fig. 4-7.

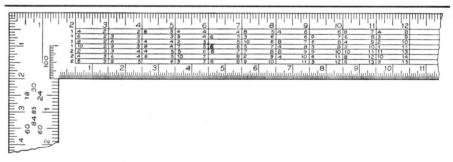

Fig. 4-7 The Essex board measure table is on the back side of the square's body.

Scales and Tables

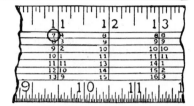

Fig. 4-8 The board measure in the example is circled.

For example: 8-3 equals 8 feet and 3 twelfths or 8′3″. If the piece being measured is thicker than one inch, multiply the figure given in the table by the thickness.

Problem: Find the board measure for a piece 8′0″ × 11″.

Solution: First find the length of the piece under the 12 column. Next, find the width on the edge of the square. This would be the 11-inch mark. The answer is found where the two come together at 7-4 or 7 feet and 4 twelfths, so 7-feet 4-inches is the number of board feet in the board. See Fig. 4-8.

If the board or timber is longer than 15 feet, let us say 29 feet, simply work the problem out twice. Figure one piece as 15-feet and the other as 14-feet, then add the two answers.

Brace Measure: This table is found at the back of the tongue along the center, Fig. 4-9. It gives the length of common braces. As an example, figures read like this: 39/39 55.15. They mean that if the rise on a post is 39 inches and the run on the beam is 39 inches, the length of the brace will be 55.15 inches or 55-1/8″, Fig. 4-10.

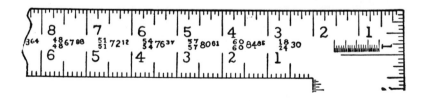

Fig. 4-9 The brace measure table is used for determining lengths of common braces.

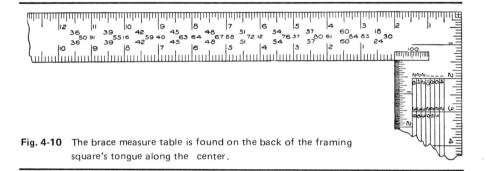

Fig. 4-10 The brace measure table is found on the back of the framing square's tongue along the center.

This is useful if the run and rise of post and beam are equal. If not, the braces are treated as common rafters. The top and bottom cuts are then determined by using the common rafter tables. That is explained in Chapter 11 on roof framing.

GLOSSARY

builder's square: another name for framing square; also called a carpenter's square.
Essex board measure: a system for determining the contents in board feet of a piece of lumber.
graduations: the markings or divisions on measuring tools and instruments.
heel: the corner of a framing square.
octagon: an eight-sided figure.
right angle: a 90-degree angle formed by two lines perpendicular to each other.

5

Construction Fasteners

The strength and stability of wood frame buildings depends on the fastenings used to join the component parts. Just as a chain is as strong as its weakest link, so is a building only as strong as its fastenings. When properly constructed with the correct fasteners, wood frame buildings can be made to withstand the forces of hurricanes and tornadoes.

Some of the fasteners commonly used in modern home construction include nails, screws, bolts, and various metal connectors. Special notched nails are made for joining truss plates to frame members as illustrated in Fig. 5-1.

Nails

Nails are the most common fastener used in frame construction. They are made in many types and sizes for specific purposes. They may

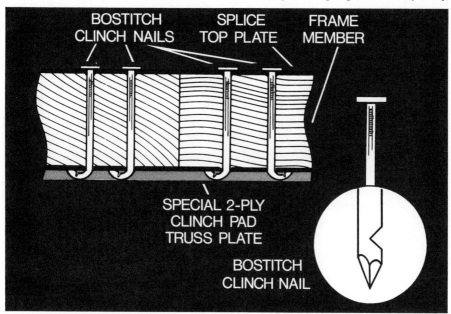

Fig. 5-1 Notched clinch-nail eliminates need for flipping trusses during assembly.

Construction Fasteners

Fig. 5-2 A machine making nails from wire. *(Keystone Steel & Wire)*

be bright, blued, cement coated, galvanized, or painted. Various metals are used in their manufacture. The shank of the nail may be smooth, barbed, or otherwise formed. Fig. 5-2 shows a nailmaking machine forming nails from steel wire.

Nails vary from one another in the design of their caps, shanks and points. The heads may be flat, round, oval, countersunk, slotted, headless, and so forth. The shanks may be smooth or threaded. The points may be regular diamond shaped, long diamond, blunt, needle pointed, sheared bevel, sheared square and chisel pointed. The nails most commonly used by the house carpenter are threaded nails, hot-dipped galvanized nails, painted nails, electroplated nails, blued nails, and cement coated nails. Fig. 5-3 shows a number of the head and point styles commonly used.

Threaded nails may have annular, spiral, screw threads, or knurled threads. The ***annular-thread nail*** is used in woodworking for its good holding power but only where there is no lateral load on the nail. It is not recommended for end-grain nailing or to edges of plywood, hardboard, or particleboard.

The ***spiral-thread nail*** has great holding power both laterally and

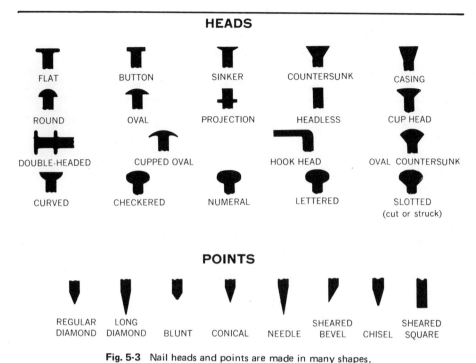

HEADS

FLAT BUTTON SINKER COUNTERSUNK CASING

ROUND OVAL PROJECTION HEADLESS CUP HEAD

DOUBLE-HEADED CUPPED OVAL HOOK HEAD OVAL COUNTERSUNK

CURVED CHECKERED NUMERAL LETTERED SLOTTED
(cut or struck)

POINTS

REGULAR DIAMOND LONG DIAMOND BLUNT CONICAL NEEDLE SHEARED BEVEL CHISEL SHEARED SQUARE

Fig. 5-3 Nail heads and points are made in many shapes.

vertically and can be used where end-grain nailing is necessary. It should not be used on plywood, but it can be used on hardboards.

The *screw-thread nail* has very good holding power for all loads. It is not recommended for nailing to edge of plywood or particleboard, or for end-grain nailing.

The *knurled-thread nail* is excellent for nailing into masonry. Wear safety eye goggles when nailing with them. Pre-drill the masonry to a 3/4″ depth to accept a nail of the same diameter as the hole diameter. Be sure to select a nail long enough to penetrate the masonry no more than 3/4 inches. Drive the nail into the masonry with a sledge hammer. For short lengths, a 2-lb. hammer is recommended; for longer lengths a 5- to 7-lb. hammer is needed; Fig. 5-4 shows a knurled masonry nail with a slight thread. (See page 60.)

Hot-dipped galvanized nails are steel nails coated with pure zinc in molten form. The nails are corrosion and stain resistant and excellent for all exterior use.

Painted nails are prime coated and finished with several coats of paint to match decor of paneling or other material being installed. To

Fig. 5-4 Knurled masonry nail with slight thread has high holding power.

protect the finish of the nail head, work with a hammer head fitted with a plastic cap.

Blued nails are steel nails that have been heat treated. The tempering process turns them blue.

Cement coated nails are nails, threaded or unthreaded, which have been given a rosin coating to increase their holding power. The adhesive coating acts as a lubricant when the nail is driven in, then it softens and establishes a super-tight bond between the nail and wood. Cement coated nails are widely used for boxing, crating, and in building trades. They are uniform in shape and gauge and ideal for use in automatic nailing machines.

The holding power of nails depends on the material being nailed as well as the penetration, diameter and surface of the nail. Nail lengths

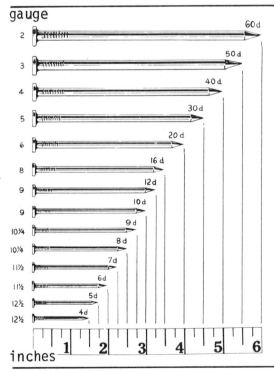

Fig. 5-5 Length and gauge of the most common wire nails.

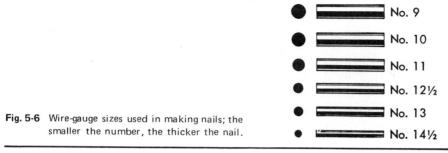

● ▬▬▬▬▬ No. 9

● ▬▬▬▬▬ No. 10

● ▬▬▬▬▬ No. 11

● ▬▬▬▬ No. 12½

● ▬▬▬ No. 13

● ▬▬ No. 14½

Fig. 5-6 Wire-gauge sizes used in making nails; the smaller the number, the thicker the nail.

may be designated in inch or in "penny" sizes. In the past, penny referred to the price of various sized nails per hundred; however it now refers only to the length of the nail, and is expressed by the small letter "d" following a numeral. For example, an eight penny nail is the same as an 8d nail and is 2-1/2″ long; Fig. 5-5 compares penny and inch sizes.

The gauge of a nail refers to the wire size used in its manufacture. The lower the gauge number, the thicker the nail, Fig. 5-6. Some of the more common sizes with the approximate count of nails per pound are shown in Fig. 5-7 below.

Common wire nails and box nails are similar in length but the box nail has a thinner shank. The common nail is used in rough framing work. The box nail is used for toenailing because its thinner shank is less likely to split the wood.

Finishing nails are made of finer wire than common nails and have smaller heads. A casing nail is similar to a finishing nail but with a conical head. Both are used in finish work since their heads can be sunk below the surface of the wood and then filled over. The casing nail holds better than the finishing nail. Like the box nail, the casing nail is made of a noncorroding metal.

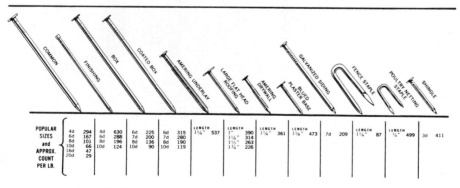

Fig. 5-7 Commonly used nails; all counts are for uncoated nails.

Construction Fasteners

Nailing

The length of a nail should be at least three times the thickness of the material being fastened, with a few exceptions. For better holding power, drive nails at a slight angle toward each other. Never drive a bent nail and avoid driving more than one nail in the same grain line near the end of a board. When toenailing, stagger opposing nails so they pass each other. For greater holding power, use annular- or spiral-shank nails. Fig. 5-8.

Common nails are driven until the head is flush with the work. Finishing nails are driven so that the head protrudes slightly; then they are driven home with a nail set. Nails for plywood are selected according to its thickness. For 3/4" plywood, use 6d casing nails or 8d finishing nails; and for 5/8" plywood, use 6d casing or 8d finishing. For 1/2" plywood, use 4d casing or 6d finishing; and for 3/8" plywood, use 3d casing or 4d finishing. For 1/4" plywood, use 3/4" or 1" brads, 3d finishing nails or 1" blue lath nails. (Brads are small finishing nails which measure anywhere from 1/2 to 1-1/2 inches in length.) Substitute casing for finishing nails when a heavier nail is needed.

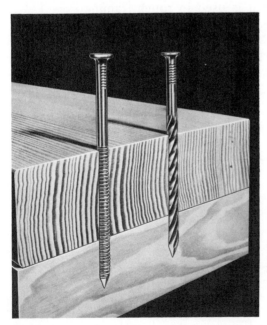

Fig. 5-8 Spiral- and annular-shank nails have high withdrawal resistance. *(Keystone Steel & Wire)*

Fig. 5-9 Proper way to drive nails into masonry with protection for head and eyes and proper hammer. *(Hand Tools Institute)*

Portable *air-* and *electric-driven nailing machines* are also made for various fastening jobs. They take much of the time and labor out of quantity nailing. Some models can drive up to 300 nails per loading and can do both flat and toe-nailing. Nailing machines are also used to secure laminated beams. The ram in the nailing head squeezes the boards together as the nails are driven into the wood.

SAFETY TIPS WHEN NAILING

* Always wear goggles when driving hardened nails.
* Always be sure to use the proper hammer for the job at hand.
* Ordinary claw hammers can chip when driving masonry nails causing injury to user; instead, use a heavy hammer as shown in Fig. 5-9.
* Hammers with broken handles or otherwise damaged, should not be used.

Screws

Screws have greater holding power than nails. But they are not generally used in construction except for special work because they are more expensive than nails. They are used for some interior work, however; and are made of steel, aluminum, brass, copper, and other materials.

Flat head, round head, and oval head screws are fairly common. Heads may be slotted or cross-slotted (Phillips). See Fig. 5-10.

Wood screws are identified by length, gauge and head type. Thus 1-1/4 - 4 FH indicates a 1-1/4-inch long screw which is 4 gauge in diameter with a flat head. They are made in sizes from 1/4 inch to 6 inches and in gauges from 0 to 24. It will be noted that each gauge rating increases or decreases by thirteen thousandths of an inch (.013). Up to one inch, screw lengths increase by eighths; from one to three inches, they increase by quarters, and from three to six inches they

Fig. 5-10 Two types of screw-head slots: *A*, slotted; *B*, cross-slotted *(Phillips)*.

Construction Fasteners

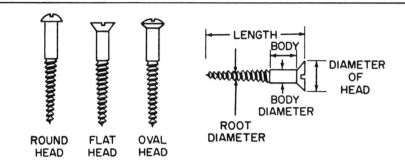

ROUND HEAD FLAT HEAD OVAL HEAD

Fig. 5-11 Three common types of wood screws used in house construction.

increase by half inches. Fig. 5-11 shows three types of screws and their nomenclature.

To properly drive a screw, drill the wood with a pilot and starter hole, as shown in Fig. 5-12. Lubricating the screw threads first with wax or soap is helpful when the screw is to go into hardwood. Make sure that the screws penetrate two-thirds into the block receiving the point, Fig. 5-13. When you choose a screwdriver, be sure it is the correct size: the length and shape should fit the work, and the tip should fit the screw slot. That is, the tip must not be wider than the screw head.

Use *sheet metal screws* for joining sheet metal to sheet metal, or sheet metal to wood. They have full-length threads. Be sure they also have a sharp point when you use them in wood, as in Fig. 5-14. The heads may be flat, oval, pan, truss or round.

When driving numerous screws, you may install a screwdriving

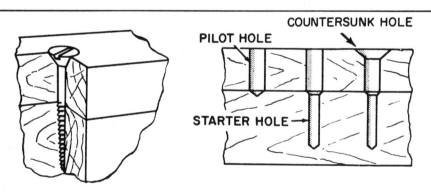

Fig. 5-12 Proper method of sinking a screw.

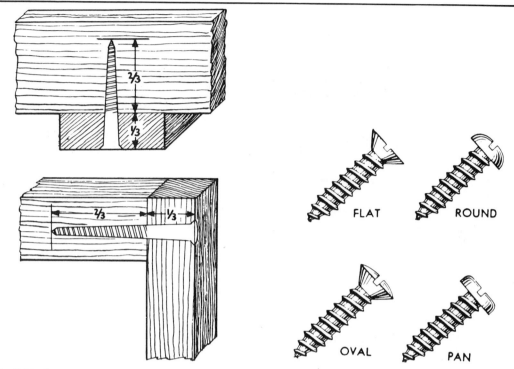

Fig. 5-13 Screws should penetrate the second block to at least two-thirds of their length.

Fig. 5-14 Sheet-metal screws with different head shapes.

attachment on the portable drill. Basically such attachments are speed reducers made to rotate either forward or in reverse. You can set a variable speed drill at slow speeds to serve the same purpose.

Bolts

Lag screws (also called lag bolts) are often used in construction. They are a heavy-duty version of a wood screw with square or hexagonal heads. Drive them with a wrench. Lag screws are available up to one inch in diameter and up to 16-inches long.

Carriage bolts have coarse machine threads and a smooth round head. The underside of the head has a square neck which keeps the bolt from turning when the nut is tightened. Other variations with ribbed and finned neck are shown in Fig. 5-15.

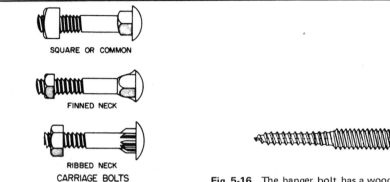

SQUARE OR COMMON

FINNED NECK

RIBBED NECK

CARRIAGE BOLTS

Fig. 5-15 Three types of carriage bolts used for fastening.

Fig. 5-16 The hanger bolt has a wood screw at one end and a machine screw at the opposite end.

Hanger bolts are threaded at each end. One end is threaded like a wood screw; the other end has a machine thread and accepts a nut, Fig. 5-16.

Masonry Fasteners

Wood may be fastened to masonry in various ways. *Hardened nails* are made with various shank styles for driving into concrete, brick, stone and other hard materials. Fig. 5-17 shows two types of masonry nails. Smooth-shank nails are for temporary fastening, while those with knurled-thread shanks provide cutting action and permanent, high holding power.

CAUTION: When driving hardened nails, you must use safety goggles. The masonry must also be predrilled to 3/4-inch depth before you drive the nails. Be sure that the nail is the proper length. It must not penetrate the concrete more than 3/4 of an inch.

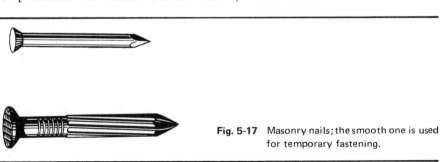

Fig. 5-17 Masonry nails; the smooth one is used for temporary fastening.

Fig. 5-18 Special tool uses explosive charge to drive masonry fastener. *(Bostitch)*

Employ ***threaded studs*** and ***drive pins*** to fasten wood to concrete. They may be hand or power driven. The power-driven units, called forced-entry tools, use 22-caliber cartridges to drive the fastener. Various power loads are available to obtain correct fastener penetration. Fig. 5-18 shows a tool being used to fasten furring to a concrete wall. Always wear goggles and a protective helmet when you work with forced-entry tools.

Framing Anchors

Metal framing anchors greatly increase the strength of structural framework. They eliminate the need for toenailing, notching, drilling and shimming. Usually made of 18-gauge zinc coated steel and in many shapes and sizes, they are used in all kinds of framing. Some are preformed, others can be shaped on the job. Fig. 5-19 shows some of the anchors employed in various stages of construction. They are widely used in post-and-beam construction.

Construction Fasteners

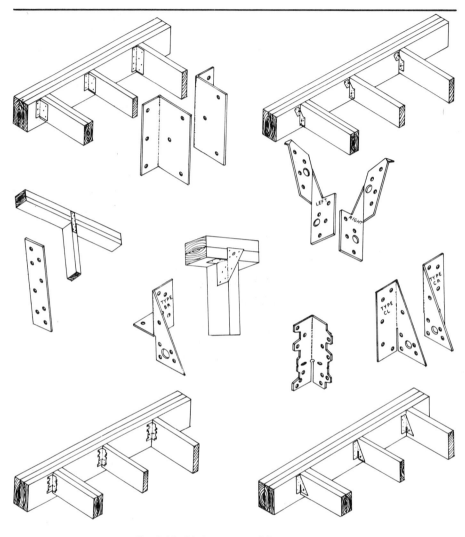

Fig. 5-19 Various types of framing anchors.

Anchor Bolts

Many types of anchoring devices are made for fastening objects to solid and hollow walls. Some are made to expand and bear against the sides of the opening; others expand after they pass through an opening. Various types are illustrated in Fig. 5-20.

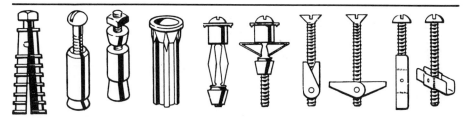

Fig. 5-20 Fasteners used for solid and hollow walls.

GLOSSARY

annular ring nail: a nail with grooved shank to increase its holding power.

casing nail: a nail similar to a finishing nail but with a conical head.

cement-coated nail: a rosin coating applied to nails to increase their holding power.

gauge: a standard of measurement for wire and sheet metal thicknesses.

hot-dipped galvanizing: a method of coating metals with zinc to prevent rust.

penny: a nail size, also expressed as the letter d; at one time it meant the cost in English pence or pennies for 100 nails of a particular size.

pilot hole: a preliminary hole drilled in wood to receive a screw or nail.

6

Scaffolding for Carpentry

Scaffolds are temporary platform structures used when the working level is above the normal reach of men on the ground. They are sometimes called stages. They may be simple structures built on the job by the carpenter, or they may be the manufactured type. Regardless of the type used, the scaffold must be safe. It should support the workers, tools and materials in a way that will permit construction to proceed in a normal and safe manner.

Made-on-Site Scaffolds

The simplest scaffold is made of a pair of sawhorses with a platform stretched between them. This is considered a *trestle-type scaffold*. The platform must be of sufficient size for proper support. Planks used for this type of scaffold should be sound and free of defects, and are generally 2 × 10's and 10-feet long. If longer boards are used, they must be supported in the center as well as at the ends.

A *double-pole wood scaffold* is shown in Fig. 6-1. It is free standing and, if used on soil, is made to rest on wood footings. Footings should be at least 2″ thick and a minimum of 6″ wide. Uprights (poles) must be clean lumber, 2 × 4's or heavier, and straight-grained. Ledgers are 2 × 6's and braces 1 × 6. The planks should be a minimum of 2 × 10, guard rails 2 × 4, and ties 2 × 6.

The *single-pole scaffold* is supported in part by the building. Ledgers rest on 2 × 6 blocks at least 12 inches long. They are notched and secured to the wall.

Swinging scaffolds are suspended from above. They hang from heavy lines and are generally used by painters.

Ready-Made Scaffolds

While many carpenters still build their own scaffolds, the trend is toward manufactured units. These *tubular metal scaffolds* are well-

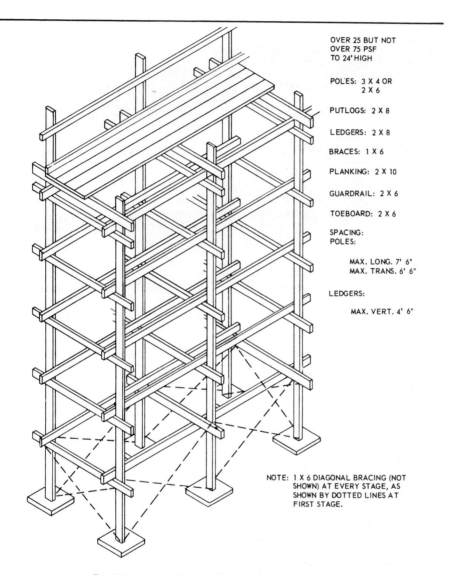

OVER 25 BUT NOT
OVER 75 PSF
TO 24' HIGH

POLES: 3 X 4 OR
        2 X 6

PUTLOGS: 2 X 8

LEDGERS: 2 X 8

BRACES: 1 X 6

PLANKING: 2 X 10

GUARDRAIL: 2 X 6

TOEBOARD: 2 X 6

SPACING:
POLES:

    MAX. LONG. 7' 6"
    MAX. TRANS. 6' 6"

LEDGERS:

    MAX. VERT. 4' 6"

NOTE: 1 X 6 DIAGONAL BRACING (NOT
SHOWN) AT EVERY STAGE, AS
SHOWN BY DOTTED LINES AT
FIRST STAGE.

Fig. 6-1 A job-built scaffold for house construction.

designed and made for maximum efficiency and safety. Fig. 6-2 shows
one type used by masons. It is fitted with sidewall brackets which
support the workers. The upper section is used to support materials.
The interlocking prefabricated sections of many ready-mades greatly
reduce assembly and dismantling time. Sections may be added vertically
or horizontally as required. They are available with adjustable legs, and

Scaffolding for Carpentry

Fig. 6-2 Standard frame fitted with brackets which supports the workers.

the frames are strengthened with diagonal braces. Sizes vary from 2 to 5 feet in width and from 3 to 10 feet in height. Special fasteners are used to secure braces to frames.

Various types of metal *sidewall bracket scaffolds* are made. They are attached to the building frame by nailing or other means. They take up little space, are quickly erected and are economical. See Fig. 6-3.

Be especially careful when you fasten this type since the bracket is held by the nail head. Therefore do not use nails with broken heads. They can be dangerous and unsafe. Another type of bracket

Fig. 6-3 Sidewall brackets are quickly attached to the building frame.

Fig. 6-4 Work platform is raised by standing on foot lever.

Fig. 6-5 Adjustable trestle jack. *(Patent Scaffold Co.)*

is held by hooking around studs; it requires that a hole be cut in the sheathing. Still others are fastened to crosspieces nailed to the inside studs. You must bore holes through the sheathing for them. Some bracket scaffolds provide for guard rails. Their installation is highly recommended.

Work platforms are supported by ***post scaffolds***. They can be raised by a foot-operated lever. (Fig. 6-4). The mechanism is supported by 4 × 4 posts and held at the top with braces.

Generally used for interior work, ***trestle jacks*** are made of steel and may or may not be adjustable. Fig. 6-5 shows an adjustable type with ledger clamps. Special locking mechanism assures safety.

SCAFFOLDING SAFETY RULES

Safety rules must be observed constantly when working on scaffolds. The following precautions should be taken:

* Inspect scaffolds daily before use. Never use equipment that is damaged.
* Follow all codes pertaining to scaffolding.
* Use adequate sills or pads under scaffold posts.
* Plumb and level the scaffold as it is being erected.
* Fasten all braces securely.
* Do not climb cross braces.
* Use guard rails and toe boards if required.
* Use caution when working near power lines. Consult power company for advice.
* Do not use ladders or other makeshift devices to increase the height of a scaffold.
* Never overload scaffolds.
* For planks, use lumber that is graded as scaffold plank.
* Planks should overlap ends by 12 inches and extend 6 inches beyond the scaffold supports.
* Secure planks to scaffolding when necessary.
* Do not ride rolling scaffolds.
* Remove all material before moving a rolling scaffold.
* Caster brakes must be applied except when moving the scaffolds.

Ladders

A number of different ladders are employed by carpenters. Some common types are shown in Fig. 6-6.

Straight ladders are sometimes built on the job using 2 × 6 and 1 × 3 stock. For safety's sake, it is best to use approved commercial ladders. Straight or single ladders are made up to 30-feet long.

Stepladders are self-supporting and usually fold flat for storage. They ordinarily have flat, broad steps instead of rungs, are made of various materials, and come in many sizes up to 20-feet high.

Extension ladders have two sliding sections which are adjustable to different heights. The maximum extension is 60 feet. The ladder is constructed so that the rungs of overlapping parts are opposite each other. On slippery surfaces install non-slipping bases to prevent slippage. Two types of non-slip bases are shown in Fig. 6-7.

Ladder jacks are available for use on straight ladders. In pairs with identical ladders you can use them to make up lightweight scaffolds. Fig. 6-8 shows one type that is adjustable to almost any slant.

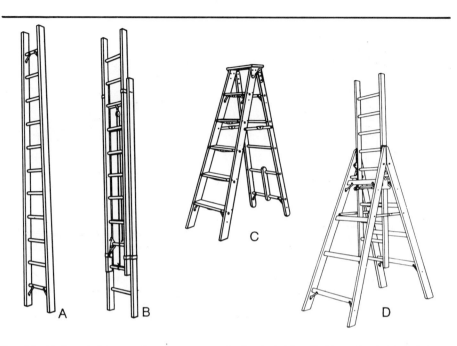

Fig. 6-6 Various ladders used by carpenters. *A*: Single ladder; *B*: Extension ladder; *C*: Step ladder; *D*: Trestle ladder. *(A. W. Flint Co.)*

Fig. 6-7 Safety shoes for ladders: **Left,** combination; **Right,** spur. *(A. W. Flint Co.)*

Fig. 6-8 Ladder jacks used in pairs are adjustable and can be used either inside or outside of ladder. *(Patent Scaffold Co.)*

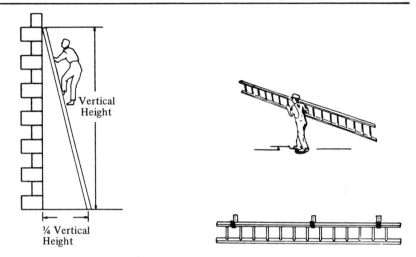

Fig. 6-9 Ladders should be placed so that the base distance to the wall is one-fourth their vertical height.

Fig. 6-10 Proper way to carry and store a ladder. *(American Ladder Institute)*

When erecting straight or extension ladders, place them at the proper angle, as explained in the safety rules that follow. Fig. 6-9 shows the proper method of setting up a ladder.

Ladders should never be painted because paint can hide defects in the wood. To protect the wood surface, apply a clear finish such as linseed oil instead. Fig. 6-10 demonstrates the proper way to carry and store a ladder.

LADDER SAFETY RULES

* Inspect all ladders carefully before each use.
* Keep all fastenings tight and oil the moving parts often.
* Be sure the ladder stands on firm ground. When needed, use non-slip bases.
* Always face ladder when ascending or descending. Place the ladder close enough to the work that you won't have to overreach.
* Keep rungs free of dirt, grease and oil.
* Never stand on the top pail-rest or rear rungs of stepladders.
* Never place a ladder in front of doors or openings unless proper precautions are taken.

* Never stand on the top three rungs of straight or extension ladders.
* Place the ladder so that the distance from its base to the wall is one-fourth of the ladder's height.
* Overlap extension ladders at least 3 feet for 36-foot lengths, 4 feet for 48-foot lengths, and 5 feet for 60-foot lengths. Install the sections so that the upper section is outermost.
* Be sure that the locks on extension ladders are hooked securely before climbing. Never extend a ladder while standing on it.
* Use extreme care when working near electric wires and equipment.
* If positioned for climbing onto a roof, a ladder must extend 3 feet above the roof.
* Never overload a ladder; and remember that standard ladders are made to support only one person. Special ladders are available for supporting two people.
* Never use a ladder in the horizontal position.
* Store ladders properly. A cool, dry, ventilated area is best. Hang the ladder so it won't sag.

GLOSSARY

guard rail: the horizontal member at the rear (outside edge) of a scaffold and designed to protect the workers from falling backwards. It is placed 42 inches above the platform.

ladder jack: a scaffold support which hooks onto a ladder; it should never be used more than 20 feet above the ground.

ledger: the horizontal support for a scaffold platform.

pad: a rectangular block placed under the posts of scaffolding to prevent the posts from sinking into the ground.

rung: the successive crosspieces of a ladder on which the foot is placed as one climbs.

scaffold: a work platform erected above ground level.

scaffold nail: a double-headed nail which can easily be removed and reused.

sidewall bracket: a platform support for a scaffold which attaches to a wall.

stage: another name for scaffolding.

stepladder: a self-supporting structure for climbing; it usually folds flat for storage.

trestle jack: a horizontal support for a platform.

7

Building Materials

The proper choice of materials for building construction is of primary importance. The quality of the materials selected helps determine the strength, durability, and appearance of the finished building. It is not enough for the modern carpenter to be skilled only in the use of tools and woodworking techniques. You must also try to be familiar with all the building products offered in today's market. You ought to know the physical properties of their materials and, in some instances, the chemical properties as well. Other factors you must carefully consider when selecting building materals are cost, availability, and appearance.

Just as the doctor must keep himself informed about new medicines, so you as a carpenter must be aware of the new building materials that are constantly being developed. It would be wise to subscribe to some trade magazines published for and by the building industry. Remember that the well-informed carpenter is a better carpenter than the uninformed one.

Wood—The Prime Building Material

Wood is probably the most important and most widely used material in the building industry. In the United States, about 25 billion board-feet of wood products are used each year by the construction industry. Most of this is used for homes and farm buildings, Fig. 7-1. Additionally, a like amount is used to produce panel products such as hardboard, particleboard, and structural insulating board, Fig. 7-2. The logs which are cut in the forests throughout the country are processed into lumber in saw mills. Huge circular saws and bandsaws are used to reduce the logs into usable lumber.

Physical Properties of Wood

Since lumber plays such an important role in the building industry, the carpenter needs a good working knowledge of the physical

Building Materials

Fig. 7-1 Wood is widely used in both home and farm building construction. *(Weyerhaeuser Co.)*

Fig. 7-2 Insulation board is manufactured from wood fibers. *(Simpson)*

characteristics of wood from its growth to its final use.

The cross section of a tree trunk reveals the familiar growth rings, Fig. 7-3. The outer layer or bark is separated from the heartwood by a light band called the sapwood. Between sapwood and bark there is a layer called the cambium. This thin tissue layer consists of minute living cells which can be seen under a microscope. The cambium continuously produces new wood and bark cells. The inner bark carries food from the leaves to the growing parts of the tree. The outer bark consists of dead cells and forms its protective layer.

The sapwood carries a watery fluid called sap from the roots to the leaves. There the sap combines with the carbon dioxide in the air.

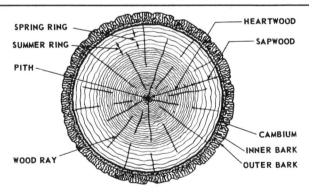

Fig. 7-3 Cross section of tree trunk.

Sunlight converts these organic elements into food which then descends to all parts of the tree. The sapwood layer usually varies from 1-1/2" to 2" thick. Some species of trees, such as yellow and ponderosa pine, may have a sapwood layer of from 3" to 6" thick. As the sapwood becomes inactive, it forms the heartwood. Sapwood made into lumber must be treated if it is to be used outdoors.

Wood rays (medular rays) connect the various layers from pith to bark. The pith is the soft center part of the tree.

Wood cells are elongated and pointed at the ends. Under a microscope hardwood cells are seen to be about 1/25th of an inch long, and softwood cells from 1/8th to 1/3rd of an inch in length.

Hardwoods and Softwoods

In forestry, the terms "hardwood" and "softwood" have no bearing on the hardness or softness of wood. Some hardwoods are softer than pine. Hardwoods are the broadleaved deciduous trees—trees which shed their leaves in the fall. Softwood trees are the evergreens, such as pine and cedar. They are also known as conifers because they are cone-bearing.

The common hardwoods include the ash, aspen, basswood, beech, birch, cherry, chestnut, elm, hickory, maple, oak, walnut, willow and poplar. The softwoods include the Douglas fir, white fir, hemlock, larch, pine, red cedar, redwood and spruce.

Identification of Wood

Some woods are easily identified because of distinctive characteristics like color, odor and grain. Black walnut is chocolate brown, often with purple streaks; cedar has a distinct odor. If possible, study actual specimens of unseasoned and seasoned woods and both as rough cut and dressed lumber.

Cutting Lumber

There are two basic ways of cutting lumber. The log can be cut with the annual growth rings forming an angle of between 45-degrees to 90-degrees with the faces of the board. This is called quartersawed if it

Building Materials

is hardwood, or edge-grained if it is softwood. The cut can also be made with the annular rings forming an angle of less than 45-degrees to the board surface. This is called plain-sawed if hardwood, or flat-grained if the lumber is softwood. See Fig. 7-4.

As a rule, plain-sawed or flat-grained wood is cheaper to produce because it is easier to cut and less waste is involved. The grain patterns are also very pleasing. It shrinks and swells less in thickness. The quartersawed lumber shrinks and swells less in the width. It is less likely to cup and twist and wears more evenly.

Moisture Content of Lumber

All wood retains a considerable amount of moisture after cutting. In order to be used commercially, that moisture content must be reduced. Generally, a moisture content of 15 percent is considered average and acceptable for framing and exterior use, and of 8 percent for interior and furniture use. The moisture content is expressed as a percentage of the oven-dry weight of the wood.

Shrinkage of Wood

Wood cut from trees contains from 30 to 300 percent water. Part of this moisture is retained as *free water* in the cell cavities and intercellular spaces of the wood. The rest of the water is absorbed into the capillaries of the wood fibers and ray cells. This *absorbed water* content is very important as it directly influences wood shrinkage and swelling. When all the free water is removed from the wood leaving only the absorbed water, the wood is said to have reached its fiber-saturation point. That is about a 30 percent moisture content for most woods. Where the moisture content is reduced below this point, shrinkage or seasoning takes place. Fig. 7-5 shows how shrinkage affects a log. If the moisture content is restored, the wood regains its original shape.

Wood shrinkage is directly proportional to the percentage loss of water below the fiber-saturation point. For every one-degree loss in moisture below 30 percent, there is a corresponding 1/30th shrinkage of the total possible shrinkage. At 15 percent moisture content, the wood has undergone 1/2 of the total shrinkage possible. Conversely, when moisture is added to wood the rate of swelling is proportional to the percentage of moisture added.

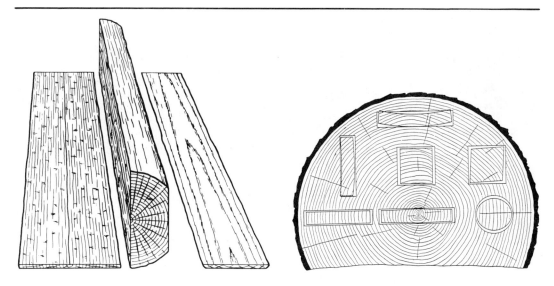

Fig. 7-4 Boards cut from a log: **Left**, quarter-sawed; **Right**, plain-sawed. *(Forest Products Lab.)*

Fig. 7-5 How wood shrinks in different parts of a log. *(Forest Products Lab.)*

Most shrinkage takes place in the direction of the annual rings; very little across them. Longitudinal shrinkage along the grain is minimal.

Determining Moisture Content

There are several ways of determining the moisture content of wood. Experienced carpenters can recognize well seasoned, dry lumber by its weight, odor and feel. Dry lumber is lighter than green wood and does not smell damp or shrink. Green wood is moist or damp to the touch. To determine exact moisture content, electrical meters are used. Fig. 7-6.

Seasoning Lumber

The excess moisture in lumber is reduced by a process called seasoning. Two seasoning methods are commonly used—air drying and kiln drying. In *air drying*, the lumber is stacked outdoors in piles. Spaces are left between layers and piles so the air can circulate freely. See Fig. 7-7. There are several disadvantages to this otherwise

Building Materials

Fig. 7-6 Precise moisture content of wood is determined with electrical meters. *(Thomasville Furniture Industries)*

Fig. 7-7 Wood being seasoned out-of-doors—spaces between layers and piles permit air to circulate freely. *(Forest Products Lab.)*

economical method. It takes a long time for the wood to dry; and it is not practical in wet, humid or cold climates.

Kiln drying is done in huge ovens where temperature, humidity and air circulation are accurately controlled. The drying is faster and the desired moisture content can be controlled. Its chief disadvantage is that it costs much more than air drying.

Defects in Wood

Lumber sometimes has certain defects and blemishes. They are usually caused during the formation or growth of the wood. In some cases either the strength or appearance (or both) may be affected. Lumber grading specifies the allowable defects. A brief description of the most common defects in lumber follows:

Knots are formed by branches growing in the body of a tree.

Checks and *splits* are separations of the wood along the grain and are usually caused in drying. Splits run through the band; checks do not.

Shakes are like splits except that they run between growth rings.

Blue stain is a bluish stain caused by a fungus growth. It does not affect the strength but may be objectionable visually.

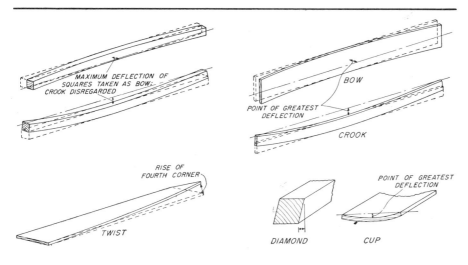

Fig. 7-8 Various kinds of warp. *(Forest Products Lab.)*

Wave is a defect, usually a depression, in the edge of a board, causing a board of timber to be less than its full width.

Warp is a distortion in a board which prevents it from lying flat or straight. It is usually caused by faulty seasoning. Several kinds of warp are shown in Fig. 7-8.

Pitch pocket is a cavity in the wood containing liquid or solidified pitch. This is commonly found in pine, Douglas fir, tamarack, and western larch.

Decay is deterioration of wood caused by fungus. The wood fibers become soft and spongy.

Lumber Grading

Lumber grading has been established by various associations of lumber manufacturers as well as by governmental agencies. The purpose of grading is to set quality control standards among mills manufacturing the same or similar woods. Member mills are closely supervised by qualified personnel. The use of official grade stamps on a piece of lumber guarantees the assigned grade. Fig. 7-9. The symbol at the lower left on this piece indicates that the lumber conforms to the Association's grading rules (Western Wood Products Association). The number at the upper left identifies the mill. Each mill has a permanent number.

Fig. 7-9 An official grade stamp used on lumber.

The symbol at the upper right identifies the graded lumber, and the species mark at the lower right identifies the wood.

Softwood Grades

The grading rules specify the poorest quality permitted in each grade of lumber. The grades for softwood are based on standards developed by the United States Department of Commerce. Grading terms and standards may vary among associations, however, depending on their interpretation of the standards. Softwood grading includes yard lumber, structural lumber and shop lumber.

Yard lumber is widely used in ordinary construction and general building purposes.

Structural lumber is used for joists, planks, beams, posts, stringers, and timbers—wherever working stress qualities are required.

Factory and shop lumber is used in factories and mills. It is graded according to the amount of clean lumber that can be used for sashes, doors, moldings and other shop uses.

Hardwood Grades

The standards for hardwood grades are established by the National Hardwood Lumber Association. The five basic grades are Firsts & Seconds (FAS), select, No. 1 common, No. 2 common, and No. 3 common. Firsts & Seconds are the best grade of hardwood boards; they must be at least 6-inches wide by 8-feet long. Selects are at least 4-inches wide by 6-feet long. The three number grades designate boards of lower quality and are rated by the number of checks, knots, and other defects present in the wood. No. 3 common is the lowest grade.

Lumber Sizes

Lumber is usually marketed by its original sawed size—(nominal size). After seasoning and surfacing, the lumber is reduced to its actual or dressed size. When 2 × 4 lumber (nominal size), is purchased at a lumber yard it actually measures 1-1/2″ × 3-1/2″ (dressed size).

Dressed lumber is designated according to the number of faces surfaced or planed smooth. For instance: S2S indicates that the lumber was surfaced on two sides. When all four surfaces are dressed (2 edges and 2 sides), the designation is S4S. Figs. 7-10 and 7-11 list the nominal and dressed sizes of various grades of lumber. Lengths are as stated.

Board Measure

Board measure is important in calculating the cost and quantity of lumber used in construction and carpentry work. Since lumber is commonly purchased at the lumber yard by its nominal size, the board measure is calculated using the nominal dimensions instead of the actual or dressed sizes.

Lumber is measured by the board-foot unit: the basic board-foot measures a piece 1-inch by 12-inches square. Measuring a 1-inch by 6-inch by 24-inch piece also makes one board foot. Likewise, a 2-inch by 6-inch by 12-inch piece of wood is one board foot. See Fig. 7-12. The following formula is used to find the amount of board feet in a piece of lumber:

$$\frac{T \times W \times L}{12} = \text{board feet}$$

T = thickness of wood in inches   L = length of wood in *feet*

W = width of wood in inches

Problem: Find the board feet in a piece of lumber measuring 1-inch × 4-inches × 12-feet

Solution: $\dfrac{1 \times 4 \times 12}{12} = \dfrac{48}{12} = 4$ board feet

Where the actual size of lumber to be used is less than one-inch thick, calculate the board feet using the one-inch nominal size. For

Building Materials

thermal conductivity

The relatively low thermal conductivity or "k", of Western Softwoods provides for significant insulation value. "k" is the amount of heat (Btu's) transferred in one hour through one square foot of material one inch thick with a difference in temperature of 1° F.

The thermal conductivity of wood increases with increased moisture content and with increased density. The "k" values for the Western Woods are shown in the table below.

Species	"k" Value*
Douglas Fir—Larch	1.05
Douglas Fir South	.96
Hem-Fir	.94
Mountain Hemlock	.94
Subalpine Fir	.74
Engelmann Spruce	.75
Lodgepole Pine	.88
Ponderosa Pine—Sugar Pine	.88
Idaho White Pine	.85
Western Cedars	.81

*"k" values shown are for wood at 12 percent moisture content. For other moisture contents, there is a change in "k" of approximately .01 for each 1 percent moisture content difference.

product classification

	thickness in.	width in.		thickness in.	width in.
board lumber	1"	2" or more	beams & stringers	5" and thicker	more than 2" greater than thickness
light framing	2" to 4"	2" to 4"	posts & timbers	5" x 5" and larger	not more than 2" greater than thickness
studs	2" to 4"	2" to 4" 10' and shorter	decking	2" to 4"	4" to 12" wide
structural light framing	2" to 4"	2" to 4"	siding		thickness expressed by dimension of butt edge
joists & planks	2" to 4"	6" and wider	mouldings		size at thickest and widest points

Standard lengths of lumber generally are 6 feet and longer in multiples of 1'

standard lumber sizes / nominal, dressed, based on WWPA 1970 rules

Product	Description	Nominal Size		Dressed Dimensions		Lengths Ft.
		Thickness In.	Width In.	Thicknesses and Widths In. Surfaced Dry	Surfaced Unseasoned	
DIMENSION	S4S	2 3 4	2 3 4 6 8 10 12 Over 12	1-1/2 2-1/2 3-1/2 5-1/2 7-1/4 9-1/4 11-1/4 Off 3/4	1-9/16 2-9/16 3-9/16 5-5/8 7-1/2 9-1/2 11-1/2 Off 1/2	6 ft. and longer in multiples of 1'
SCAFFOLD PLANK	Rough Full Sawn or S4S	1 1/4 & Thicker	8 and Wider	Same	Same	6 ft. and longer in multiples of 1'
TIMBERS	Rough or S4S	5 and Larger		Thickness In. 1/2 Off Nominal	Width In.	6 ft. and longer in multiples of 1'
		Nominal Size		Dressed Dimensions		
		Thickness In.	Width In.	Thickness In.	Width In.	Lengths Ft.
DECKING Decking is usually surfaced to single T&G in 2" thickness and double T&G in 3" and 4" thicknesses	2" Single T&G	2	6 8 10 12	1 1/2	5 6 3/4 8 3/4 10 3/4	6 ft. and longer in multiples of 1'
	3" and 4" Double T&G	3 4	6	2 1/2 3 1/2	5 1/4	
FLOORING	(D & M), (S2S & CM)	3/8 1/2 5/8 1 1/4 1 1/2	2 3 4 5 6	5/16 7/16 9/16 3/4 1 1 1/4	1 1/8 2 1/8 3 1/8 4 1/8 5 1/8	4 ft. and longer in multiples of 1'
CEILING AND PARTITION	(S2S & CM)	3/8 1/2 5/8 3/4	3 4 5 6	5/16 7/16 9/16 11/16	2 1/8 3 1/8 4 1/8 5 1/8	4 ft. and longer in multiples of 1'
FACTORY AND SHOP LUMBER	S2S	1 (4/4) 1 1/4 (5/4) 1 1/2 (6/4) 1 3/4 (7/4) 2 (8/4) 2 1/2 (10/4) 3 (12/4) 4 (16/4)	5 and wider (4" and wider in 4/4 No. 1 Shop and 4/4 No. 2 Shop)	25/32 (4/4) 1 5/32 (5/4) 1 13/32 (6/4) 1 19/32 (7/4) 1 13/16 (8/4) 2 3/8 (10/4) 2 3/4 (12/4) 3 3/4 (16/4)	Usually sold random width	4 ft. and longer in multiples of 1'

ABBREVIATIONS
Abbreviated descriptions appearing in the size table are explained below.
S1S — Surfaced one side.
S2S — Surfaced two sides.
S4S — Surfaced four sides.
S1S1E — Surfaced one side, one edge.
S1S2E — Surfaced one side, two edges.
CM — Center matched.
D & M — Dressed and matched.
T & G — Tongue and grooved.
EV1S — Edge vee on one side.
S1E — Surfaced one edge.

Fig. 7-10 Lumber sizes.

coverage estimator

The following estimator provides factors for determining the exact amount of material needed for the five basic types of wood paneling. Multiply square footage to be covered by factor (length x width x factor).

	Nominal Size	WIDTH Overall	WIDTH Face	AREA FACTOR*
SHIPLAP	1 x 6	5½	5⅛	1.17
	1 x 8	7¼	6⅞	1.16
	1 x 10	9¼	8⅞	1.13
	1 x 12	11¼	10⅞	1.10
TONGUE AND GROOVE	1 x 4	3⅜	3⅛	1.28
	1 x 6	5⅜	5⅛	1.17
	1 x 8	7⅛	6⅞	1.16
	1 x 10	9⅛	8⅞	1.13
	1 x 12	11⅛	10⅞	1.10
S4S	1 x 4	3½	3½	1.14
	1 x 6	5½	5½	1.09
	1 x 8	7¼	7¼	1.10
	1 x 10	9¼	9¼	1.08
	1 x 12	11¼	11¼	1.07

	Nominal Size	WIDTH Overall	WIDTH Face	AREA FACTOR*
PANELING PATTERNS	1 x 6	5 7/16	5 1/16	1.19
	1 x 8	7⅛	6¾	1.19
	1 x 10	9⅛	8¾	1.14
	1 x 12	11⅛	10¾	1.12
BEVEL SIDING	1 x 4	3½	3½	1.60
	1 x 6	5½	5½	1.33
	1 x 8	7¼	7¼	1.28
	1 x 10	9¼	9¼	1.21
	1 x 12	11¼	11¼	1.17

*Allowance for trim and waste should be added.

Product	Description	Nominal Size Thickness In.	Nominal Size Width In.	Dressed Dimensions Thickness In.	Dressed Dimensions Width In.	Lengths Ft.
SELECTS AND COMMONS S-DRY	S1S, S2S, S4S, S1S1E, S1S2E...	4/4 5/4 6/4 7/4 8/4 9/4 10/4 11/4 12/4 16/4	2 3 4 5 6 7 8 and wider	¾ 1 5/32 1 13/32 1 19/32 1 13/16 2 3/32 2⅜ 2 9/16 2¾ 3¾	1½ 2½ 3½ 4½ 5½ 6½ ¾ Off nominal	6 ft. and longer in multiples of 1′
FINISH AND BOARDS S-DRY	S1S, S2S, S4S, S1S1E, S1S2E ...	⅜ ½ ⅝ ¾ 1 1¼ 1½ 1¾ 2 2½ 3 3½ 4	2 3 4 5 6 7 8 and wider	5/16 7/16 9/16 ⅝ ¾ 1 1¼ 1⅜ 1½ 2 2½ 3 3½	1½ 2½ 3½ 4½ 5½ 6½ ¾ off nominal	3′ and longer. In Superior grade, 3% of 3′ and 4′ and 7% of 5′ and 6′ are permitted. In Prime grade, 20% of 3′ to 6′ is permitted.
RUSTIC AND DROP SIDING	(D & M) .... If ⅜″ or ½″ T & G specified, same over-all widths apply. (Shiplapped, ⅜-in. or ½-in. lap) ..	1	6 8 10 12	23/32	5⅜ 7⅛ 9⅛ 11⅛	4 ft. and longer in multiples of 1′
PANELING AND SIDING	T&G or Shiplap.................	1	6 8 10 12	23/32	5 7/16 7⅛ 9⅛ 11⅛	4 ft. and longer in multiples of 1′
CEILING AND PARTITION	T&G	⅝ 1	4 6	9/16 23/32	3⅜ 5⅜	4 ft. and longer in multiples of 1′
BEVEL SIDING	Bevel or Bungalow Siding....... **Western Red Cedar Bevel Siding available in ½″, ⅝″, ¾″-nominal thickness. Corresponding thick edge is 15/32″, 9/16″ and ¾″. Widths for 8″ and wider, ½″ off nominal.**	½ ¾	4 5 6 8 10 12	15/32 butt, 3/16 tip ¾ butt, 3/16 tip	3½ 4½ 5½ 7¼ 9¼ 11¼	3 ft. and longer in multiples of 1′ 3 ft. and longer in multiples of 1′
				Surfaced Dry / Green	Surfaced Dry / Green	
STRESS RATED BOARDS	S1S, S2S, S4S, S1S1E, S1S2E...	1 1¼ 1½	2 3 4 5 6 7 8 and Wider	¾ / 25/32 1 / 1 1/32 1¼ / 1 9/32	1½ / 1 9/16 2½ / 2 9/16 3½ / 3 9/16 4½ / 4⅝ 5½ / 5⅝ 6½ / 6⅝ Off ¾ / Off ½	6 ft. and longer in multiples of 1′

See coverage estimator chart above for dressed Shiplap and Tongue and Groove (T&G) widths.

MINIMUM ROUGH SIZES Thicknesses and Widths Dry or Unseasoned All Lumber (S1E, S2E, S1S, S2S)
80% of the pieces in a shipment shall be at least ⅛″ thicker than the standard surfaced size, the remaining 20% at least 3/32″ thicker than the surfaced size. Widths shall be at least ⅛″ wider than standard surfaced widths.
When specified to be full sawn, lumber may not be manufactured to a size less than the size specified.

Fig. 7-11 Lumber sizes.

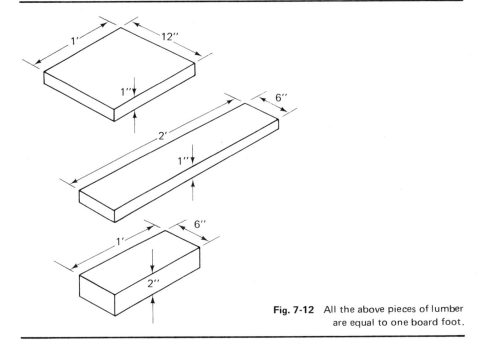

Fig. 7-12 All the above pieces of lumber
are equal to one board foot.

lumber more than one-inch which involves a fraction use the next larger
nominal size for board-feet calculations.

Plywood: Types & Use

Plywood consists of layers of veneer crossbanded and glued, and the
layers may be of any thickness up to 1/4 of an inch. The number of
layers or plies is always odd—3, 5, 7 and so on. The grain direction of
the alternate layers is laid at right angles. The grain direction of the
outer faces (front and back) are parallel to each other. The layers under
the face pieces are called crossbands; all other inside layers are called
the core.

In *veneer plywood* all the inner layers are generally as thick as the
face plies. In *lumber core plywood* the center ply is thicker than the
face plies and the core material is usually basswood or poplar. Other
core materials are also used in its construction, however, Fig. 7-13.

Plywood: Types & Use

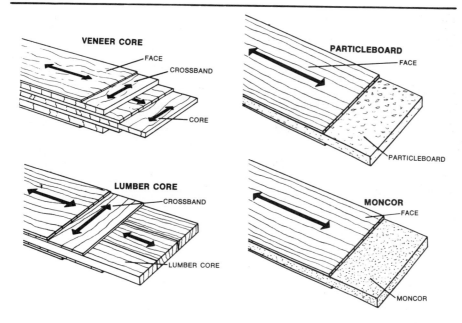

Fig. 7-13 Plywood is made with various core materials. *(Evans Products Co.)*

There are three principal methods of making plywood: the sheets may be rotary cut, flat sliced, or quarter sliced from the logs, Fig. 7-14.

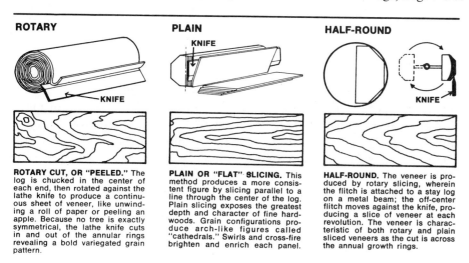

ROTARY

KNIFE

ROTARY CUT, OR "PEELED." The log is chucked in the center of each end, then rotated against the lathe knife to produce a continuous sheet of veneer, like unwinding a roll of paper or peeling an apple. Because no tree is exactly symmetrical, the lathe knife cuts in and out of the annular rings revealing a bold variegated grain pattern.

PLAIN

KNIFE

PLAIN OR "FLAT" SLICING. This method produces a more consistent figure by slicing parallel to a line through the center of the log. Plain slicing exposes the greatest depth and character of fine hardwoods. Grain configurations produce arch-like figures called "cathedrals." Swirls and cross-fire brighten and enrich each panel.

HALF-ROUND

KNIFE

HALF-ROUND. The veneer is produced by rotary slicing, wherein the flitch is attached to a stay log on a metal beam; the off-center flitch moves against the knife, producing a slice of veneer at each revolution. The veneer is characteristic of both rotary and plain sliced veneers as the cut is across the annual growth rings.

Fig. 7-14 Three methods of making plywood. *(Evans Products Co.)*

Building Materials

Fig. 7-15 Special machines repair defects in plywood veneer with patches as strong as the original wood. *(American Plywood Assoc.)*

Fig. 7-16 Veneers are glued and "layered up" to assemble a panel of proper thickness. *(American Plywood Assoc.)*

shows the three methods and the resulting grain patterns. Figs. 7-15 through 7-17 show various stages in the manufacture of plywood.

Plywood is made for both interior and exterior use. The glue used for interior plywood is usually only moisture-resistant. Consequently such plywood should be used where it will not be exposed to high moisture and humidity. Exterior plywood has a 100 percent waterproof glue line and can be used for all exposed applications. The veneers used for the inner plies of interior plywood may be of lower quality than those used in exterior products. Structural grades are always made with a waterproof glue line.

Standard Plywood Sizes: Plywood is available in widths of 36, 48, and 60 inches and in lengths from 60 to 144 inches. Other sizes are made on special order. Thicknesses range from 1/4" to 1-1/4" or more.

Softwood Plywood: It is widely used in the construction industry. Softwood plywood is manufactured according to standards set by the United States Department of Commerce which are described fully in U.S. Product Standard PS 1-66. It provides a uniform standard for designating types, grades, and sizes of construction and industrial plywood. The guide is available from the U.S. Government Printing Office and from the American Plywood Association.

Plywood Grades: In each type of plywood there are various appearance grades which are rated according to the veneers used for the face and back of the panel. They are identified by the letters N, A, B, C and D. Plywoods are classified structurally by groups. The strongest are

Fig. 7-17 Once veneer panels are assembled and cut to size, they are fed into a press where great heat pressure-bonds them. *(American Plywood Assoc.)*

Group I; the lower the group number, the stronger and stiffer the panel. Fig. 7-18 shows the appearance grade stamp usually found on the back of the panel. Also shown is an edge brand. Both markings indicate the plywood structural group and the recommended use. The index number to the left of the slash mark refers to the maximum allowable rafter spacing when the panel is to be used for roof decking; the number to

Designates the type of plywood
Exterior or Interior

Product Standard governing
manufacture

S T A N D A R D

32 / 16
INTERIOR
PS 1-66
000

TESTED
DFPA
QUALITY

Mill number

The sign of a quality tested
and inspected product

A-A · G-1 · INT-DFPA · PS 1-66

Fig. 7-18 Examples of typical plywood back and edge stamps.

the right of the slash specifies the floor-joint spacing when the panel is to be used for subflooring. If a zero appears to the right, the panel is *not* to be used for subflooring. These index numbers do not apply when the plywood is to be used for wall sheathing.

Hardwood Plywood: Hardwood plywoods are made in four types and are used mostly in the manufacture of furniture. Type I is made with waterproof adhesives and is not affected by exposure to water. Type II is made with a water-resistant glue line and while it is not waterproof, it can be used where high dampness and humidity are present. Type III is made with moisture-resistant glue. It can be used in areas of limited dampness and excessive humidity. Technical Type is similar in quality to Type I but differs in thickness and ply arrangement.

Some of the woods used in hardwood plywoods are birch, black walnut, cherry, gum, Lauan mahogany, maple and teak. The sheets are generally stocked in 4' X 8' panels but other sizes are available, Thicknesses range from 1/8 to one inch.

Veneer core plywood is made with 3, 5, 7, or 9 plies, with the layers glued as for softwood plywood.

Lumber core plywood is used widely for doors and in furniture manufacturing. The solid core is easier to work with, holds screws better and simplifies finishing. Fig. 7-19 shows cross section of veneer and lumber core plywood for comparison.

Particleboard-core plywood is similar to lumber core (review Fig. 7-13), except that the core is made with particleboard. It is highly stable and resists warping.

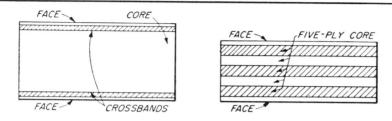

Fig. 7-19 **Left**, 5-ply lumber core plywood; **Right**, 7-ply veneer core plywood.

Overlay Plywood: Medium density overlay or MDO plywood panels are made with a resin-treated fiber facing. The smooth surface is an ideal base for painting. It is recommended for siding, signs, displays, and other outdoor uses.

High density overlay plywood or HDO is similar to MDO but the surface is very hard and smooth. It can be used as is, without painting. It has high resistance to chemicals and abrasion. It is used for concrete forms, signs, cabinets, counter tops and other hard-usage applications.

Hardboard

Hardboard is made of refined wood fibers which have been formed into sheets under heat and pressure. The two basic types are tempered and untempered.

The *tempered hardboard* contains oils and resins which make it hard and moisture resistant. It is made in various thicknesses from 1/12 inch to one inch and over. The face dimensions are either 4- or 5-foot widths; the lengths are 4, 5, 6, 7, 8, 9, 10, 11, 12, 16 and 18 feet. It is made with one side smooth or both sides smooth. It is available plain, patterned, embossed, perforated, or grooved. Decorative hardboard panels are available in 16-inch widths and 8-foot lengths. Installation is with concealed metal clips and adhesive. Hardboard is also used for siding, forms, and underlayment, cabinet backs and drawer bottoms. It is easily cut with woodworking tools.

Untempered hardboard is similar to the tempered material, except that it is softer and not as moisture resistant. Sizes are similar. It is widely used for paneling, underlayment, door panels, drawer bottoms and cabinet dividers. It has many other applications in homes and industry.

Building Materials

Particleboard

Particleboard is a construction material made into sheets by combining wood particles with resin under pressure. It is made in thicknesses of from 1/4" to over 1-7/8". It is usually supplied in 4' X 8' panels. In home construction it is used mainly for interior wall coverings and as an underlayer for resilient floor coverings. It is also used widely as a core for cabinet doors and countertops.

Sheet Metal

Sheet metal is used in house construction mainly for flashing, gutters and downspouts. The most common metals used are galvanized metal, aluminum, copper, stainless steel and terneplate.

Galvanized metal is sheet steel coated with zinc, either by electroplating or by hot-dipping. Zinc gives the steel a protective rust-proof coating. Two weights of galvanized metal are generally used; 26- and 28-gauge. Gutters should be 26 gauge. For other metal work, the 28 gauge is satisfactory.

Aluminum used for flashing should be a minimum of .019 in thickness and a minimum of .027 for gutters. Copper should be a minimum of .020 in thickness. Fasteners employed with the various kinds of sheet metal should be made of the same material. When dissimilar metals are used, galvanic action takes place and causes corrosion.

Adhesives

Modern adhesives are available for all types of gluing jobs, Fig. 7-20. Some adhesives are waterproof, others are highly heat resistant. Some require clamping while setting, others do not. For best results, select the most appropriate glue for the job at hand, and be sure to follow the manufacturer's instructions for application.

Some adhesives contain solvents which may be injurious to your health if improperly handled. If the instructions recommend using the product with adequate ventilation, *be sure you do so.*

Some adhesives are flammable and must be handled very carefully. For instance, vapors from contact cement are highly explosive. Use great care with any of these materials. All flames must be extin-

Fig. 7-20 Some adhesives used in construction work.

guished—pilot lights as well. Observe no smoking rules completely. Do not use electrical equipment while adhesive vapors are present since the spark from an electric switch can be enough to cause an explosion.

Brief descriptions of the more common adhesives used by the carpenter follow:

Contact cement is a liquid of creamy consistency. Apply it with a brush or spreader to both surfaces to be glued. Next allow it to air dry for about 15 minutes; then bring both surfaces together precisely since it bonds on contact. This cement is widely used for bonding laminates to countertops.

Liquid resin is a polyvinyl emulsion glue better known as "white" glue. It comes ready to use in squeeze bottles and in larger economical containers. It is a fast-setting non-waterproof glue suitable for wood, leather, cork and similar materials. It is not recommended in areas of high humidity or high temperature.

Plastic resin comes in powdered form and is mixed with water prior to use. It is water-, mold-, and rot-resistant. Use it for wood, particleboard, veneer and the like—always with clamps.

Powdered casein is a strong, water-resistant glue ideally suited for use on oily woods such as teak. It is mixed with water and sets at low temperatures.

Resorcinal is a waterproof glue used when structural strength is required. It is generally supplied in two parts which must be combined before use. It has a dark red color and requires clamping while setting.

Building Materials

Liquid glues are made from heads, skins and bones of fish. They provide a good strong joint suitable for furniture making. They are not recommended for use in areas of high dampness.

Mastics are tacky adhesives widely used for installing ceiling tiles, floor tiles, subflooring, wallboard, panelling and other building materials. Depending on the area of application, they may be troweled, brushed or applied in ribbon-like strips. The two basic types are latex and rubber resin. The resin type is usually furnished in tubes which fit standard caulking guns, Fig. 7-21.

Other building materials include glass, plastics, ceramics, asbestos, asphalt and steel. They are discussed in later chapters. Clear plastic sheets are widely used as glazing because they are practically unbreakable.

Fig. 7-21 Adhesive being used for laying a subfloor. *(Western Wood Products)*

GLOSSARY

bark: the outer layer of a tree trunk.

cambium: the growing part of a tree trunk.

checks: lengthwise cracks in a piece of wood.

contact cement: a liquid glue which bonds without clamping pressure; used widely with plastic laminates.

edge-grain: the grain produced when softwood is cut so that the annular rings form an angle greater than 45 degrees with the surface of the board.

flat-grain: the grain pattern formed when softwood is cut so that the annular rings form an angle of 45 degrees or less with the surface of the board.

hardwood: wood from a tree which has broad leaves; most are harder than softwoods, but not all.

heartwood: the inner part of a tree trunk, just inside the sapwood.

knots: defects caused by branches: small knots under 1/2 inch in diameter are known as pin knots; medium knots are 1/2 to 3/4 inches in diameter; large knots are over 1-1/2 inches in diameter.

overlay plywood: plywood with a resin-treated fiber facing.

particleboard: a highly compressed panel made of wood chips and resin.

pitch pocket: a cavity in wood containing pitch (resin).

pith: soft spongy wood at the center of a tree.

plain-sawed: hardwood cut with annular rings at an angle of 45 degrees or less with the surface of the board.

plies: the layers of wood in a sheet of plywood.

plywood: a sheet construction material made of three or more layers of wood with the adjacent layers crossgrained; the numbers of layers or plies is always an odd number.

quarter-sawed: hardwood cut so that annular rings form an angle greater than 45 degrees with the face of the board.

sapwood: layer of wood between the heartwood and bark of a tree.

shakes: not to be confused with shingles and shakes they are defects in wood, consisting of tiny splits between growth rings.

seasoned wood: wood which has had its moisture content reduced so that it is suitable for carpentry; standards for seasoning may vary according to kind of construction.

softwood: wood from a tree with needle-like leaves and cones; such as pine and spruce; all evergreens are softwood trees.

warp: a twist in a piece of wood.

Part Two

PRACTICAL CARPENTRY

8

Building Foundations

The foundation of a structure must support the weight of floor, wall and roof loads. Therefore a sound sturdy foundation is essential to good quality building construction. Although many types of foundations are in use in the building industry, we shall concern ourselves only with the most common types which are concrete, concrete block and wood.

Except when small amounts are required, concrete is generally supplied by ready-mix plants as in Fig. 8-1. The concrete is delivered to the job site in huge trucks, ready to pour.

Foundation walls rest on a footing of poured concrete which must be placed below the frost line in areas subject to freezing temperatures. The wall may enclose a full basement, or a crawl space, or it may support a slab on grade. Fig. 8-2 shows a typical foundation. The footing transmits the load to the soil.

Fig. 8-1 Concrete mixed in modern plants has exact proportions of aggregate, cement and water.

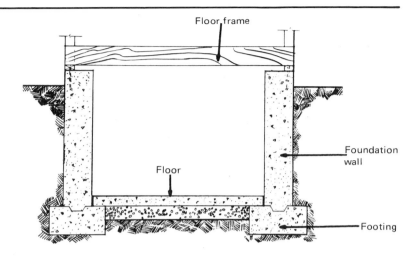

Floor frame

Foundation wall

Floor

Footing

Fig. 8-2 Cross section of a typical foundation.

While the excavating and concrete work is often done by craftsmen in other trades, the carpenter may be called upon to layout or assist in laying out the building lines, to set up batters and to build the necessary forms. It is therefore important that the carpenter have a thorough knowledge of foundation construction, the ability to read architectural drawings and to use layout instruments. Building codes dictate wall thicknesses and types of construction.

Architectural Plans

Before starting any actual construction or foundation work, the carpenter must study and familiarize himself with the architects' plans or blueprints. Plans are working drawings prepared by the architect. They show in great detail exactly how a building is put together. They also contain a certain amount of written material or specifications. The plans together with the specifications enable the carpenter to build the house. The architects' drawings (sometimes in the form of blueprints) will include the site plan, floor plans, elevations, framing, foundations and construction details. They may also include plumbing, heating and electrical work; otherwise separate utility plans must be made.

You will find that dimensions over one foot are always given in feet and inches, such as 6'-3". If there are no inches, this must be specified by adding a zero to the measure, instead of leaving the space blank; for example, a six-foot measurement will appear as 6′-0″ on the plan.

Architectural Symbols

Symbols are used in plans to simplify these detailed drawings. They represent various materials and items graphically and help eliminate written labels. Some of the most common symbols are shown in the Appendices. The carpenter should be able to recognize each of these.

Scale in Drawings

It would be very impractical to make architectural drawings full size. Instead, they are drawn at greatly reduced size but to scale, thus retaining their exact proportions. The reduction is determined by the size of the building and by the size of and detail in the drawings required. Usually, architectural drawings are made to a scale of 1/8 inch or 1/4 inch to the foot. This means that for a 1/4-inch scale, every foot of measurement in a structure will be shown as 1/4″ on the drawing. For example, a sheet of plywood 4′0″ × 8′0″ would be drawn 1″ × 2″ on the plan.

When greater detail must be shown, a larger scale may be used. If a drawing or section is not drawn to scale, that will be noted on the drawing. The note will read "not to scale," "do not scale," or "no scale."

The architect uses one of the several kinds of architect's rules when making his drawings. They are calibrated or marked off in various graduated scales. Typically, a graduation represents a fraction of an inch to the foot.

There are three general types of architectural drawings: floor plans, elevations, and sections. The floor plans are views seen directly from above and show the floor and room layouts, doors, windows, stairs, etc., as though the building had been "sliced through" horizontally. The elevation drawings are head-on views of vertical surfaces such as the interior and exterior walls. The section drawings are "sliced open" views of parts of the building. In addition, detail drawings of certain parts to show construction specifications more clearly, may be made on floor plans, or section or elevation drawings.

Building Foundations

The carpenter must be able to read these drawings and to take appropriate measurements from them, before proceeding with his work.

Plot Layout

The layout of the building lines is done after the site is cleared. Usually the corners of the house are plotted by a surveyor. Before the first stake is driven, however, the carpenter should know that all legal requirements are satisfied. Local building codes must be consulted regarding minimum setback and side yard requirements.

When the building location has been determined, building lines are laid out to indicate the outside lines of the foundation wall. Set corner stakes with nails at the top to mark the corners of the structure. Lay them out accurately with transits and levels as outlined in Chapter 3. However, you may utilize lot markers, curbs and property lines to establish building lines without the use of instruments. It is important that you keep corners square and all lines parallel. To make sure that the corners are square, measure the diagonals. They will be equal in length if the corners are square, Fig. 8-3.

Another way to check for squareness is shown in Fig. 8-4. This is sometimes called the 3-4-5 method. Multiples of these figures are used to measure the sides and diagonal. For example, if one side of a right triangle measures 3 feet and the adjacent side measures 4 feet, the diagonal will measure 5 feet. For greater accuracy, multiples of these figures are used as shown in the drawing. This method can also be used to layout square corners in the first place.

Batter boards are located at each corner and three to four feet away

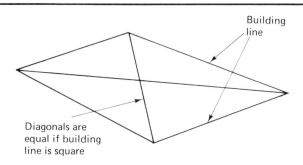

Building
line

Diagonals are
equal if building
line is square

Fig. 8-3 If diagonals are equal, the layout will be square.

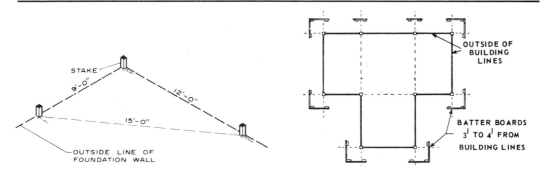

Fig. 8-4 A method for laying out and checking square corners.

Fig. 8-5 Layouts with abutments require additional batters.

from the excavation line. The batters are made with 2 × 2 stakes and 1 × 6 lumber. In a simple rectangular layout four batters will be needed, one at each corner. A more complex layout with abutments or L-shaped outlines will require more. See Fig. 8-5.

Stretch lines across opposite batter boards and adjust them so they align with the nails of the corner stakes. Use a plumb bob for exact placement of this line. Then mark the location of the string on the ledger board and make a small kerf cut. Tie a weight (stone or brick) at the ends of the line to keep it taut. Repeat this procedure on all batter boards. See Fig. 8-6. Check your diagonals for accuracy by the method mentioned above.

The height of the foundation walls is dependent on the design of the structure, the grade, and the location. Normally, wall height is governed by the highest point or elevation of the excavation. Build the top of the foundation wall at least 8 inches above the finished grade. This assures good drainage and moisture protection of the framing members, Fig. 8-7.

If rough grading is necessary because of steep terrain, have the top soil removed and piled separately so that it can be reused later when the area is graded.

Excavation Requirements

The width and depth of the excavation is determined by the type of foundation. If drain tile is necessary because of poor drainage, allow for

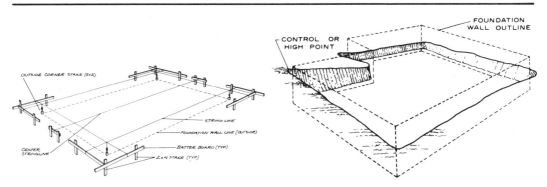

Fig. 8-6 Method of extending chalk lines over batter boards.

Fig. 8-7 Foundation should be at least 8 inches above finished grade. *(Forest Products Lab.)*

it; if clay or the like makes the soil stable, you will need little back slope. If you are excavating in sandy soil you must provide ample back sloping to prevent cave-ins.

Footings

Footings are generally made of poured concrete and support the foundation wall. They should be made wide enough to accept the load above without settling. In normal construction the thickness or depth of the footing is equal to the thickness of the wall; the width is generally twice that thickness. In poor soil, these figures may need to be increased.

In northern climates subject to freezing temperatures, the footings must be placed below the frost line. In most cases this is four feet; however, local building codes should be consulted. Footings must be placed on undisturbed soil—never on loose or replaced soil. If the soil is suitably firm, footings can be poured directly into a trench without the use of forms.

Footings for concrete block walls are generally left flat at the top, Fig. 8-8. For poured walls, it is desirable to key the top of the footing. Keyways add strength by improving the bond between footing and the wall joints. It also prevents water from entering the wall at this point. The keyway is made by inserting a wood strip at the top of the footing before the concrete sets. When the strip is removed after the concrete has set, an interlocking key or groove will be formed in the concrete as shown in Fig. 8-9.

Fig. 8-8 Footings for block walls are generally left flat on top.

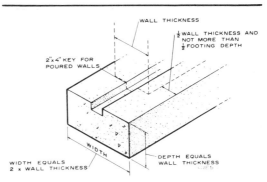

WALL THICKNESS

½ WALL THICKNESS AND NOT MORE THAN ½ FOOTING DEPTH

2"x 4" KEY FOR POURED WALLS

WIDTH

WIDTH EQUALS 2 x WALL THICKNESS

DEPTH EQUALS WALL THICKNESS

Fig. 8-9 Key in footing adds strength to foundation.

Footing forms are made of one-inch or two-inch lumber, and the *stakes* should be made of 2 × 4's, Fig. 8-10. Start by restringing the chalk lines. Locate the outer edges and corners of the footing. Drive stakes at the corners and intermediate points along the footing. Be sure to allow for the thickness of the form boards when placing the stakes. The inside edge of the form board will represent the outside edge of the footing. Drive all stakes so that their tops are slightly below the form boards. This facilitates leveling the concrete when poured. Use grade stakes to level the form boards.

Use double-headed nails for assembling the forms. Place them so they won't be locked in after the concrete sets. Drive them from the stake side, not from the inside. If a footing passes over a pipe trench or other obstruction, you must use reinforcing rods.

The inner wall of the form is assembled next. Take measurements from the outer form and repeat the procedure. Use temporary spacers cut from 1 × 2's to keep the form spread evenly. Remove all loose dirt, debris and grade stakes before pouring the concrete. Remove the spacers as the concrete is poured.

Step footings are necessary on sloping ground to prevent slippage of the foundation. Such footings are used when either other parts of a house or the garage are at different levels from the main part. The vertical step should be at least six inches thick and no more than two feet high. Check the local building code for specifications. Fig. 8-11 shows a stepped footing.

Columns and posts should have square-shaped footings with a pedestal. The pedestal usually contains a protruding steel pin to which the wood post or column is anchored. The footing may be

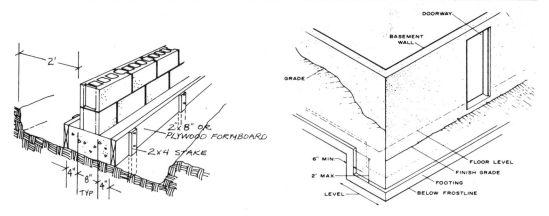

Fig. 8-10 Typical footing for a concrete block foundation.

Fig. 8-11 Details of a stepped footing.

24" × 24" × 12" or 30" × 30" × 12". The pedestal should be 3 inches above the finish basement floor and 12 inches above the finish grade in a crawl-space area.

Foundation Walls

Concrete weighs about 150 pounds per cubic foot. The carpenter must consider this when he builds forms for the foundation walls. The forms must be strongly built and must be smooth and free of defects; and all joints must be tight. Forms are usually made of one inch boards or of 3/4-inch plywood. You may use other suitable materials.

Reusable forms which can be assembled and disassembled quickly are widely used, Fig. 8-12. They can be made by the carpenter or purchased ready-made. Many types are available; most have locking devices which hold the sections securely. Some are held with nails, Fig. 8-13.

Formwork must be constructed for each wall face of poured concrete foundation walls. To prevent the forms from bulging outward due to the pressure exerted by the impact of the wet concrete, strong and proper bracing is needed. For walls up to three feet high, studs spaced two feet apart will suffice. Walls higher than three feet require studs braced with walers, Fig. 8-14. Use spreaders to keep the form

Fig. 8-12 Reusable forms for foundation walls are widely used in modern construction.

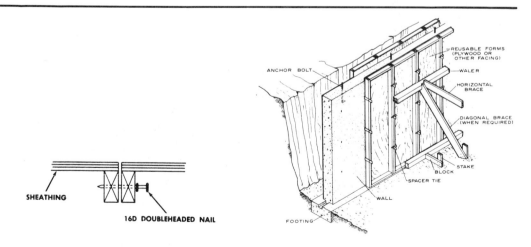

Fig. 8-13 A method of joining wall forms in line.

Fig. 8-14 Method of bracing forms.

Building Foundations

Fig. 8-15 A spreader is used to support upper part of form; wall is low at this point.

walls properly spaced. A 2 × 6 spreader is being used to space the upper part of a form shown in Fig. 8-15. The bottom edge is just above the wall line. Cut simple spreaders from 2 × 4 lumber.

Ties also prevent the walls from bulging. They can take many forms; perhaps the simplest is the wire tie with clamp, Fig. 8-16. A three-section threaded tie is shown in Fig. 8-17. The outer sections are removed when the forms are stripped.

Fig. 8-16 Clamps hold wire ties secure and are easily removed.

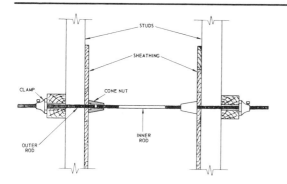

Fig. 8-17 A three-piece threaded wall tie.

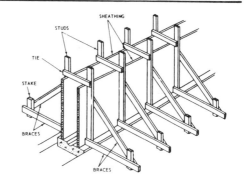

Fig. 8-18 A method of forming and bracing a low foundation wall.

Ties are usually designed with a break-off feature which allows the tie to break below the surface of the concrete. That is especially important in basement walls where protruding metal ties would be objectionable. The small holes left in the wall after the break-off are patched with mortar cement.

Wood ties may be used at the top of a low foundation wall, Fig. 8-18.

Openings for windows and doors in foundation walls are built into the formwork. They are called bucks or core boxes in poured concrete construction. Nailing strips may be added to the form so that they become an integral part of the wall when cast in concrete. Fig. 8-19 shows a basement window frame embedded in concrete. In cement block construction, window and door frames are set in place and mortar is troweled into the grooves provided. This forms a keyway which

Fig. 8-19 A basement window frame set into concrete wall; treated wood should be used.

Fig. 8-20 In some construction, openings are left for windows which will be installed later.

Building Foundations

strengthens and supports the frame permanently. In some construction, window frames may be installed after the foundation is completed, Fig. 8-20.

Lintels are used over window and door openings when the openings are below the top of a wall. These reinforced horizontal supports carry the load above the opening. In poured concrete walls, they are made with steel reinforcing bars imbedded in the concrete, as in Fig. 8-21. They are held in place with wire and should be at least two inches above the opening. In concrete block construction, lintels may be made several ways; for example, either reinforced pre-cast lintels can be put in place or steel angle-iron may be used. Another popular method is to use special lintel blocks supported by the vertical members of the door or window frames. These blocks which are open at the top are then filled with concrete after steel bars have been inserted in them.

Masonry steps are usually poured with the foundation if they are included in the plans. Forms for such steps are built with the main form. If they are to be poured separately for any reason, connect them to the main framework with steel bars.

Notches for beams and *girders* are provided by inserting blocks in the formwork. If wooden beams and girders are to be used, the notches must be made enough larger than the beams to give 1/2″ clearance at the ends and sides. This provides sufficient ventilation for moisture to escape. See Fig. 8-22.

When an outside wall is to have a masonry veneer finish, you must make the foundation wall with a ledge wide enough to support the

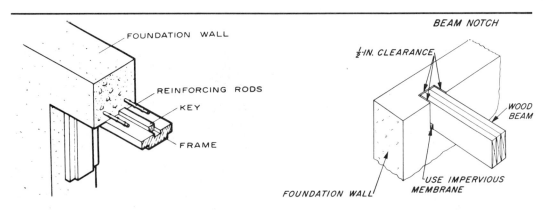

Fig. 8-21 Reinforcing rods are used over door and window openings. *(Forest Products Lab.)*

Fig. 8-22 Clearance for beams must be provided in foundation wall. *(Forest Products Lab.)*

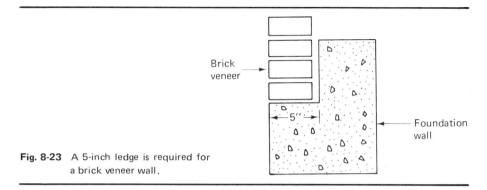

Brick
veneer

5"

Foundation
wall

Fig. 8-23 A 5-inch ledge is required for
a brick veneer wall.

masonry. For a brick veneer, place the ledge below the sill and be sure it
measures 5″ wide. See Fig. 8-23.

Before *pouring concrete* check all the forms for accuracy and
stability. Be sure that nail markers are installed on the perimeter of the
form at the exact height the wall is to be. Use the level to assure that
the nails are placed accurately since they mark the correct height that
the concrete is to reach.

Figure 8-24 shows wales being installed on a special form made with
high-density overlaid plywood.

If wheelbarrows must be used to carry concrete to some parts of
the form, provide easy and safe access. In most cases, however,
ready-mix trucks will be able to pour directly into the forms, Fig. 8-25.
Oil or coat the forms with a suitable compound so they will not stick to
the hardened concrete; and dampen the soil beneath if the concrete is
poured in hot dry weather. This will prevent the soil from absorbing the

Fig. 8-24 Wales are being installed on a special form.

Fig. 8-25 Concrete is being poured into oiled forms.

Building Foundations

Fig. 8-26 Concrete is spaded to prevent honeycombing.

water in the concrete. If the soil is muddy or frozen, the pouring should be postponed.

Pour concrete as close as possible to the place in which it is needed. Never put it at one spot and simply allow it to flow around the form since that causes the aggregate to collect in too-large amounts. Instead, pour the concrete in 8- to 12-inch layers successively around the entire wall. Honeycombing or pitting occurs when too-large air pockets form around large bits of aggregate collected too near the face of the wall. See Fig. 8-26. To prevent air pockets, spade the concrete with a scraper or other flat tool.

Sill anchors are set into the top of the foundation wall when the concrete becomes plastic enough to support them. After these anchors have been set at the proper depth, be sure to jiggle them up and down in order to work out any air pockets that may have formed. For concrete walls, sill anchors should be 1/2 inch in diameter and 6 inches

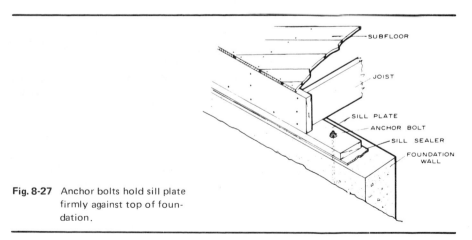

Fig. 8-27 Anchor bolts hold sill plate firmly against top of foundation.

SUBFLOOR

JOIST

SILL PLATE
ANCHOR BOLT
SILL SEALER
FOUNDATION WALL

long. For cement block walls, they must be 1/2 inch in diameter and 15 inches long. For either type of foundation they should be placed at 6-foot intervals. Fig. 8-27 shows an anchor bolt in a concrete wall. In masonry-block construction the anchors are placed at regular intervals in openings in the blocks which have been filled with concrete. A piece of lath at the second joint below the top keeps the concrete in place.

After the concrete has set, remove the forms carefully. If necessary, brace high walls temporarily as shown in Fig. 8-28. Remove all braces when you install floor framing.

Masonry blocks or *concrete blocks* may be used to build foundation walls instead of poured concrete. Such block walls do not require forms. The two types commonly used are the concrete block and the cinder block.

The concrete block is made of cement, sand and crushed rock or gravel. Cinder blocks are similarly made except that they contain volcanic cinders or slag instead of crushed rock. Concrete blocks weigh about 50 pounds each, each cinder block about 30 pounds. They are available in many shapes.

The standard size for masonry blocks is 8 × 8 × 16, but their actual size is 3/8 of an inch less in each dimension to allow for the mortar joint. An 8 × 8 × 16-inch block will therefore actually measure 7-5/8 × 7-5/8 × 15-5/8 inches.

Blocks are also available in widths of 4, 6, 8, 10, and 12 inches and heights of 4 and 8 inches. Mortar joints in a block wall are tooled smooth to prevent water seepage and for improved appearance, Fig. 8-29. If necessary, the cores of the blocks can be filled with insulating material.

Fig. 8-28 Braces to temporarily support a high foundation wall.

Fig. 8-29 Tooling of block walls improves the appearance and prevents seepage.

Building Foundations

Fig. 8-30 A typical pilaster in a block wall.

Fig. 8-31 Metal ties protrude from a block wall.

To prevent cracking of block walls, place control joints at 20 foot intervals and fill them with a special caulking compound. Also use control joints at doorways and windows and at the junction of walls and partitions.

Pilaster blocks add lateral strength to a wall. They also provide a larger bearing surface for beam ends. Fig. 8-30 shows a pilaster in a wall section. Reinforcing rods will be placed in the opening together with concrete or mortar fill.

If the wall is to be faced with brick, imbed metal ties throughout the wall. They are made to protrude from the mortar joint (Fig. 8-31); and the opposite ends are bent down into the opening in the block.

Blocks are used for decorative as well as structural purposes. Those used for foundations and other load-bearing applications must meet the American Society for Testing Materials (ASTM) standards.

Waterproofing foundation walls prevents water from entering the foundation of a house through cracks in the concrete or at bad joints in the core. To prevent seepage below, place drain tiles around the base of the foundation as well.

Waterproof concrete block walls with a coat of portland cement plaster. Apply it in an even coat about 3/8 of an inch thick; and do the entire wall in one operation. Coat the plaster with two applications of hot asphalt brushed on at right angles to each other to ensure additional waterproofing.

Waterproof poured concrete walls with bituminous waterproofing, as in Fig. 8-32. Note that the waterproofing is applied along the intended grade line.

Drain tiles are pipes made of clay, concrete or other material and are generally about 4 inches in diameter. They are sometimes

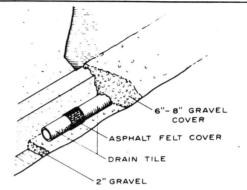

Fig. 8-32 Waterproofing of concrete wall follows grade line.

Fig. 8-33 Drain tile carries away ground water.

6"– 8" GRAVEL COVER

ASPHALT FELT COVER

DRAIN TILE

2" GRAVEL

perforated. These tiles are placed in a bed of gravel alongside the footing to carry away ground water. If they are not perforated, lay them 1/4 of an inch apart. Then cover the joint over with asphalt or roofing paper to prevent dirt washing into the pipe as the water enters. Before backfilling, you must cover the drain tiles carefully with coarse gravel. See Fig. 8-33.

Concrete Basement Floors

Concrete floors for a basement are not poured until the house is framed and the roof covered. If the floor were poured before the finished roof and walls go up, rain water could accumulate and remain since there would be no way for it to drain off easily.

Slab Construction

In homes without a basement, the foundation walls rest on a concrete slab directly on the ground. Poured properly, a slab makes a satisfactory floor. It is important that the ground around the slab be graded to slope away from the floor on all sides, at least one foot in every 25. The slab itself should be at least eight inches above the surrounding grade.

The soil beneath the slab (the subgrade) is cleared of roots, grass and debris, and given a granular fill such as gravel or crushed stone. The fill should be well-compacted and at least 4 inches thick. This subbase usually assures a dry floor slab; a vapor barrier must be used between the subbase and the slab, however. The barrier may be made from 6-mil polyethylene, 55 pound roll-roofing or various asphaltic materials manufactured for the purpose.

Perimeter insulation is installed where the slab meets the foundation wall. In warm climates, the insulation prevents excessive heat gain in the living quarters; in cold climates, it reduces excessive heat loss to the outside.

The insulation must meet certain specifications: it must be rigid enough to support the weight of the slab; it must also be waterproof and vermin proof; and it should not absorb moisture. Chapter 18 on insulation covers the various types of insulation. Note that the insulation is placed above the vapor barrier and between the slab and foundation wall, Fig. 8-34.

Pouring the slab is timed so that all necessary apertures or connections for utilities are provided for or first installed. Sleeves are

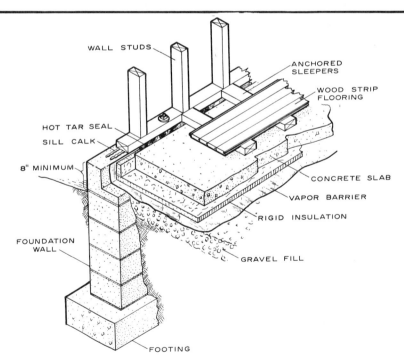

Fig. 8-34 Proper insulation of the slab is important. (*Forest Products Lab.*)

Fig. 8-35 Reinforced wire mesh, furnished in coils, is cut to size on the job. *(Keystone Steel & Wire)*

used to protect any electric lines and water pipes that must pass through the concrete.

Six-inch reinforcing wire mesh is embedded in the first layer of concrete above the granular fill, Fig. 8-35. Supported thus, the wire mesh rests 1-1/2 inches below the surface of the completed poured slab.

After the concrete is poured and leveled or screeded, the surface is troweled to a smooth finish. Most troweling is now done with powered trowels since they do better finishing than hand trowels and are much faster.

Building Foundations

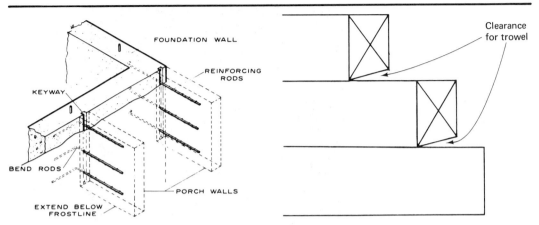

Fig. 8-36 Method of connecting porch walls to the foundation; stair walls can also be done this way.

Fig. 8-37 Riser bottoms are beveled to allow clearance for troweling.

Concrete Steps

Porticos, steps and platforms, if made of concrete, should be an integral part of the foundation and poured when the foundation is—if possible. If they are added later, they should be anchored to the main structure. Fig. 8-36 shows how porch walls are tied to the foundation. Steps can be joined this way also; they are generally poured as a part of the platform. Riser form boards are nailed to the side-wall sections and their bottoms are beveled, as shown in Fig. 8-37. This is done so the treads can be troweled up to the inside corners. Fig. 8-38 shows details of a form for simple platform steps.

Steps should have a 7-1/2″ rise and a tread or depth of about 10 inches. The tread is the horizontal part of the step on which the foot is placed. If the width of the steps is more than 3 feet, the form boards for the risers should be made from 2 inch stock. Use steel rods, 1/2 inch in diameter for reinforcement. The concrete ought to be troweled with a wood float to give the surface "tooth" since a steel trowel produces a smooth finish that is not safe for steps.

Walks and Driveways

If the soil for a walk is well-compacted, the concrete can be laid directly on the ground. A better method, especially on poorly drained

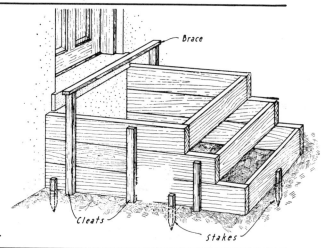

Fig. 8-38 Form construction for platform with two steps.

soil, is to use a well-compacted fill of crushed stone or gravel. The fill should be from 4- to 6-inches thick.

Forms for walks and driveways are made of 2 × 4 or 2 × 6 lumber and are placed so they slope away from the house. Make the slope 1/4 inch per foot and place control joints at four-foot intervals. Make them by using a groover and a straight edge. Round all edges with an edging tool.

To make simple poured concrete "flagstones" for walks, make forms as shown in Fig. 8-39. Assemble with screws and oil them well before use.

Treated Wood Foundations

A wood foundation system has been developed with pressure-treated wood for single dwellings, garden apartments and small office buildings. The system provides a new way to install house foundations quickly and easily. Basically a stud wall below grade, it has many advantages. Costs are lower than for other types of foundations; work can be done in any weather; and there is no waiting for ground to dry or thaw. Also, construction time is cut considerably. You can install panels in a few hours.

Although this All Weather Wood Foundation (AWWF) is a recent addition to residential construction, it is not an untried method. The United States Forest Products Laboratory in Madison, Wisconsin, built

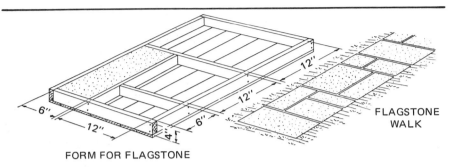

FORM FOR FLAGSTONE

FLAGSTONE WALK

Fig. 8-39 Form construction for making flagstones for a walk.

an experimental home with a pressure-treated wood foundation more than three decades ago. The structure was analyzed after thirty years, and found to be in excellent condition.

AWWF has been approved by the FHA for federal mortgage insurance. It is also being included in some state and local codes.

AWWF fabrication: The system is made with pressure-treated lumber and plywood. Foundation sections made of nominal 2" lumber are either shop-built or made on the job. Footing plates rest on a 4" bed of gravel or crushed stone. The exterior of the foundation wall is covered with a polyethylene film held in place with a waterproofing compound. Details of the system are shown in Fig. 8-40. For crawl space construction maintain space between floor joints and ground; it should be a minimum of 8 inches. Then it is recommended that a vapor barrier be placed over the soil in order to prevent decay-causing condensation on the wood framing members.

Site preparation: for basement construction the site is excavated and leveled. Then plumbing lines are installed and provision is made for drainage. A sump is recommended if the soil drains poorly, Fig. 8-41.

After plumbing and drainage are all installed, the basement site is covered with gravel to a minimum depth of 4 inches.

Footings: The width of the footing plates is determined by the vertical loads on the foundation wall and the bearing capacity of both soil and gravel bed. Normally, 2 × 6 or 2 × 8 footings are adequate. For brick veneer construction, footing plates should be 2 × 10 or 2 × 12.

A load-bearing partition may be used as a center support, Fig. 8-42. It is constructed in the same manner as the exterior foundation walls.

Walls: The framing members of the foundation are designed to resist the lateral pressure of the fill and the vertical forces of live and dead loads. Pour the concrete slab after the walls are erected and

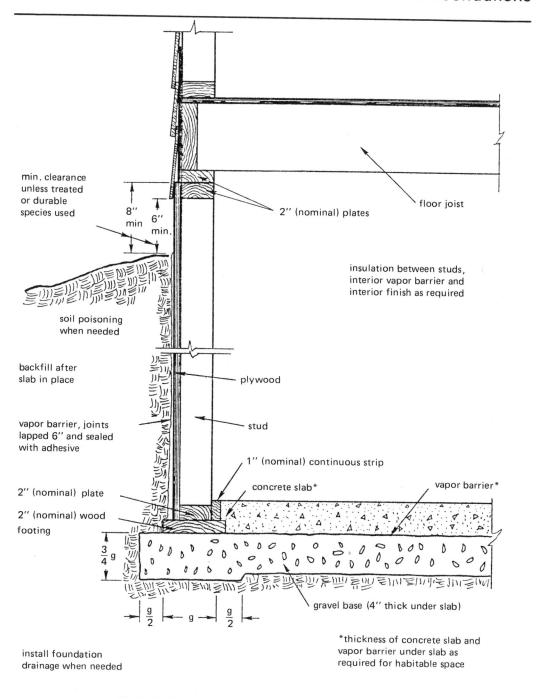

min. clearance
unless treated
or durable
species used

8"
min

6"
min.

2" (nominal) plates

floor joist

insulation between studs,
interior vapor barrier and
interior finish as required

soil poisoning
when needed

backfill after
slab in place

plywood

vapor barrier, joints
lapped 6" and sealed
with adhesive

stud

1" (nominal) continuous strip

2" (nominal) plate

2" (nominal) wood
footing

concrete slab*

vapor barrier*

$\frac{3}{4}$ g

$\frac{g}{2}$ g $\frac{g}{2}$

gravel base (4" thick under slab)

*thickness of concrete slab and
vapor barrier under slab as
required for habitable space

install foundation
drainage when needed

Fig. 8-40 Basement construction details for wood foundation.

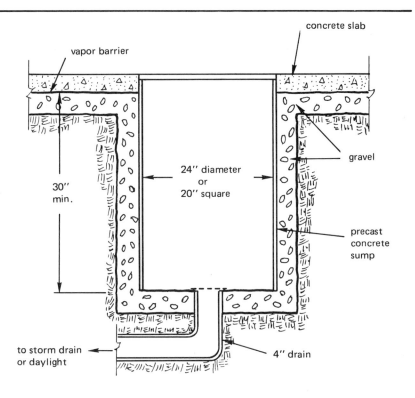

vapor barrier

concrete slab

24" diameter
or
20" square

30"
min.

gravel

precast
concrete
sump

to storm drain
or daylight

4" drain

Fig. 8-41 Details of basement sump for an all wood foundation.

anytime before backfilling. The concrete slab also braces the bottom of the walls.

End-nail the treated studs to the top and bottom plates with two 16d nails per joint. Attach treated plywood panels 1/2 inch or thicker with 8d nails spaced 6 inches on center along edges and 12 inches on center at intermediate supports. You may apply panels either horizontally or vertically. Locate vertical joints between panels over a stud.

Attach foundation sections to the wood footings and adjacent sections with 10d nails 12" OC and face-nail them through the bottom plate and mating studs. Then face-nail a field-installed untreated top plate to the top plate of the foundation wall with 10d nails, 16 inches or less on center. Stagger joints in the untreated top plate with those of the treated wall panels. A typical corner detail is shown in Fig. 8-43.

All nails used in treated wood foundations should be made of hot-dipped galvanized steel.

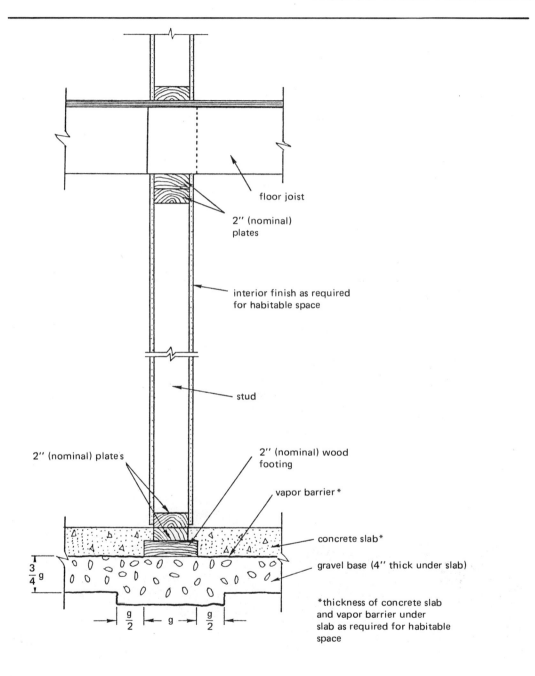

floor joist

2″ (nominal) plates

interior finish as required for habitable space

stud

2″ (nominal) plates

2″ (nominal) wood footing

vapor barrier *

concrete slab*

gravel base (4″ thick under slab)

$\frac{3}{4}$ g

$\frac{g}{2}$ ⊢ g ⊣ $\frac{g}{2}$

*thickness of concrete slab and vapor barrier under slab as required for habitable space

Fig. 8-42 Details for load-bearing wall.

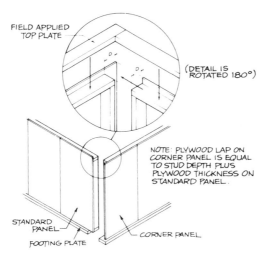

FIELD APPLIED
TOP PLATE

(DETAIL IS
ROTATED 180°)

NOTE: PLYWOOD LAP ON
CORNER PANEL IS EQUAL
TO STUD DEPTH PLUS
PLYWOOD THICKNESS ON
STANDARD PANEL.

STANDARD
PANEL

FOOTING PLATE

CORNER PANEL

Fig. 8-43 Typical corner joint detail.

Cover the wall frames with a polyethylene film either before or after they are erected. Hold the polyethylene in place by mopping it with a coal-tar waterproofing compound. The erected walls are temporarily held with braces as shown in Fig. 8-44. As already described, the concrete slab is poured on the 4″ gravel bed after the walls are in place. After the concrete sets, add backfill and frame the rest of the house, as in Fig. 8-45. Brace the top of the wall adequately before backfilling, of course.

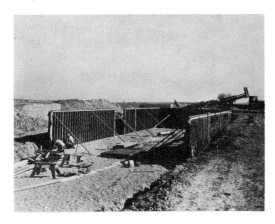

Fig. 8-44 Erected wall sections are temporarily braced. (*Koppers Co., Inc.*)

Fig. 8-45 Completed structure ready for interior and exterior finish work. (*Koppers Co., Inc.*)

Curbs

Two-inch forms are used to make curbs, and they are installed in a trench with stakes, spreaders and ties. See Fig. 8-46. Place expansion strips, made from 1/2-inch material at intervals of 20 feet.

Fig. 8-46 Forms for curbing; the line assures accuracy.

Building Foundations

GLOSSARY

aggregate: sized bits of hard material used for making concrete; it includes sand, crushed rock and gravel.

backfilling: replacing soil around a foundation.

batter boards: horizontal boards on corner posts which assist in the accurate layout of foundation and excavation lines.

bituminous waterproofing: done with an asphalt or tar-based compound.

building lines: the outside lines of a building; the perimeter of the structure.

cement blocks: also called concrete blocks; made of cement, sand, crushed rock and gravel in graduated fragments.

cinder blocks: similar to cement blocks, but made with volcanic ash instead of crushed rock and lighter in weight.

control joint: expansion joint in masonry; it allows movement due to expansion and contraction.

drain tiles: cylindrical tiles, usually perforated, are used at the base of foundations to carry away ground water.

footing: the base or support for a foundation wall, designed to spread the force of concentrated loads.

foundation: the support upon which a structure rests; it carries the weight of the floors, walls and roof.

frost line: the maximum depth at which soil is frozen; that point below which the soil does not freeze.

honeycombing: a pitting in poured concrete caused by formation of air pockets along the form faces; it can be prevented by spading so that the aggregate does not collect too near the surface.

keyway: an interlocking groove or channel made in wood or concrete joints for reinforcement.

lintel: a horizontal support used over doors, windows, and fireplace openings; often made of wood, steel or reinforced concrete.

pilaster: an upright pier or column which projects from a wall and adds strength to the structure; if exposed, it may be used decoratively as well.

screeding: leveling up poured concrete surfaces with a board or plank.

sill anchor: a fastener projecting from a foundation wall or slab and used for securing the sill.

slab: a concrete floor poured on a prepared ground site.

spreader: in form work, a wood or steel member temporarily inserted between form walls to keep them apart evenly.

stakes: pieces of wood or metal pointed at one end so that they can be driven into the ground; often used as corner or boundary markers.

vapor barrier: sheeted layers or a membrane, usually plastic; often used between subbase and floor slab to prevent moisture penetration from below.

waler (also whaler): a horizontal reinforcement used to keep newly poured concrete forms from bulging outward.

9

Floor Framing

Floor framing may be called the superstructure of a frame house since it is built over the foundation as the platform on which the entire house rests. The floor framing members include the sill, posts, joists, beams and subfloor. The framing for walls and partitions rest in turn on this platform. Fig. 9-1 shows a floor frame near completion. Note the opening for basement stairs.

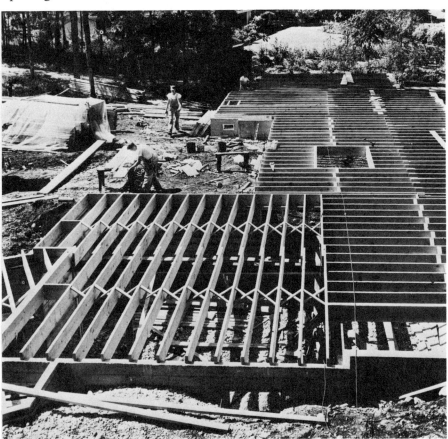

Fig. 9-1 Floor frame being completed; subfloor will be added next.

Floor framing methods are pretty much standardized throughout the country. Variations occur mostly because of the type of framing used. The two basic types are platform framing and balloon framing. These are covered in detail in Chapter 10. The differences in framing are clearly shown in Fig. 9-2 and Fig. 9-3.

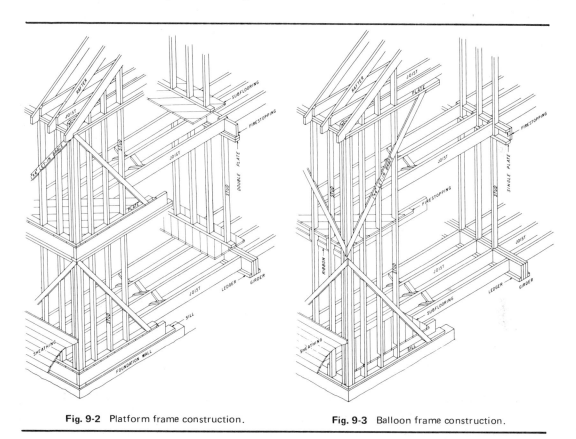

Fig. 9-2 Platform frame construction.   **Fig. 9-3** Balloon frame construction.

The sill is the horizontal timber which rests on the foundation wall and forms the bearing surface for all construction above the foundation. It is used in frame and brick veneer structures. It is not used when exterior walls are brick or concrete.

The sill is usually 2 × 6 lumber but the size may vary, depending on local building codes. All sills should be foundation grade cedar, redwood, or other preservative-treated wood. If termites are a threat, sills should be installed with termite shields. These are metal strips with edges turned down at a 45-degree angle. They are placed between the foundation wall and the sill. How a wall and post are shielded from termites is illustrated in Fig. 9-4 below.

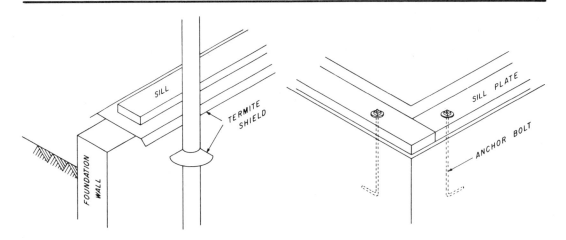

Fig. 9-4 Installation of termite shield on exterior wall.   **Fig. 9-5** Anchor bolts hold sill to foundation.
(*National Forest Products Assoc.*)

The sill is mounted on the foundation wall and held with the anchor bolts which were embedded earlier when the foundation was made. Locate the sill so that its outer edge aligns with the outer edge of the foundation, less the thickness of the sheathing. The bolt holes should be bored a bit larger than the bolt diameter to allow for adjustment, Fig. 9-5. Use a sealer strip, usually made of fiberglass between foundation and sill to keep out drafts and insects. Be sure the sill is perfectly level and exactly aligned since it forms the permanent base for the walls and flooring. Fill any low spots and voids at the top of the foundation wall with mortar.

Girders

Girders are load-bearing structural members. They support floor joists where the span or distance between walls is great enough to require such support. A short span between walls does not require girders. The girder ends rest on the foundation wall or on piers. Wood or steel posts often support timber girders at intermediate points.

Girders may be made of wood or steel. Wooden girders are either built-up or solid. Steel girders are generally I-beams, so called because the cross section resembles the capital letter I. Fig. 9-6 illustrates a steel I-beam supported by a Lally (pipe) column.

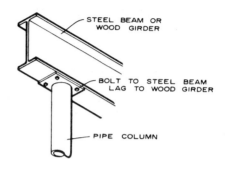

STEEL BEAM OR WOOD GIRDER

BOLT TO STEEL BEAM LAG TO WOOD GIRDER

PIPE COLUMN

Fig. 9-6 An I-beam supported by a pipe or Lally column.

Girder size is determined by the load to be carried. Construction plans will specify both the cross section and length of the girders. Built-up wood girders will support slightly less than solid girders of the same dimensions. However, built-up girders of kiln dried lumber undergo less shrinkage than solid girders; so they are more widely used.

Generally, built-up girders or beams consist of two or more pieces of 2" lumber. Two-piece girders are nailed from one side with tenpenny nails spaced 16 inches apart and staggered. Three-piece girders are nailed from both sides with twentypenny nails spaced 30 inches apart and also staggered. See Fig. 9-7. Note how the end joint is staggered.

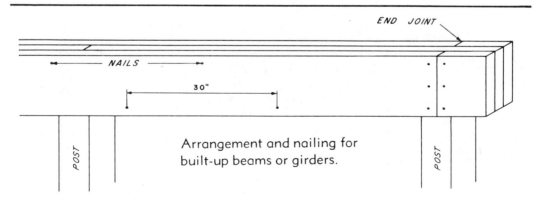

END JOINT

NAILS

30"

POST

POST

Arrangement and nailing for built-up beams or girders.

Fig. 9-7 Nailing pattern for built-up beams.

Floor Framing

Girder Installation: The ends of a girder must bear on at least four inches of the foundation wall. If the wall is notched, maintain at least 1/2″ air space at the sides and end of wood girders, as shown below in Fig. 9-8.

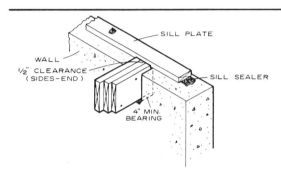

Fig. 9-8 A minimum 4″ bearing is required for wood girders.

To align the tops of the girder and the sill, nail a wood block to the top of the sill. See Fig. 9-9.

Fig. 9-9 Temporary stop block is used to align top of girder with sill.

Posts which support girders and beams are either wood, steel or masonry. Masonry piers are generally used in crawl-space construction.

Wood Posts

Wood posts are usually solid timbers, but you can build them up by spiking several pieces of 2 inch stock. If you use built-up posts, use pieces flat and free of knots or other imperfections. Rest the posts on footings above the floor line to keep moisture from the post bottom, Fig. 9-10. It is also advisable to place a piece of asphalt-impregnated felt between the footing and the bottom of the post, in addition to treating the end of the post. The posts are kept from shifting by steel pins which project from the footing. You can also use special adjustable fasteners to keep the post off ground level. Wood posts ought to be at least 6" X 6" but they can be only 4" X 6" if you are using them as part of a framed wall or partition.

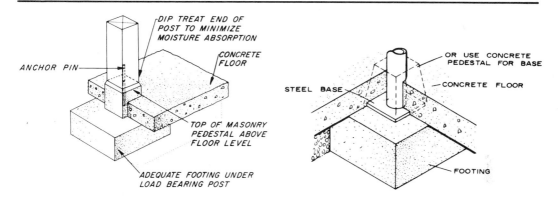

Fig. 9-10 Posts should rest on supports above the floor level. (*Forest Products Lab.*)

Fig. 9-11 Method of anchoring the bottom of the Lally column.

Steel Posts

Tubular steel posts called Lally columns are used for supporting both wood girders and steel beams. Bearing plates at each end allow them to be anchored to the floor and beam. Fig. 9-11 shows two methods of anchoring them. Some steel columns are adjustable in height by means of a threaded section. This permits the carpenter to readjust them after shrinkage of wood members has taken place. For added strength, Lally columns are usually filled with concrete.

Floor Framing

Joists

Joists are main supporting members which carry floor loads to the sills and girders. They are usually made of 2 or 3 inch lumber and placed on edge. Generally floor boards are attached to them. Spacing is normally 16 inches on center (16" OC) but other spacings may also be used. Generally joists run the short way of the structure. The size depends on the floor load and span. You must use joists capable of withstanding the intended load with a minimum of deflection. Permissible deflections will not crack plaster. Consult local building codes for information on specific joist sizes.

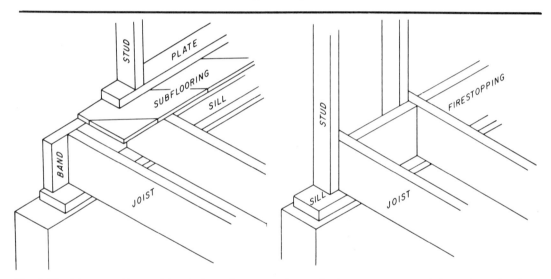

Fig. 9-12 Joist assembly used in platform framing.   **Fig. 9-13** Joist assembly used in balloon framing.

If a joist has an edge bow, install it with the crown on top. Also, place any lumber with edge knots so that the largest knots are up.

A study of Figs. 9-12 and 9-13 clearly shows the difference between platform and balloon framing with regard to joist assembly. In platform framing, the joist ends are nailed to the sill and header joist. In the balloon frame, the joist is nailed only to the sill.

You can make the joist layout on the sill, but for platform construction it is more practical to locate the joist markings directly on the header joist. Do this with a measuring tape or a joist rod. A joist rod is simply a long piece of wood with joist spacings marked on it.

If joists are butted or in line, the markings on opposite header joists will be the same. When joists are lapped, the markings should be offset accordingly. Make the markings with a square. Placing an "X" mark next to the line indicates the joist location and prevents the possibility of errors when nailing.

For 16" OC spacing, the first joist is 16 inches, measured from its own center to the outside edge of the end joist. All other markings are placed 16 inches apart. If the joists are lapped, then your first measurement on the opposite header will be 14-1/2 inches from edge to center. See Fig. 9-14. The joists are toenailed to the sill. Place the joists so the crown, if any, is on the top side; they will straighten out when loaded.

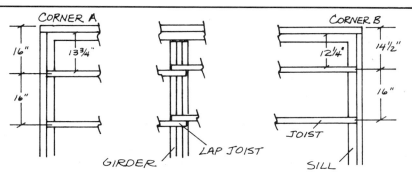

Fig. 9-14 Spacing of joists differs at corners *A* & *B* when joists are lapped.

The nailing pattern is important. Use three 16d nails if you are nailing through header joists; for toenailing joist to sill, use two 10d nails. *Note*: Codes vary regarding nailing requirements, so check your local building codes.

There are several ways to fasten the joists to the beam; Fig. 9-15 shows the two generally used for wood beams. Two 10d nails are used when the joists rest on a ledger. Special nails are required for the metal hanger.

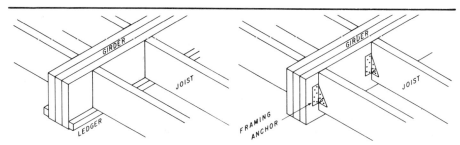

Fig. 9-15 Two methods of fastening joists to beams: **Left**, on ledger; **Right**, with framing anchors.

Floor Framing

To gain headroom space in a basement, the wood beam supporting the joists is sometimes set higher than the sill. The joists are then notched and made to rest on ledger strips nailed to the lower part of the beam, as shown in Fig. 9-16. When joists rest on a girder, the ends should overlap, Fig. 9-17. Bridging is placed between the joists. Figure 9-18 shows various methods to fasten joists on steel beams. If the joists rest on the steel beam, place a nailer on top of the beam, Fig. 9-19.

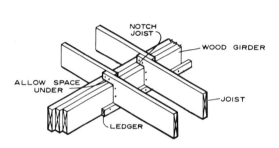

Fig. 9-16 Method of gaining headroom in a basement.

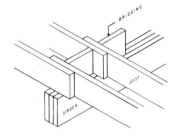

Fig. 9-17 Overlap is required when joists rest on girder.

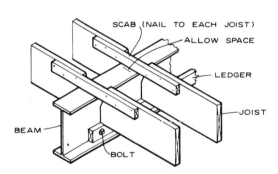

Fig. 9-18 Method of fastening joists to metal beams.

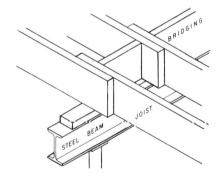

Fig. 9-19 The joists rest on a nailer, bolted to the steel beam.

Double Joists

Joists are doubled for additional strength. This is necessary when openings are made for chimneys, fireplaces, and stairways. Doubling up is also necessary when a partition above runs parallel to the joists. If the partition wall is to carry heating ducts or pipes, the doubled joists may be spaced, Fig. 9-20.

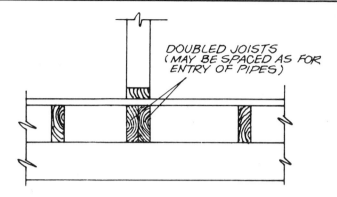

Fig. 9-20 Joists are doubled under partition walls.

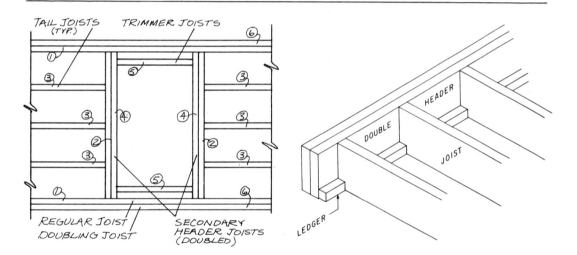

Fig. 9-21 Framing of floor opening: numbers indicate order of installation of the various members.

Fig. 9-22 Ledger can be used to fasten joist to double header.

When a joist is doubled up around an opening, it is called a trimmer, otherwise it is simply a doubled joist. A very important rule to remember is that a joist location is never altered to accommodate an opening. Instead, cut through the joists to make the required opening. See Fig. 9-21. The remaining joists are called tail joists. Tail joists are supported by headers which in turn are supported by trimmers; Fig. 9-22 shows how tail joists are fastened to double headers.

When laying out an opening in the floor frame, always work from a centerline. This will ensure accurate placement of inside edges. The nail sequence for floor openings is important. Following an established

Floor Framing

procedure makes end nailing possible, thus eliminating toe-nailing. The proper sequence is shown in Fig. 9-23. First, install the trimmers, follow with the first headers; next, the tail joists, and followed with the

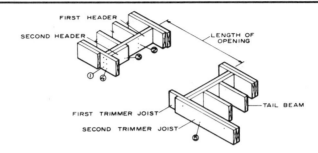

Fig. 9-23 When framing floor openings, the proper nailing sequence can eliminate toe-nailing.

second header. Install the second trimmers last. The numbers indicate the proper sequence. For greater support it is possible to use joist hangers. Use them either alone or in conjunction with end nailing.

Floor Projections

Overhangs and bay windows may require special framing. In simple projections, under two feet, the joists are simply extended. For longer projections, special anchoring may be needed. When the projection is parallel to the joists, construct the joists as shown in Fig. 9-24, left.

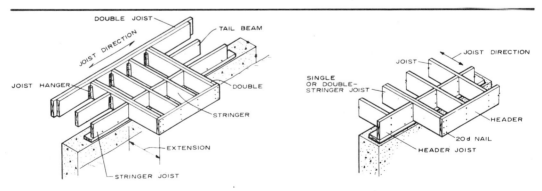

Fig. 9-24 Projection framing: **left,** parallel to joists; **right,** perpendicular to joists.

Bridging

Bridging is used to reinforce and stiffen joists. Several types of bridging are possible: cross, solid, and metal. Cross bridging, also called herringbone, utilizes diagonally placed lumber, either 1 × 3's or 1 × 4's, crossed to form an "X". The tops are nailed but the bottoms are left loose and nailed only after the flooring is in place and shrinkage is no longer a problem. See Fig. 9-25.

Fig. 9-25 Bottom of bridging is not nailed until flooring is installed.

Fig. 9-26 Staggered solid bridging permits the end-nailing of joists.

Solid bridging is sometimes used. The objection to this is that shrinkage can cause snug-fitting bridges to work loose. By staggering solid bridging, each piece can be end nailed, Fig. 9-26.

Metal bridging is available in several configurations. Some types have serrated ends which eliminate the need for nails. The advantage with these is that they can easily be tightened after shrinkage takes place in the joists.

Subflooring

The subfloor is placed directly over the joists and serves as the base surface for the finish floor. Subfloor material may be common boards, shiplap, T & G or plywood. Except for plywood, the boards are laid out at right angles or at 45-degree angle to the joists. All end joints must fall

Floor Framing

on a joist. However, if tongue-and-groove subflooring with grooved ends is used, the end joints need not fall on a joist. Fig. 9-27 shows the subfloor of a house ready for the wall framing.

Fig. 9-27 The subfloor of a platform construction. (*Weyerhaeuser*)

Laying the subfloor diagonally or straight is mostly a matter of choice. If the boards are applied diagonally, the finish floor can be laid out parallel to or across the joists. If the layout of the subfloor is straight, the finish floor must be perpendicular to the subflooring.

For accuracy in laying a diagonal subfloor, start the first piece about ten feet from the corner. Measure and mark a point 10 feet from the corner along the header and end joist. This establishes a 45-degree starting line. Figure 9-28 shows how diagonal subflooring is laid. Use two 8d nails for each bearing, three nails if the boards are 8-inches wide. For diagonal subflooring on balloon framing, you must use a nailing strip for the flooring ends. See Fig. 9-29.

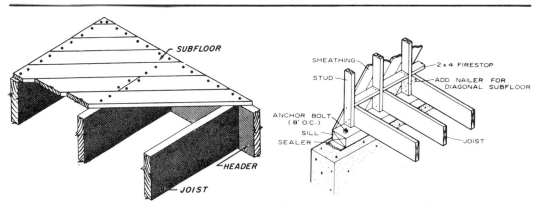

Fig. 9-28 Detail for a diagonal subfloor.

Fig. 9-29 A nailer strip for subfloor must be added to balloon frame.

Plywood Subfloors

Because of the size, plywood subflooring is more economical to install than narrow boards. In addition to reducing installation time, it also provides a strong, smooth surface for any type of finish floor.

Generally C-D interior grade plywood is used for subflooring. Panels for this purpose are marked with identifying marks such as 32/16. This means that the panel may be used for roof sheathing with rafter spacings of 32 inches or as flooring with joist spacings of 16 inches.

When installing plywood, place the better face up. Lay the panels with the 8-foot (long) dimension across the joists. Do not butt panels tightly. Leave a 1/16" space at the ends and 1/8" at the sides. To assure exact spacing for end joints (Fig. 9-30), use a 1/16" spacer. You can

Fig. 9-30 A spacer assures proper spacing of subflooring. (*American Plywood*)

Floor Framing

make it easily from sheet metal or any rigid material 1/16″ thick. To hold 1/8″ side joints, simply use two layers of spacer material. Double all joint spacings between panels in wet or humid areas. Be sure to stagger the joints in each row. If you start the first row with a 4′ × 8′ panel, start the second with a 4′ × 4′ panel. See Fig. 9-31. Nails for 1/2″ plywood should be 6d or 8d common for 5/8″ and 3/4″ plywood. Space the nails 6 inches apart around the perimeter and 10 inches apart along the intermediate members. Snap chalk lines on the plywood to locate the joist centers. This speeds up and simplifies nailing. Charts are available from plywood manufacturers detailing correct nail sizes for various plywood thicknesses.

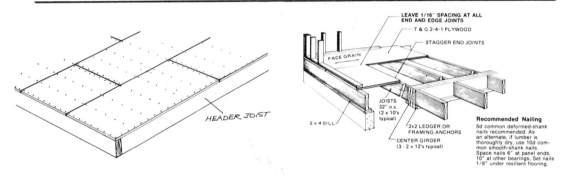

Fig. 9-31 Joints in plywood subflooring must be staggered.

Fig. 9-32 Floor framing details for 32″ joist spacing.

When you lay 1-1/8″ tongue-and-groove plywood as a combined subfloor and underlayment surface, use 1/16″ spacing at ends and edges of joints. The standard panel sizes are 4′ × 8′ but 4′ × 10′ panels are also available on special order. Fig. 9-32 shows framing detail when joists are spaced 32 inches on center.

Glued Floor System

Glued plywood floors permit the use of a one-layer floor instead of the conventional two-layer. This system results in savings in both labor and material. In addition, squeaking and nail-popping are reduced and floor stiffness is increased. The success of the system depends on the adhesive used and its application. In effect, the glued floor acts as a one-sided stressed skin panel. The various steps involved in the installation of a glued floor are shown in Figs. 9-33 and 9-34. See Fig. 9-35 for the nailing details of a typical glued floor.

Fig. 9-33 Spread enough glue to lay one or two panels at a time. (*American Plywood Assoc.*)

Fig. 9-34 Stagger all joints, leaving 1/16″ space between each.

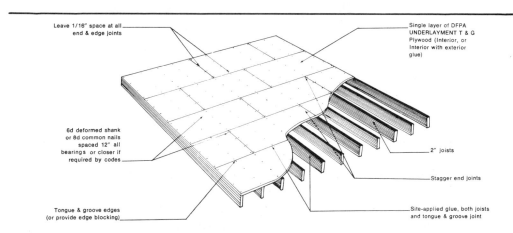

Leave 1/16″ space at all end & edge joints

Single layer of DFPA UNDERLAYMENT T & G Plywood (Interior, or Interior with exterior glue)

6d deformed shank or 8d common nails spaced 12″ all bearings or closer if required by codes

2″ joists

Stagger end joints

Tongue & groove edges (or provide edge blocking)

Site-applied glue, both joists and tongue & groove joint

Fig. 9-35 A typical glued-floor nailing schedule.

Termite Control

The best and easiest time to take preventive steps against termite entry into a wood building is during the laying of the foundation and subflooring. The damage caused by termites in the United States is quite extensive, running into hundreds of millions of dollars yearly.

Floor Framing

Subterranean and dry-wood termites are the two types that affect wood structures in the United States. Of the two, the subterranean termite causes the most destruction. The damp-wood termite, a third type, causes relatively little damage.

Subterranean Termites live underground where they build tunnels in soil in order to reach their food supply which is the cellulose of wood. They live in the dark, but they must have moisture to live. At certain times of the year, males and females swarm the area, fly for a short while, then lose their wings and mate. They resemble flying ants except that they have straight bodies compared to the ant which has a narrow waist. See Fig. 9-36.

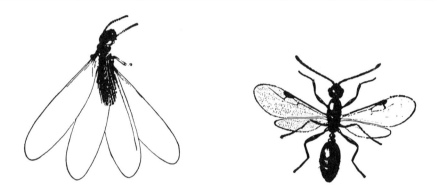

Fig. 9-36   Left, winged termite; **Right,** winged ant.

Dry-Wood Termites do not build tunnels to reach their food supply. They fly directly to the wood and bore into it from the surface. They cut across the grain, leaving large spaces supported by the tunnel walls. Dry-wood termites exist mostly in the southern states, and the damage they cause is not as extensive as that done by subterranean termites.

Unfortunately, termites eat the wood fibers from the inside, so that destruction may be well under way before it is discovered. Redwood, heartwood, bald cypress and very resinous southern yellow pine are some of the woods generally immune to attack by termites.

The best way to prevent termite entry is to poison the soil around and under the foundation. In addition, treated wood should be used for the lower members of the frame. Where dry-wood termites are known to do damage, posts and poles should have full-length treatment with wood preservative.

In order to be effective, wood should be pressure-treated, Fig. 9-37. Brushing or dipping wood is not effective because only the surface is protected that way. The preservatives used in treating wood are water soluble salts, creosote and pentachlorophenol (penta). When impregnated with any of these chemicals, the wood fibers are rendered useless as food for fungi and insects.

Fig. 9-37 Lumber entering pressure-treating tank. (*Koppers Co., Inc.*)

Not all wood in a house requires treating. Sills, joists, and subfloors are danger zones because of their location, near or below grade.

Various products are made for poisoning the soil. Some of these are dangerous to animals and humans and should be used with great care. Follow the manufacturer's recommendations and observe all safety rules.

To minimize termite infestation, good drainage away from the foundation is recommended. All tree stumps, roots, and untreated construction lumber scraps should be removed from the site. The temptation is great to bury all debris when backfilling, but this almost guarantees termite infestation later on.

Floor Framing

GLOSSARY

beam: a structural member of wood, steel or concrete used to support loads between walls or posts.

bearing: the part of a building member resting on its supports.

bridging: wood or metal pieces placed between floor joists or wall studs to stiffen them.

crawl space: the space between the first floor and ground in houses without basements.

girder: a large supporting beam used to support concentrated loads along its length.

header: a horizontal load-bearing support over an opening such as a door or window; also a supporting member placed at the head and foot of an opening.

headroom: the vertical height in a doorway; also, the clear height between tread and ceiling in a stairway.

joist: a horizontal timber laid on edge to support floor and ceiling loads.

joist hanger: a metal bracket used to support joists.

Lally column: a metal column or post used to support beams and girders, and usually filled with concrete.

overhang: a projection beyond a wall.

pier: a masonry pillar used for supporting structural members.

sealer strip: the layer of material, usually fiberglass, placed between the sill and top of the foundation wall; it keeps out drafts and insects.

sill: the lowest part of a structure resting on the foundation; also, the lowest member of a window or exterior door.

stressed skin: a panel consisting of two wood faces glued to a framework.

subfloor: the rough flooring laid directly over joints.

termite shield: a thin metal strip placed between foundation wall and sill to prevent the entry of termites; also used around pipes and other projections.

tongue and groove (T & G): shapes cut into the edge of boards. The tongue is a projection which fits into the groove or rectangular channel of the mating piece.

trimmer: a beam or joist into which a header is framed.

underlayment: the material on which a finished floor is placed.

10

Wall Framing

Generally, wood frame construction is either of two types: balloon or platform. Both methods are used extensively and each has its advantages and disadvantages.

Balloon Framing

In balloon framing, the wall studs extend in one piece from the sill to the top plate at the roof line. The studs and floor joists rest on the sill and are anchored to it, Fig. 10-1.

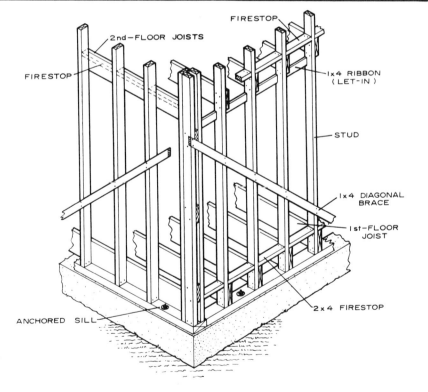

Fig. 10-1 Balloon framing.

The second floor joists rest on a ribbon which is set into the studs. Study the drawing, and note that it shows that subflooring cannot be installed until the studs are erected. This is a distinct disadvantage. Also notice the firestops which are essential in balloon framing to retard the spread of fire.

One advantage of balloon construction is the reduced shrinkage that occurs in its lengthwise framing members. (Most shrinkage occurs across the grain in lumber, not lengthwise.) Because of this low shrinkage, balloon framing works well in construction where the outside walls are masonry, such as brick veneer, stone or stucco, There is also less chance of plaster cracking at ceiling and floor lines.

Another worthwhile feature of balloon construction is that ducts, plumbing and electrical lines are installed with ease. Because of the limitation of studding lengths, however, balloon framing is limited to two-story structures. Also, it is impractical to work with studs more than 18 feet long.

Platform Framing

In platform framing the wall sections are erected above the subfloor one story at a time, 10-2. This greatly simplifies construction. Sections can be constructed flat on the subfloor, then tilted up into position, Fig. 10-3. The second floor in a two-story structure is erected in the

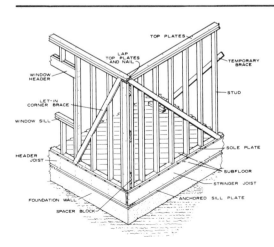

Fig. 10-2 Platform framing.

Fig. 10-3 In platform construction, wall sections made on the floor can then be erected. (*National Forest Products Assoc.*)

Wall Framing

same manner. Firestops are not needed as the top plates of each wall serve this purpose.

In some construction, as in a single story house, you may use a combination of balloon and platform framing. End-wall studs may run to the end of the rafters of a gable, while the side walls may be of platform design.

Wall Frame Members

The wall frame consists of vertical and horizontal members which make up the exterior walls and interior partitions. In addition to forming the walls, they also serve to support the ceiling and roof of a structure.

The main parts of a wall are the studs and the plates. The studs are the vertical members; the plates run horizontally. In addition, a wall may contain headers, trimmers, rough sills, cripples, firestops and ribbons.

Studs and plates are usually made from 2 X 4's, but 2 X 6 stock may also be used, especially in a wall or partition containing bathroom plumbing or duct work.

Sole Plate

The sole plate is the lowest horizontal member of a wall. It rests on the subfloor and is nailed into the bottom of the wall studs and into the floor frame. Because the plate is not a load-bearing member, the lumber you use for it need not be as good as that you use for studs and top plates. Large or numerous knots will also not affect the usefulness of this lumber.

Studs

The studs are the vertical support members of a wall. Like the sole plate, they are generally made of 2 X 4's. Spacing is usually 16″ OC, but 24-inch spacing may be permitted in some localities. In dwellings with more than one story, the upper-story studs must be spaced not more than 16″ OC. The "on-center" measurements in house framing are very important as they allow the use of standard size panels and wall covering materials, Fig. 10-4.

Fig. 10-4 Modular stud spacing permits use of standard size wall covering.

The lumber used for the wall frame must be strong and straight. Moisture content of the lumber should be no greater than 19 percent. The wood used depends on the availability in a particular locality. Some of the more common species used for framing are spruce, Douglas fir, yellow pine and hemlock.

Stud grade lumber is available in 8' lengths or as precut 7'9" lumber. Precut lumber saves time and labor for the carpenter and is used when the ceiling height is 8 feet. The precut size permits the use of 8' drywall or 16" rock lath with allowance for plaster grounds at the floor level.

Top Plate

The top plate is the horizontal member above the studs. It is supported by the studs and it in turn supports the ceiling joists and roof. The plate is usually made up of two pieces of 2 × 4, Fig. 10-5.

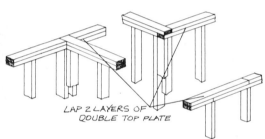

Fig. 10-5 Left The top plate supports the ceiling joists and roof.

Fig. 10-6 Right The upper section of the top plate is lapped to tie the wall section.

Normally, the second plate is installed after the wall is erected in platform construction. This permits lapping of the corners, thus tying the wall sections. See Fig. 10-6.

Fire Stops

Fire stops are barriers made of either noncombustible material or of wood at least 2" thick, placed in concealed air spaces to prevent drafts that would encourage the spread of fire in a building. They may be used in walls, floors, or roofs. In a wall, the fire stops are placed horizontally between studs. Fire stops may be used in platform construction but they are a necessity in balloon-type frame construction.

Bracing

In wall construction, bracing may be used permanently to stiffen wall sections, or only temporarily until the framing is completed.

Bracing may be applied straight or diagonally. The lumber used is generally 1 × 6 or 1 × 4 stock, but 2-inch stock of the same thickness as the studs may also be used. One-inch stock is "let-in"; that is, the studs are notched to permit the brace to lay flush with the studs, as illustrated in Fig. 10-7.

Fig. 10-7 Let-in bracing of a stud wall.

If stock 2-inches or heavier is used, it is "cut-in." This means that every block must be cut to fit between the studs. The blocks are usually cut diagonally and installed between plates. When plywood sheathing, 5/8" or thicker, is used, bracing may be eliminated.

Wall Framing

Door and Window Framing

Door and window openings in wall framing are referred to as rough openings (RO). The size of these rough openings depends on door and window frame sizes. Usually the architect's plans will indicate the RO sizes. As a rule, for double hung windows the rough opening width is equal to the glass size plus 6 inches. The RO height is equal to the glass height plus 10 inches.

To figure the RO size for casement windows add 11-1/4 inches to the glass width, and 6-3/8 inches to the glass height. For exterior doors, add 2-1/2 inches to the width and 3 inches to the height for the rough opening size.

When doors and windows are installed, regular studs must be cut. They are replaced by shorter studs called cripples. The sides of an opening are reinforced with trimmers, or supporting studs. These shortened studs also support the header (or lintel).

Headers span the opening of a door or window and support the load over them. The size of a header is determined by the length of the span and the ceiling plus roof loads above it. As the span increases, so also should the depth of the header. See Fig. 10-8.

Headers are made from two pieces of 2-inch stock spiked together with a spacer in between. The spacer can be plywood or any other suitable material of the proper thickness. The combined thickness of the double header with filler should equal the stud thickness. If the depth of the header leaves little room for cripples, the carpenter may increase the header size to occupy the entire opening.

The RO sizes for doors and windows are determined by the millwork used. Rough opening sizes for various stock units are listed by manufacturers.

Corner Posts

Corner posts join two intersecting walls. They also provide nailing surfaces for inside and outside wall covering materials. They may be erected as a separate unit as in balloon construction, or they may be formed when two sections of a platform wall are raised into position.

Some carpenters prefer to erect the corner posts first. The posts can then be used as a guide for plumbing the preassembled wall sections

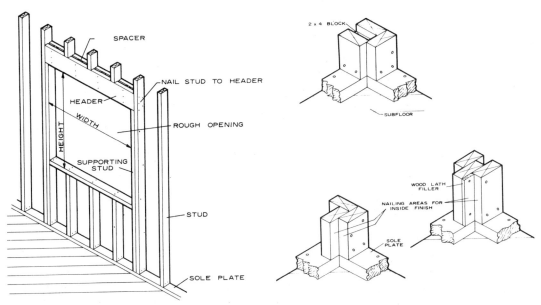

Fig. 10-8 Header spans the opening of a window.

Fig. 10-9 Three methods of constructing corner posts. (*Forest Products Lab.*)

when they are raised into position. However, this method does not permit the sheathing of the sections before they are erected.

There are several ways of constructing a corner post. They are usually made with three or more regular studs. The lumber used for corners must be straight and free of defects; Fig. 10-9 shows some of the more common arrangements used in building corners. As shown in the drawing, in platform construction the end wall stud is butted against two or more of the side wall pieces. These are nailed with 10d nails. If the corner is formed with blocking, use at least three pieces.

Intersecting Walls

When a partition meets an outside wall special construction is needed. The partition must have a solid base to nail into. In addition, corners must be formed to provide a nailing surface for the inside wallcovering material. Several methods can be used to construct these intersecting walls. See Fig. 10-10 at the top of the next page.

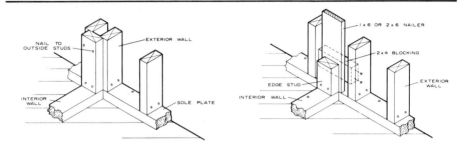

Fig. 10-10 Construction details for a partition wall intersecting an exterior wall.

Sole Plate Layout

The sole plate or bottom plate layout is the first step taken in framing the walls. Partitions, windows and doors are located on the subfloor. Chalk lines are struck to layout the partitions. Doors and windows are indicated with center lines. Lumber for the sole plates (2 X 4's) is placed on the subfloor in proper position around the perimeter. Choose straight pieces and tack them into place so they won't move. Side wall plates along the longest wall should run to the end of the floor frame. Butt the end wall plates to them.

After deciding on the corner post method to be used, start laying out the stud locations. Begin with the side wall and note that the centerline of the first stud is 16 inches from the outside corner of the frame, Fig. 10-11. For example, assume a 16" stud spacing. Measure 3/4 of an inch to either side of the center line and draw a pencil line.

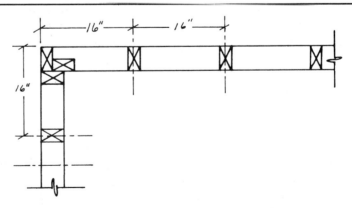

Fig. 10-11 Stud layout on sole plate; for intermediate studs use 16" spacing, center to center.

This represents the first regular stud. Now, measuring from the center of the mark, pick off a mark every 16 inches along the length of the wall. Do *not* alter the regular stud locations for doors and windows. Some carpenters like to use the 16" leg of the framing square for this. The problem with that method is that cumulative errors may develop. A more accurate method is to use the steel tape, most of which have prominent figures marked off for stud spacing.

After regular stud centers are laid out, use a short piece of 2 × 4 as a template to mark the thickness of each stud. Next, locate the corners, partitions, doors and windows. For the windows and doors, mark off the rough openings by making equal measurements on both sides of the door or window center line. These lines represent the inside edge of the trimmer stud. Again, using the short template stud, mark off the thickness (1-1/2 inches) of these studs. Next to these and away from the opening add another stud mark which will represent full studs. Mark the full studs with an "X". Trimmer studs are identified with a "T". Lay out the partition studs, using one of the methods shown. Since these are full studs, mark them with an "X". Cripple studs are identified with a "C". The markings are now transferred to a second 2 × 4, similar to the one just marked. This will be used for the top plate. This procedure will be repeated for the balance of the sole plates. See Fig. 10-12.

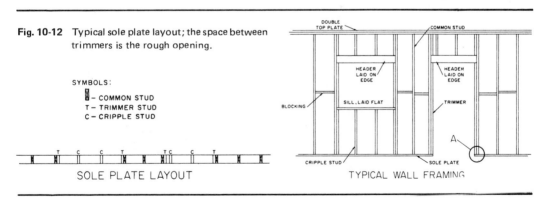

Fig. 10-12 Typical sole plate layout; the space between trimmers is the rough opening.

SYMBOLS:

X – COMMON STUD
T – TRIMMER STUD
C – CRIPPLE STUD

SOLE PLATE LAYOUT

DOUBLE TOP PLATE

COMMON STUD

HEADER LAID ON EDGE

HEADER LAID ON EDGE

BLOCKING

SILL, LAID FLAT

TRIMMER

A

CRIPPLE STUD

SOLE PLATE

TYPICAL WALL FRAMING

The Story Pole: After the sole plate layouts are completed, the stud sections must be cut and the wall assembled. In order to lessen the chance of error, the carpenter makes and uses a story pole or rod. This is simply a suitable piece of lumber marked off with the sizes of studs, trimmers, headers, cripples and sills. Since all markings are laid out full size, there is little chance of error. Choose lumber for the pole that is straight and the same thickness as the studs. Make it a little longer than

Wall Framing

the height of the wall. Ordinarily, a ten-foot length of 2 × 4 will suffice. Your markings should include sole, rough sill, header and top plates. If some markings are too close to each other, use the other side of the "pole." Fig. 10-13 shows story pole markings for a header and double plate.

Fig. 10-13 Story pole markings for a header and top (double) plates.

The story pole for balloon-frame construction must be long enough to extend from the sill to the top plate which may be two or more stories high. Floor joists and ribbons of the second story are also marked off on the pole.

Wall Assembly

The actual assembly of the wall can now be started. Cut the various stud lengths as indicated by the story pole. You can save considerable time by using precut studs. Since they are factory trimmed each piece

Fig. 10-14 Place studs and plates on edge during assembly—a flat and clean surface is a must.

will be identical in length. Place the sole and top plates on edge, far enough apart so that the studs can be installed, Fig. 10-14. Be sure the floor is swept clean for this operation. Nail the top and bottom plates to the regular studs. Use 16d nails. Before nailing the door studs, lay them flat on the floor and add the trimmers. Use 10d nails and be sure the bottom edges are aligned.

After the regular and RO studs are in place, install the headers, rough sills and cripples. It will be necessary to toenail the cripple studs which butt the headers. Use 8d box nails for toenailing. Box nails are thinner than common nails and have less tendency to split the wood.

When all members have been installed, check the section for squareness. Measure the diagonals which should be equal, Fig. 10-15. If the wall is not to be pre-sheathed, install temporary diagonal bracing to keep the assembly square while erecting. If sheathing is to be applied, the bracing is not required. Tack the assembly to the floor, however, to prevent movement while sheathing.

The sheathing may be plywood, composition panels or boards. If bracing is needed, you must install it before applying the sheathing. Consult building codes regarding the bracing. Normally, a wall sheathed with 5/8" plywood will not have to be braced.

Fig. 10-15 Wall frame is checked; if diagonals are equal, the section is square. (*National Forest Products Assoc.*)

Fig. 10-16 A wall section is raised or "walked" upright while a helper stands by to secure it.

Raising the Wall is not difficult. Sufficient help should be on hand to support the weight and one person should be free to attach temporary bracing. Fig. 10-16 shows a sheathed section being erected. Be sure to leave an opening in the double plate wherever a partition intersects the wall. End plates are half-lapped at the corners to tie them in. Nail the lower plate to the floor frame with 16d nails spaced 16 inches apart. Secure the corner posts with 10d nails 12 inches apart.

Wall Framing

Partition walls are assembled after the outer walls have been erected. They may be bearing or non-bearing. Load-bearing partitions run perpendicular (at right angles) to the ceiling joists. Non-bearing partitions run parallel to the ceiling joists. If the partition is load-bearing, door and window openings must be doubled just as they are for outer walls. Otherwise, single studs may be used. To keep costs down, some carpenters build non-load bearing partitions with 2 × 3's.

Partition walls are usually constructed perpendicular to adjacent walls. There are times however when plans call for diagonal walls. These must be laid out carefully so their intersections provide a good nailing surface for the wallcovering. Do enough nailing to keep the joint from opening. In Fig. 10-17 a diagonal wall is seen at the floor line.

After walls and partitions have been erected, add the finishing touches. Remove sole plate bridging the door openings. Cut alongside the trimmer studs with a handsaw. If the section has been inadvertently nailed, remove the nails and discard the piece.

Fig. 10-17 Floor-line construction of diagonal partition.

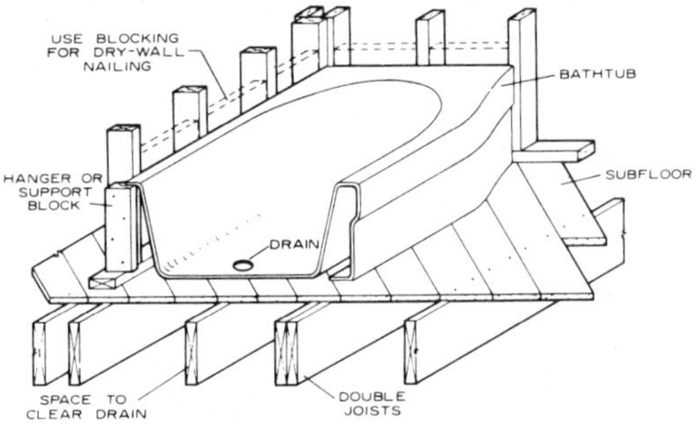

Fig. 10-18 Blocks nailed to wall studs support tub; their lower ends rest on the sole plate.

Framing for Utilities

Wood backing must be installed for plumbing and heating units during the rough framing stage. Also add corner blocks to the framing intersections wherever they are needed—to provide a nailing surface for baseboards, drywall and other materials; Fig. 10-18 shows the installation of bathtub supports and the drywall backing around the tub.

Most medicine chests are designed to fit between studs. Some, however, may require cutting through one or more studs. These openings require headers.

Framing for Utilities

Some of the items that require backing: towel bars, plumbing units, hot air registers, hand rail brackets, shelves, valances, fans, vents, air conditioners, attic access doors, clothes chutes and central vacuum outlets. Also install backing for electrical outlets, toilet tanks, wash basins, soap dishes, tissue dispensers, shower doors and the like; Fig. 10-19 shows how a wall can be framed for recessing a convector.

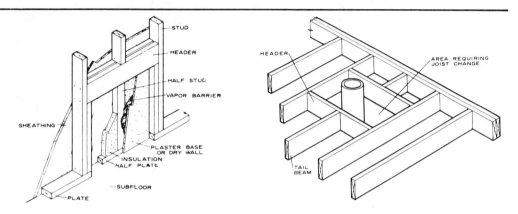

Fig. 10-19 Framing layout for recessing a convector.   **Fig. 10-20** Layout showing headers installed to support a joist.

Cut or notch framing members only if it is absolutely necessary. When floor joists are cut, drilled or notched, their strength may be affected. If it is essential that they be cut, reinforce them by nailing scabs on each side of the joist.

To eliminate the cutting of joists, headers may be installed as in Fig. 10-20. Figure 10-21 shows two methods of framing vent pipes.

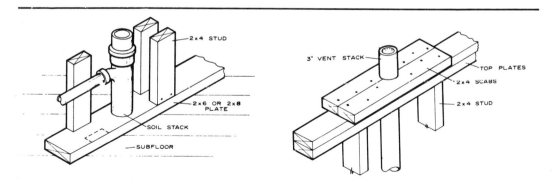

Fig. 10-21 **Left,** framing layout for 4″ soil stack; and **Right,** for 3″ vent stack.

Fig. 10-22 A header supports wall above brick fireplace.

Fireplace Wall Frame: The wall above a fireplace should never rest on the masonry. Instead, a header should be installed as shown in Fig. 10-22. The header size depends on the span.

Ceiling Joists

Ceiling joists serve several purposes. They form a nailing surface for the ceiling finish material. They also tie together opposite walls and they resist the thrust of rafters. They may also serve as an upper story or attic floor.

Ceiling joists are installed after the wall and partition framing is completed. Installation is somewhat similar to the procedure for floor framing. The main differences are that there are no header joists and the lumber used is not the same size. If the ceiling is going to support a floor, the joists will be heavier than those used in an unused attic. Consideration must be given to the fact that an unused attic may be converted into living quarters at a later date. If this is likely, ceiling joists should definitely be heavier.

Ceiling joists usually run the short way of the structure, the same as the floor joists. There are times, however, when they may run in two directions, as in Fig. 10-23. This is done to reduce the length of the span, especially in L- or T-shaped dwellings. Joists that project above the rafters in shallow pitched roofs are another reason for changing the direction of the joists.

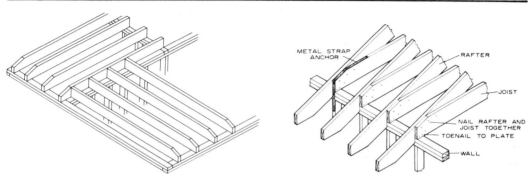

Fig. 10-23 Ceiling joists running in two directions.

Fig. 10-24 Ceiling joists connected at outside wall.

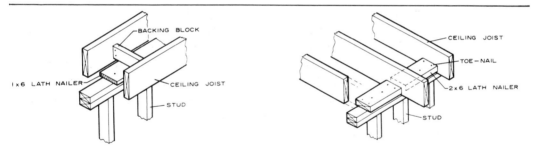

Fig. 10-25 Method of anchoring partition to joist.

Fig. 10-26 Nailers between joists are 2 X 6 stock.

If a wall is constructed with a double top plate, the ceiling joists need not be placed over the studs. The double plate provides sufficient rigidity. If possible, space the ceiling joists the same as the rafters. This makes for a more rigid structure since the joists and rafters can be nailed into each other as well as the plate, Fig. 10-24. In high wind areas, metal strap anchors are recommended. The procedure is to toenail the joists to the plate with 10d nails, then nail the rafters to the joists. Joists are also nailed to load-bearing partitions. If the joists run parallel to the partition, you must provide blocking to anchor the partition to the joist. A nailer for ceiling material is also needed. See Fig. 10-25. Nailers consisting of short 2 X 6's are installed between joists when the joists are perpendicular to the partition, Fig. 10-26.

Wall Framing

When the joists cannot span the walls in one piece, they are joined either by lapping or butting. If lapped, a filler block should be used. That allows the placing of joists on both sides of the rafters, Fig. 10-27.

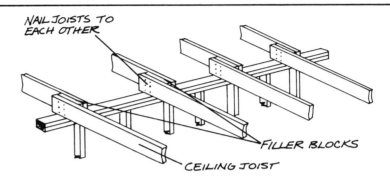

Fig. 10-27 Spacer (filler) blocks lap joists too short to reach across span of wall rafters.

When plans call for a large ceiling without a load-bearing partition to help support it, you may use a flush beam. The beam must be of sufficient size to support the joists. Anchor the joists to it by one of several methods. Joists may be toenailed or they may be held with metal hangers. Steel straps are also used for added strength. Another method is to make use of a ledger strip as is done in floor framing.

Openings in ceilings may be needed for a hatchway or for disappearing stairways. Hatchway sizes are determined by local codes. Rough-opening sizes for disappearing stairways are furnished by the manufacturer of the unit.

Low Cost Module Framing

A low cost module-framing system has been devised for single and multi-family dwellings. It utilizes plywood-over-lumber framing spaced on 24" centers. The result is a considerable saving in both labor and materials.

The two-story house shown in Fig. 10-28 was constructed with universal 24" framing for floors, walls, and roof.

Called the MOD 24 Framing System, this system is recognized and approved by the FHA and five major model codes. In effect, the system eliminates one stud for each four feet of wall, thus saving raw material, time and labor, Fig. 10-29.

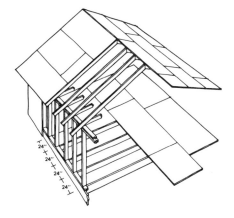

Fig. 10-28 Two-story house built in 24" modules. (*American Plywood Assoc.*)

Fig. 10-29 All framing members align in the 24" module system. (*American Plywood Assoc.*)

Floors in Mod 24 System: To better utilize materials, single floor construction is recommended. Space joists 24" OC and use tongue and groove 3/4" plywood for the subfloor, as shown in Fig. 10-30. Glued flooring is recommended because it increases stiffness and eliminates loose nails and squeaks.

Fig. 10-30 Glued floor construction increases stiffness. (*American Plywood Assoc.*)

Wall Framing

Walls in Mod 24 System: To obtain the full benefit of modular spacing, plan the wall openings carefully. The comparison of the two wall sections in Fig. 10-31 illustrates how pre-planning can result in a substantial saving of material. Studs up to 10-feet long can be used in single-story construction. In two-story construction, limit the first floor studs to 8 feet.

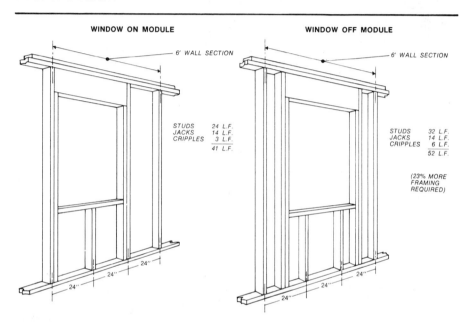

Fig. 10-31 Proper design can cut cost and labor; on-module construction at the left requires 23 percent less framing than unit on the right.

Exterior plywood siding should be at least 1/2" thick when single wall construction is used. The plywood in this type of construction serves as sheathing and siding. If two-layer construction is used, 3/8" plywood is acceptable under lumber or other types of siding. Gypsum wallboard 1/2" thick is acceptable for interior wall lining.

Roof Framing in Mod 24 System: Trussed rafters are commonly used to support roof and ceiling loads in light frame construction. Such trusses are designed and fabricated to meet the requirements of various building codes. They are usually spaced on 24" centers, Fig. 10-32. Truss manufacturers have tables available which show allowable clear spans.

Fig. 10-32 Interior view of framed and sheathed house; note in-line placement of studs and trusses. (*American Plywood Assoc.*)

GLOSSARY

balloon framing: a method of construction in which the wall studs extend in one piece from the foundation to the roof.

brace: a piece of lumber used to reinforce a wall or other framing.

cripple studs: shortened studs cut to fit above and below openings in a wall frame.

end wall: the short wall of a structure.

firestop: cross blocking in a frame wall to retard the spread of fire.

nailer: a strip of wood or blocking which serves as a backing into which nails can be driven.

partition wall: an interior wall separating one area of a house from another.

plaster ground: a strip of wood used as a thickness gauge when plastering a wall; usually placed at the floor and around windows and doors.

plate (also top plate): a horizontal framing member nailed to the top of wall studs; it supports the ceiling joists and roof rafters.

platform framing: a method of construction in which each floor is framed independently; the joists of the floor above rest on the top plate of the floor below.

Wall Framing

ribbon: a horizontal timber nailed to the face of the studs; it usually supports floor joists.

rough opening (RO): the opening in a framed structure.

side walls: the long walls of a structure.

sole plate: the bottom horizontal member of a wall.

stud: the upright member of a wall frame.

11

Roof Framing

The roof frame forms the base to which roofing materials are applied. Its function is to keep out the weather and to protect the occupants. It must also provide proper drainage of water and be strong enough to withstand high winds and, in cold climates, heavy snow loads. In addition, the roof must be pleasing to the eye and compliment the structure it adorns, Fig. 11-1.

Roof framing is an art. It is perhaps the most complicated task the carpenter will encounter in house construction. However, if he learns to use the framing square accurately and has some knowledge of geometry, he should have little difficulty in mastering this rewarding craft.

Fig. 11-1 A well-designed roof on a house. (*Certain-Teed Products Corp.*)

Roof Framing

Types of Roofs

Roofs vary in their intricacy. A flat roof is the simplest to build while the intersecting hip roof represents one of the more complex types. The most commonly used roof types and a description of each is included in this chapter. Some are shown in Fig. 11-2.

Flat Roof: In it, the ceiling joists also serve as the rafters on which the roofing material is applied. A slight slope is usually incorporated for drainage. Flat roofs must be of sturdy construction because they carry roof and ceiling loads. They do not contain attic spaces and they are usually insulated. If the roof overhangs on four sides, lookout rafters are employed, Fig. 11-3.

Shed Roof: Also called a *lean-to,* it is pitched in a single plane and

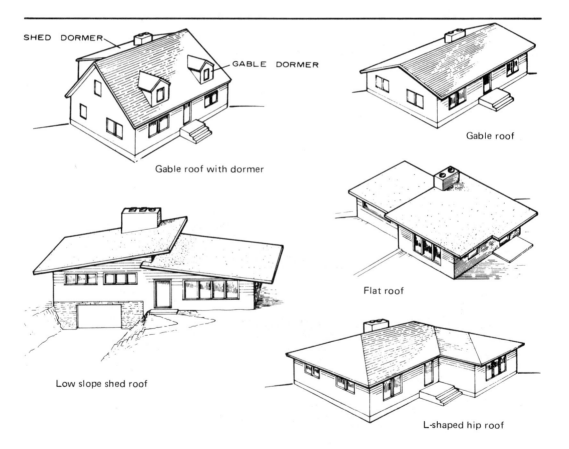

SHED DORMER

GABLE DORMER

Gable roof with dormer

Gable roof

Low slope shed roof

Flat roof

L-shaped hip roof

Fig. 11-2 Roof types used in residential construction.

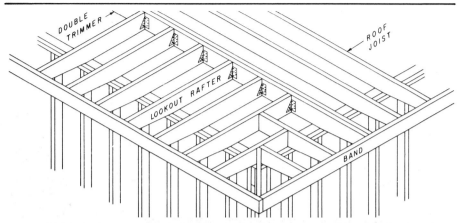

Fig. 11-3 Construction detail for flat roof. (*National Forest Products Assoc.*)

may or may not lean against another building. It is supported by walls or posts which are higher on one side than the other.

Gable Roof: This roof is pitched with two slopes meeting at the ridge. It is of simple design, economical to construct, and one of the most common roofs used in residential homes. It is sometimes combined with a shed roof.

In effect, the gable roof is like two shed roofs back to back. By reversing the slopes, the roof becomes a **butterfly** with the low point at the center.

Hip Roof: It has four sloping sides which rise to meet at the ridge. The lower edges of this roof form an overhang on all four sides of the structure. The hip roof is used mostly on low silhouette ranch type homes. When the plan of the structure is square, all four sides meet at a center point and the ridge is eliminated. It is then called a **pyramid roof**.

Gambrel Roof: This roof has two sloping sides stepped at different angles on each side of the center ridge, Fig. 11-4. The upper slope is

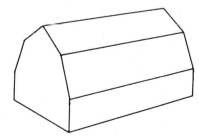

Fig. 11-4 A gambrel roof.

Roof Framing

Fig. 11-5 A mansard roof.

slight while the lower is quite steep. This design gives greater headroom than a gable or hip roof.

Mansard Roof: It may be likened to the gambrel except that it has four sides, each with a double slope. The upper slopes in this roof are almost flat. It, too, gives more headroom on the second story than with a gable or hip roof. See Fig. 11-5.

Roof Framing Members

The basic roof frame consists of three main members—the plate, common rafters, and ridge board. In a gable roof these are the only members employed. As the roof becomes more complex, hip, valley and jack rafters are employed. Rafters support the roof in much the same manner as studs support the walls and joists support the floors. They are the structural members of the frame and their size and spacing is determined by the load placed upon them. Generally, rafters are made with 2 × 6 stock and the ridge board is usually 2 × 8 lumber. The plans you use will specify the size. See Fig. 11-6 which illustrates the various roof frame parts described on the next page.

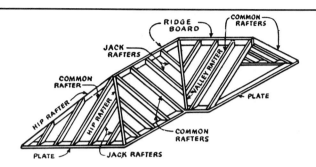

Fig. 11-6 Roof frame parts. *(Stanley Tools)*

Common rafters are those members extending from the plate to the ridge board at right angles to both.

Hip rafters extend diagonally from an outside corner plate to the ridge board.

Valley rafters extend diagonally from an inside corner plate to the ridge board at the intersection of two roof surfaces.

Jack rafters may be classified as hip jacks, valley jacks and cripple jacks, depending on where they are used.

Hip jacks extend from the plate to the hip rafter.

Valley jacks extend from valley rafters to the ridge board.

Cripple jacks extend from hip jacks to valley jacks.

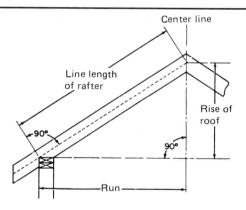

Fig. 11-7 Roof frame terms.

All the terms used in roof framing must also be fully understood. Refer to the diagram above in Fig. 11-7 and learn and memorize each term on the following page:

Roof Framing

Span is the distance between two opposite walls, measured from the outside of the plates.

Run is the horizontal distance from the outside of the plate to the center of the ridge. Generally it is one-half the distance of the span.

Rise is the vertical distance from the top of the plate to the center line of the ridge.

Rise in inches is the number of inches the roof rises for every 12 inches of run.

Total rise is the distance from the top of the plate to the top of the ridge.

Pitch is the amount of slope in a roof from the ridge to the plate. It is usually expressed as a fraction such as 1/3, 1/2, or the like. It is determined by dividing the total rise by the span.

For example: A roof with a 6 foot rise and a 24 foot span will have a 1/4 pitch. (6/24 = 1/4). See Fig. 11-8.

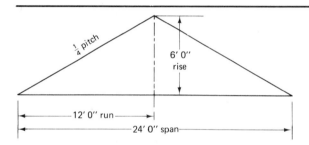

Fig. 11-8   The rise divided by the span equals the pitch. It is usually expressed as a fraction.

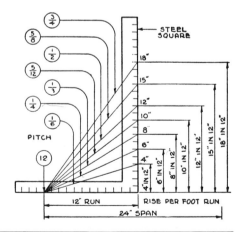

Fig. 11-9   Common roof pitches used in house construction.

Pitch may also be expressed as 6 in 12, 4 in 12, and so on. The first number is the vertical rise in inches per foot of horizontal run, and the second is the unit of run, per foot, expressed in inches. See Fig. 11-9.

Layouts for Common Rafters

The common rafter may be considered the imaginary hypotenuse or slant of a right triangle (the long side opposite to its right angle). An imaginary line, called the *measuring line* is used when taking measurements. This measuring line passes through the center of the rafter and the top outer corner of the plate, Fig. 11-10 below.

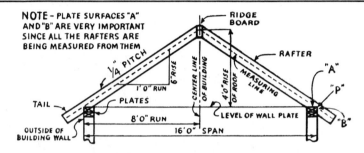

Fig. 11-10 Rafter measurements are taken along imaginary "measuring" line. (*Stanley Tools*)

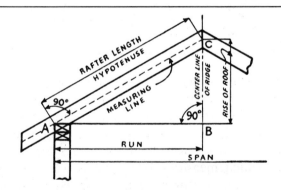

Fig. 11-11 Rafter length is calculated from outer corner of plate to center of ridge, along the measuring line. (*Stanley Tools*)

The total length of the rafter or *rafter length* is measured along its center (the measuring line) from the outer edge of the top plate to the center of the ridge. See Fig. 11-11 above.

The *tail of the rafter* is that portion which projects beyond the wall, measured along the rafter. The distance from the end of the rafter to the wall is the overhang. It is measured horizontally.

The principal cuts in a common rafter are shown in Fig. 11-12. The *plumb cut* (top cut) is the edge which rests against the ridge board. The *bottom* or *heel cut* rests on the plate. Because of its shape, this interior cut or notch is often referred to as the **bird's mouth.** Hip, valley and jack rafters have **side cuts** in addition to the top and bottom cuts. Side cuts are also referred to as **cheek cuts.**

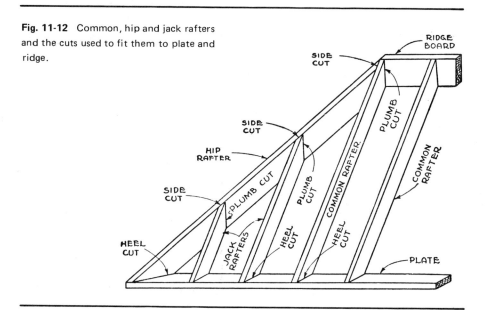

Fig. 11-12 Common, hip and jack rafters and the cuts used to fit them to plate and ridge.

The framing square plays an important part in roof frame layout and in measuring rafter cuts. It is recommended that you review the section in Chapter 4 on the framing square before you proceed.

You will recall that the framing square has a body and a tongue: the body is a blade 2 inches wide and 24 inches long; the tongue is a narrower blade which measures 1-1/2 inches wide by 16 inches long; the side with the manufacturer's name stamped on it is its face; the opposite side is its back. The scale divisions marked on the square include eighths, tenths, twelfths and sixteenths. See Fig. 11-13.

The rafter tables are on the face of the square's body. They are used to determine the lengths of common rafters. The inch marks on the outer edge of the body face represent the **rise in inches per foot run.** See Fig. 11-14. The first row of figures below the edge indicates the actual length per foot of the common rafter. For a rafter with a rise of 8″ refer to the rafter table. Under the 8″ mark is the figure 14.42. This means that for every foot of run, the rafter length is 14.42 inches. Therefore, to find the length of the common rafter, simply multiply the figure in the table by the number of feet of run.

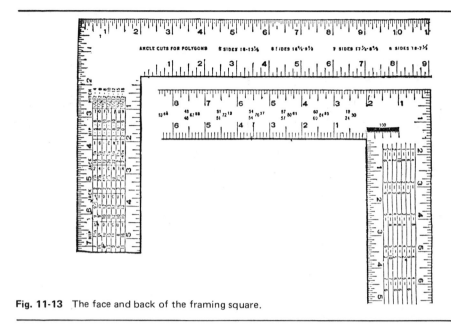

Fig. 11-13 The face and back of the framing square.

LENGTH	COMMON		RAFTERS	PER FOOT	RUN	21 63		15 00	14 42	13	
1 1	HIP	OR	VALLEY	1 1	1 1	24 74		19 21	18 76	18	
DIFF	IN LENGTH		OF JACKS	16 INCHES	CENTERS	28 84		20	19 23		
1 1	1 1		1 1	2 FEET	1 1	43 27		30	28 84		
SIDE	CUT		OF	JACKS	USE	6 11/16		9 5/8	10		
1 1	1 1		HIP	OR	VALLEY	1 1	8 1/4	10 6/8	10 7/8		6

Fig. 11-14 Rafter tables on the face of the square's body. *(Stanley Tools)*

Example: Run of rafter is 10 feet, rise is 8 inches. The table shows 14.42 under the 8" mark. Multiply 14.42 by 10. The answer is 144.20 inches. Divide 144.20 by 12 to get the answer in feet. The answer is 12.01 feet, but for practical purposes drop the .01. The length of the rafter on the measure line is 12 feet. See Fig. 11-15 on page 182.

The rafter table actually represents the hypotenuse of a right triangle. The inch mark is the rise and the 12 represents the unit run. Check this out by placing a ruler so that it crosses any one of the inch marks on the body and the 12" mark on the tongue. This diagonal will measure the same as the figure printed under that same inch mark.

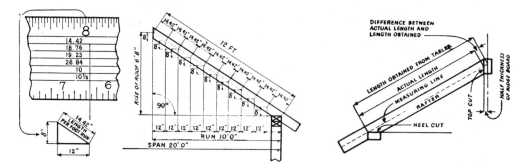

Fig. 11-15 The rafter table in use for measuring rafter Fig. 11-16 Rafter cutting lines.
lengths.

The measure line obtained from the table is to the center of the ridge. Therefore, half the thickness of the ridge board must be deducted to obtain the actual cutting line. The deduction is measured perpendicular to the center line as shown in Fig. 11-16. To make the actual cut lines, place the square as shown in Fig. 11-17. The body now

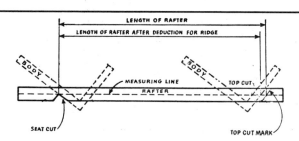

Fig. 11-17 The square used to reckon cutting lines.

represents the unit run (12 inches) and the tongue represents the rise per foot run. Deduct for the ridge thickness and draw a line. That will be the top cut. Now move the square to the bottom of the rafter with the rise and run marks again aligned with the edge of the rafter. Allow the body to coincide with the plate mark and draw a level line from plate mark to bottom edge of rafter to represent the seat cut. Draw a line perpendicular to the seat. That is the bottom plumb line and when both those lines are cut, the "bird's mouth" will be formed.

NOTE: the term plumb and level as used above refer only to the direction of the lines as they will appear on the rafter when it is in its proper position on the roof.

Layouts for Common Rafters

The framing square is not scaled for odd inches, so if the run of a building has an odd number of inches it is handled thus:

Example: The building span is 24'-10"

Answer: That means a run of 12'-5", but the scale is used only for the 12 feet. The additional 5 inches are added to the last plumb line. See Fig. 11-18.

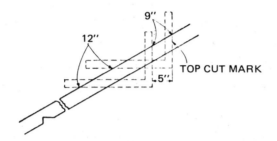

Fig. 11-18 Method of adding odd inches to length of rafter.

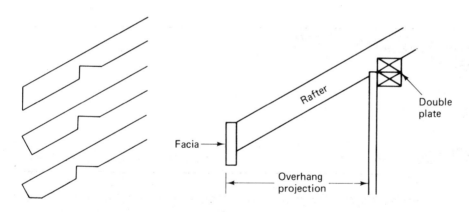

Fig. 11-19 End of a rafter tail—a combination of plumb and level.

Fig. 11-20 Method for calculating layouts.

The *tail* or *eave measurement* is added to the plumb line of the bird's mouth next. The end of the tail may be square, plumb, level, or a combination of plumb and level as shown in Fig. 11-19. The working plans specify the shape of the end as well as the amount of overhang. The simplest way to layout the tail is to make a full size layout of the overhang. Actual length of the tail can then be added to the rafter template. Be sure to allow for facia thickness, Fig. 11-20.

Roof Framing

There are other ways of laying out rafters. One method used is called **step-off**. This is a step and repeat process in which the square is stepped successively to calculate the desired length. The number of steps used is equal to the number of feet in the run. The square is placed with the unit and rise figures on the rafter edge. The unit run or 12 inches is set on the body and the unit rise on the tongue. Use a sharp pencil or knife and mark each step starting with the ridge line. Keep the blade aligned with the edge marks, then draw the line. This will be the line length of the rafter. The tail is treated as outlined above for the rafter tables. Fig. 11-21 shows the various steps involved.

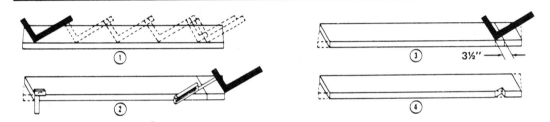

Fig. 11-21   Rafter layout by step-off method.

Fig. 11-22   Cutting rafters on the job site.

Cutting Rafter Templates: A template may now be cut and used to trace the balance of the rafters. However, the template should be checked first for accuracy. Trace and cut two rafters and assemble them on the roof. If they are correct, the rest may be cut. Use *only* the master template for tracing. For greater accuracy and economy, the rafters should be cut with power tools. A radial arm saw is most practical for cutting rafters. Work can be ganged, and special cutters are available for making seat cuts in one pass. When rafters have to be cut on the job, however, stationary tools are not usually available. Most carpenters set the rafters on sawhorses, then proceed to cut them as shown in Fig. 11-22. After cutting, be sure to place the rafters against the frame of the structure.

Erecting the Gable Roof

The gable roof is the simplest of the pitched roofs to assemble. First, ceiling joists should be put in place; then rafter locations should be marked on the top wall plates as well as on the ridge board, Fig. 11-23. Normally, the ridge board will have been laid out during the layout of the joists on the wall plates. Spacing of the rafters may vary. If the spacing is the same as the ceiling joists, the rafters will be installed adjacent to each joist. When the spacing differs, only some of the rafters will coincide with the joists.

Fig. 11-23 Rafter installation can start when ceiling joists are in place.

Fig. 11-24 Method of joining short sections of ridge board with plywood splices.

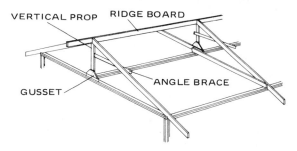

Fig. 11-25 Temporary bracing is used when erecting ridge board. (*American Plywood Assoc.*)

Select flat and straight rafters for the end pieces. Temporarily nail the first rafter to the plate while an assistant holds the free end. Install the mating rafter, then temporarily nail both to the ridge board. The first pair should be flush with the outside edge of the end wall. If the ridge board must be made of more than one piece because of its length, do not join the sections beforehand. It will be easier to work with the shorter lengths. Joining can be done with plywood splices, Fig. 11-24.

Install two more rafters, also temporarily; these should be toward the opposite end. Plumb the rafters, and temporarily brace the ridge board. See Fig. 11-25. Use a plumb line or level to check for squareness.

Roof Framing

If everything is true, nail permanently. Continue to install the intermediate rafters in pairs. One rafter in each pair is face nailed through the ridge board. Others are toenailed. When a rafter is adjacent to a joist, nail it to the joist as well as the plate. Use 16d nails through the ridge board and 8d box nails when toenailing. Add the balance of the rafters until all have been installed. The ridge board may be spliced to increase its length; as each section of the ridge board is installed, check carefully to make sure it is straight. Fig. 11-26 shows several types of roof anchors commonly used. In high-wind areas, special metal straps may be required.

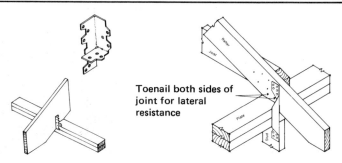

Toenail both sides of joint for lateral resistance

Fig. 11-26 Several types of metal rafter anchors.

In hurricane-prone areas, rafter bracing is often used. The braces act as trusses and effectively reduce the span of the rafters. The braces also tie the rafters to the ceiling joists. See Fig. 11-27.

Collar-beam ties are also used to reinforce the roof frame. See Fig. 11-28. In an attic crawl space, they are placed horizontally in the upper

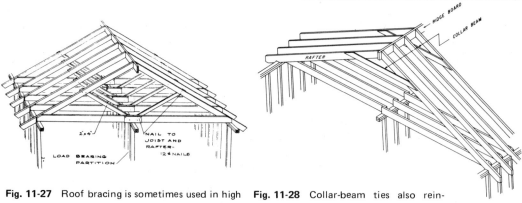

Fig. 11-27 Roof bracing is sometimes used in high wind areas.

Fig. 11-28 Collar-beam ties also reinforce the roof frame.

third of the crawl space. They are usually made of 1 × 6 stock nailed at every fourth rafter. If the attic is to contain a finished room, the collar beams are made of 2 × 4's or 2 × 6's and placed on every rafter; they also serve as joists for the ceiling.

The gable studs are installed on the end wall plates in line with the wall studs. Notch the tops to fit under the rafters, Fig. 11-29. The diagonal line may be traced directly from the rafters. Hold the studs plumb when marking the cutting line. Be sure that the end rafters are not crowned. If necessary, install a temporary brace to straighten the rafter while the gable studs are being installed.

To layout the cut line without tracing each one, mark the first two studs, noting the difference in length between them. Increase or decrease each succeeding stud by this same amount.

Toenail into plate and rafters with 8d nails, and spike the top into the rafter with 16d nails. If a vent is to be installed in the gable end, space the center studs to accommodate it, Fig. 11-30.

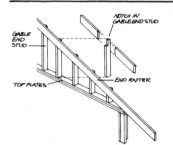

Fig. 11-29, Left: End wall studs are notched to fit under the rafters.

Framing for vent in the gable end of a wall.

If the roof has an overhanging gable end (extended rake), short lookouts are used to provide the necessary support, Fig. 11-31. Another method of framing the overhang is shown in Fig. 11-32. The intermediate rafters are notched to accept 2 × 4's. If the gable wall overhangs brace it temporarily.

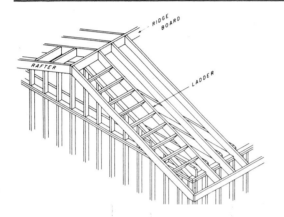

Fig. 11-31 Short lookouts support overhanging gable. *(National Forest Prod. Assoc.)*

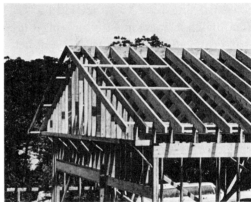

Fig. 11-32 Another method of supporting an overhanging gable end.

Erecting the Hip Roof

If the ends of a gable roof are sliced diagonally, a hip roof results. The corner rafters of such a roof are called hips or hip rafters. Generally they extend from the plate to the ridge. When two gable roofs intersect, a valley is formed, Fig. 11-33. The corner rafter at the point of intersection is called a valley rafter. Like the hip rafter, it also extends from the plate to the ridge. Thus the total rise of hip and valley rafters is the same as the common rafter. See Fig. 11-34 at the top of the next page.

Fig. 11-33 A valley is formed where two gable roofs intersect on a house roof.

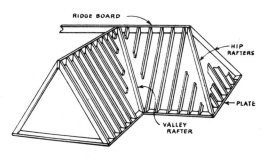

Fig. 11-34 Hip and valley rafters in position.

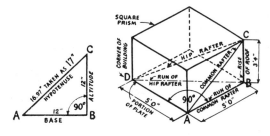

Fig. 11-35 Relative position of the hip rafter can be studied in the diagrams above.

The hip rafter can be compared to the diagonal of a square prism, Fig. 11-35. The figure DAB forms a right triangle. If lines AD and AB are each 12 inches, then line DB (the hypotenuse) equals 16.97 inches. The figure 16.97 is so close to 17 that for all practical purposes, 17″ is used. It will be noted that line DB represents the run of the hip rafter and line AB the run of the common rafter. Thus for every foot run of the common rafter, the run of the hip rafter will be 16.97 (or 17″) Fig. 11-36. The letters A, B, C and D refer to those shown in Fig. 11-35. To find the total run of the hip rafter multiply the total run of the common rafter by 16.97.

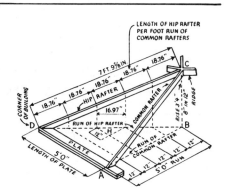

Fig. 11-36 Comparison of layouts and dimension for both hip and common rafters.

When laying out common rafters, use the 12″ mark on the body of the square, and for laying out hip rafters, use the 17″ mark.

To lay out the top and bottom cuts of a hip rafter, use the 17″ mark on the body of the square and the rise per foot on the tongue. The diagonal on the body will give the seat or bottom cut; the diagonal at the tongue will be the top cut. For extra inches under 12, let the odd inches equal the sides of a square. Measure the diagonal and add the sum to the length. See Fig. 11-37.

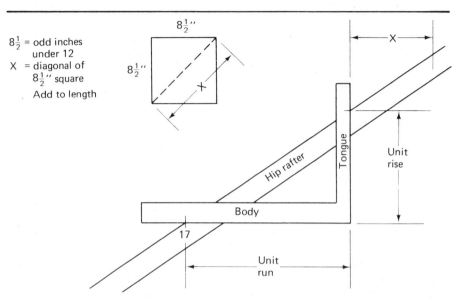

$8\frac{1}{2}$ = odd inches under 12

X = diagonal of $8\frac{1}{2}''$ square

Add to length

Fig. 11-37 Hip rafter layout: the 17″ mark on the body of the framing square gives the seat cut; the diagonal at the tongue gives the top cut.

Rafter Table for Hips and Valleys: The second line of the rafter table on the square reads "Length of hip or valley per foot run". See Fig. 11-38. It means that the figures in the table indicate the length of hip and valley rafters per foot run of common rafters. To find the length of a hip or valley rafter, multiply the length given in the table by the number of feet of run of the common rafter.

LENGTH	COMMON	RAFTERS	PER FOOT	RUN	21 63	15 00	14 42
HIP	OR	VALLEY			24 74	19 21	18 76
DIFF	IN LENGTH	OF JACKS	16 INCHES	CENTERS	28 84	20	19 23
			2 FEET		43 27	30	28 84
SIDE	CUT	OF	JACKS	USE	6 11/16	9 5/8	10
		HIP OR	VALLEY		8 1/4	10 5/8	10 7/8

Fig. 11-38 Second line of rafter table gives lengths of hip and valley rafters.

Example: Find the length of a hip or of a valley rafter where the rise per foot is 8 inches and the span of the building is 10 feet.

Solution: Find the inch line on the square's body which corresponds to the rise of the roof. Under the 8 mark on the second line is the figure 18.76. Multiply 18.76 by 5 (the run of common rafters for 10-foot span). Therefore, $18.76 \times 5 = 93.80$ inches. Divide 93.80 by 12 to obtain the answer in feet.

Answer: 7.81 or 7' 9-13/16".

The measurements for hip and valley rafters are taken along the center of the top edge, Fig. 11-39. The measurements are made to the center point of the common rafters, Fig. 11-40. In order to fit the intersection, hip and valley rafters must be shortened. The amount of

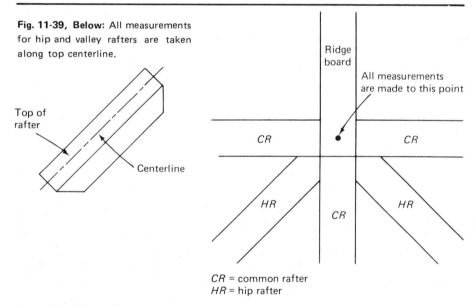

Fig. 11-39, Below: All measurements for hip and valley rafters are taken along top centerline.

Top of rafter

Centerline

Ridge board

All measurements are made to this point

CR

CR

HR

HR

CR

CR = common rafter
HR = hip rafter

Fig. 11-40, Above: Measurements for hip and valley rafters are made to the center of the common rafters.

the shortening is equal to one half the diagonal thickness of the common rafter. See Fig. 11-41. The measurement is made parallel to the plumb top cut. In addition, side cuts must be made as shown at Y in the figure. The side cuts are based on figures taken from the sixth

Roof Framing

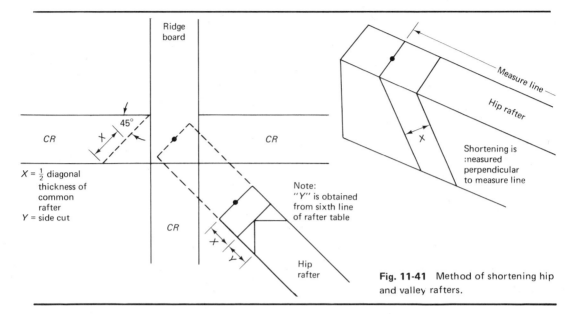

Fig. 11-41 Method of shortening hip and valley rafters.

line of the table on the blade of the square, Fig. 11-42. Using an 8" rise as an example, locate the figure 8 on the body of the square. On the sixth line marked "side cut hip or valley", you will find the figure 10-7/8 directly under the 8. Take this figure on the body and 12 inches on the tongue. Apply the square to the back edge of the hip rafter and draw the diagonal from the center of the rafter out. Flop the square,

Fig. 11-42 Sixth line of rafter table gives side cuts for hip and valley rafters.

and repeat for the second cut, Fig. 11-43. Hip and valley rafters are treated similarly except for the tail cut. Fig. 11-44 shows the difference between the two.

Dropping Hip Rafters: If the hip rafter were installed "as is" after making the top, bottom and side cuts, its top edges would protrude above the line of common and jack rafters. That in turn would prevent

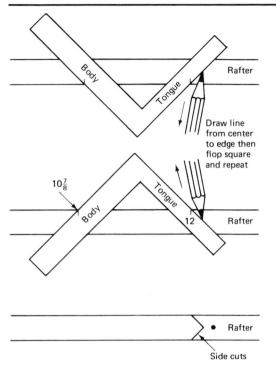

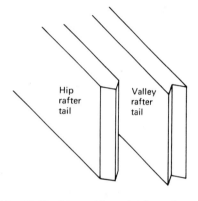

Fig. 11-43, Left: Marking side cuts on hip and valley rafters.

Fig. 11-44, Above: Hip and valley rafter tails.

the roof sheathing from laying flat. To eliminate the problem, the hip can be either backed or dropped. In backing, the edge is beveled as shown. The bevel lines can be obtained by placing the hip rafters in place and marking with a pencil, Fig. 11-45-A. The shaded area is then beveled with a plane. To drop the hip rafter, cut away a portion of the seat equal to the amount it is to be backed, Fig. 11-45-B.

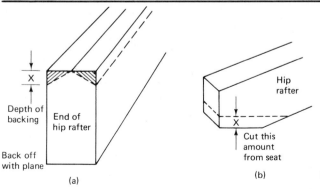

Fig. 11-45-A Left, Bevelling the hip rafter.

Fig. 11-45-B Right, Dropping the hip rafter.

Roof Framing

The square may also be used to determine the amount of drop or backing. Set the square on the stock with the rise on the tongue and the figure 17 on the body. Draw a diagonal then measure along this line a distance one half the thickness of the rafter. Make a mark at this point. The distance from the rafter edge to this mark is the amount of backing or drop required.

Layouts for Jack Rafters: Jack rafters can be classified as shortened common rafters. If they run from the plate to a hip rafter, they are called "hip jacks." Those between the ridge and valley rafters are valley jacks and those between valley and hip rafters are cripple jacks, Fig. 11-46.

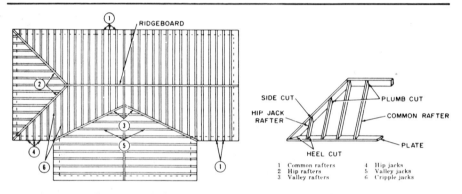

Fig. 11-46 Rafter terms and their locations in roof layouts.

Since all three jacks lie in the same plane as common rafters, they have the same run as common rafters. Generally, they have the same spacing as common rafters, either 16 or 24 inches. Equally spaced, they will also have a common difference in length as they rest against hip or valley rafters. The second jack will be twice as long as the first. The third will be three times as long as the first, and so on.

The framing square has two tables which can be used to determine the difference in length of the rafters. The third line gives the difference in length for rafters 16" OC. The fourth line is for 24" spacings, Fig. 11-47. To use the table, locate the rise-per-foot figure on the body of the square. The third line (16" spacings) under the figure will give the length of the first jack rafter. (This is the same as the difference in length.) If the rise was 8 inches, the figure will be 19-1/4 (some tables show the actual figure 19.23, but 19-1/4 is close enough). Thus 19-1/4"

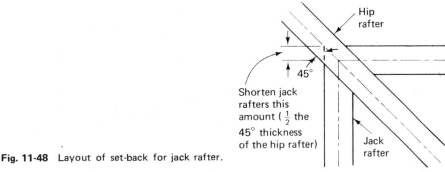

23	22	21	20	19	18	1
LENGTH	COMMON	RAFTERS	PER FOOT	RUN	21 63	20
'' ''	HIP OR	VALLEY	'' ''	''	24 74	24
DIFF	IN LENGTH	OF JACKS	16 INCHES	CENTERS	28 7/8	27 3/
'' ''	'' ''	'' ''	2 FEET	'' ''	43 1/4	41 3/
SIDE	CUT	OF	JACKS	USE	6 11/16	6 15/16
'' ''	'' ''	HIP OR	VALLEY	'' ''	8 1/4	8 1/2

22 21 20 19 18 17 16

Fig. 11-47 Third and fourth lines on the framing square's rafter table give difference in lengths of jacks.

is the length of the first jack; the second will be 2 × 19-1/4 or 38-1/2 inches; the third will be 3 × 19-1/4 or 57-3/4 inches. From these figures deduct half of the diagonal thickness of the hip or valley rafter on the plan. See Fig. 11-48. The top and bottom cuts are the same as for the common rafters.

Hip
rafter

45°

Shorten jack
rafters this
amount ($\frac{1}{2}$ the
45° thickness
of the hip rafter)

Jack
rafter

Fig. 11-48 Layout of set-back for jack rafter.

To simplify the layout of jack rafters, carpenters often use a jack rafter pattern. This is a master pattern similar to the common rafter pattern described earlier. The lumber for hip and valley rafters should be at least 2 inches wider than for common rafters to provide full bearing surface for the angular cuts of the jack rafters.

Start from the ridge, reading along the measure line. Place a mark at a point equal to the common difference taken from the rafter table. Deduct half the diagonal thickness of the hip rafter. Square this line to the top edge of the rafter and mark its center. Use the side cut (line 5 on rafter table) figure on the body of the square and the 12" mark on the tongue. Draw this angle through the center mark and repeat the markings for the rest of the jacks. The same procedure is used for the valley jacks.

Side Cuts are required at the end where the jack rafters butt the hip or valley rafters. The side cuts for jacks are found on the fifth line of the rafter table and is marked "side cut of jacks use". Again see Fig. 11-47. To obtain the side cut for a jack rafter, take the figure shown in the table on the body of the square and 12 inches on the tongue. Mark along the tongue for side cut.

Example: Find the side cut for jack rafters of a roof having 8-inch rise per foot run or 1/3 pitch. Fig. 11-49 shows side cuts for hip and valley jacks.

Solution: Under figure 8 in fifth line of rafter table you will find "10". This figure taken on the outside edge of the body and 12 inches along the tongue will give the required side cut.

Answer: Place framing square on edges of jack rafter with figures 12 on tongue and 10 on body coinciding with edge of rafter toward you. (Remember: Keep tongue to your right.) The side cut is marked along the tongue edge.

Figure 11-50 shows a method of laying out side cuts for jacks with an 8 to 12 slope.

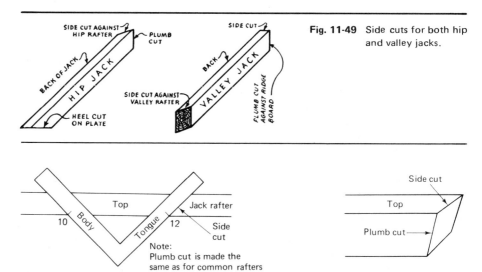

Fig. 11-49 Side cuts for both hip and valley jacks.

Fig. 11-50 Method for laying out side cuts on jack rafters.

After the jacks are laid out, cut them out carefully. Hand saws may be used, but for greater accuracy and uniformity, power tools are recommended. The radial arm saw is especially suitable as it can be set for cutting compound miters.

Installing Jacks: The jack rafters are installed in pairs. If necessary, install bracing to prevent movement of the hip and valley rafters during erection. Use 10d nails when installing jacks.

Cripple Jacks extend from hip to valley rafters. Since the hip and valley rafters are parallel to each other, the cripple jacks spanning them will be of equal length, Fig. 11-51.

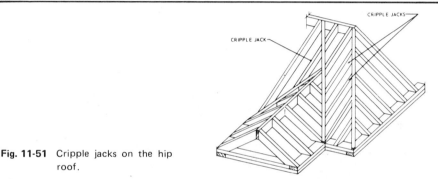

Fig. 11-51 Cripple jacks on the hip roof.

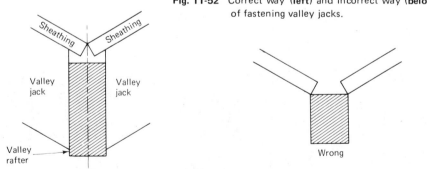

Fig. 11-52 Correct way **(left)** and incorrect way **(below)** of fastening valley jacks.

The run of a cripple jack is equal to the distance from the center of the hip to the center of the valley rafter, measured along the plate line. To obtain the line length, multiply the unit length of the common rafter by the number of units of run covered by the cripple jack.

Valley jacks are nailed flush at the ridge board but slightly above the top edge of the valley rafter, Fig. 11-52. This prevents a gap at the joint when the sheathing is applied.

Roof Framing

The Gambrel Roof

This roof has a single ridge but two sets of common rafters. The intersecting rafters are supported by purlins which are horizontal members resting on partitions. Collar beams spanning the rafters provide additional support.

The lower slope of the gambrel roof is steeper than the upper. When the rafters are laid out, treat them as two separate pitches. Fig. 11-53 shows the framing for a gambrel roof. Fig. 11-54 shows a gambrel roof with dormers; and Fig. 11-55 the layout of a gambrel roof.

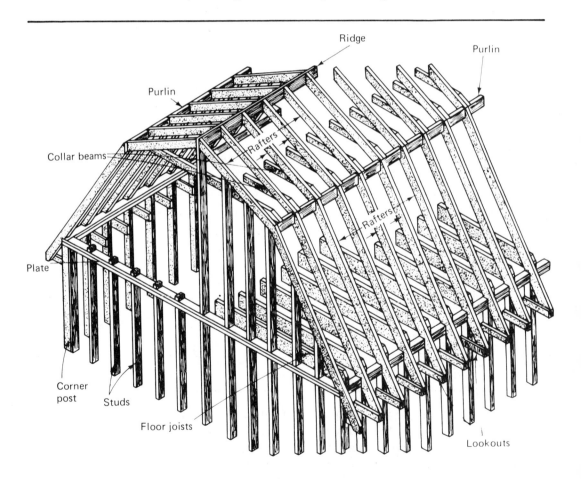

Fig. 11-53 Framing for a house with a gambrel roof.

Fig. 11-54 Gambrel roof with dormers.

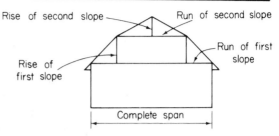

Fig. 11-55 Method of laying out rise and run of gambrel roof.

Roof Trusses

Trussed rafters are used when plans call for a wide span without intermediate supports. This permits the design of large spacious rooms without load bearing partitions. For special design, the truss beam carries the roof loads to the bearing walls or other supports.

Trusses can be purchased ready-made or they may be constructed on site. Their use permits rapid assembly of the roof frame, thus the structure can be enclosed and protected in less time.

Basically, the truss consists of upper and lower chords and diagonals. The upper chords represent the rafters while the lower chords correspond to the ceiling joists. The diagonals serve as braces.

Many truss designs are available. The most commonly used in house construction are shown in Fig. 11-56. Widely used in house construction, the W-truss is economical and easy to assemble. The King-post truss is the simplest of all trusses. Because of its lack of diagonals, however, it cannot span as great a distance as the W type, even when made of the same size members. The scissor truss is ideally suited for houses with sloping "cathedral" ceilings.

Fig. 11-56 Common truss designs used in house construction: **Left,** W-type; **Center,** King-post; **Right,** scissor. *(Forest Products Lab.)*

Roof Framing

Several important factors must be considered in truss design. Wind and snow loads must be taken into account, as well as the slope and weight of the roof itself. These and other factors determine the size and spacing of the trusses. Normally, the spacing is 24" on center.

Trusses are usually assembled with plywood gussets or metal truss plates, but special split-ring and other connectors are available.

Trusses are usually made using a master template as a guide. Using straight sound lumber, layout the template carefully. Check all measurements before and after assembly. Some carpenters prefer to make a chalk-line layout on the floor boards. They then lay out and cut the various members to match the "drawing". Obviously the chalk-line layout must be done with great care and precision.

If wood gussets are used, these should be cut from 3/8" or 1/2" exterior grade plywood and assembled with glue and nails. In areas where moisture is a problem, a waterproof glue should be used. Nailing pattern and face grain direction of the plywood are important. Fig. 11-57 shows a typical W-truss with nail and grain details: for 3/8" plywood gussets, use 4d nails; for 1/2" and 3/4" plywood gussets, use 6d nails.

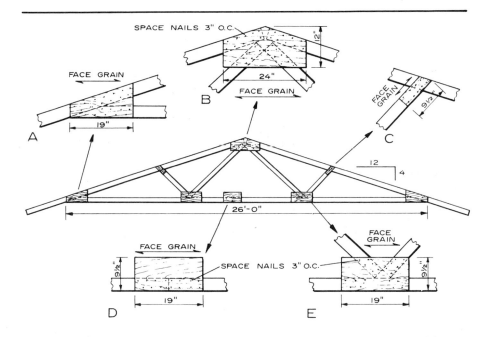

Fig. 11-57 Construction details for a W-truss with plywood gussets. *(Forest Products Lab.)*

To check the accuracy of the master template after assembly, flop it on the layout. If the template is accurately constructed it should still match the layout perfectly.

When large quantities of trusses are required, a jig saw may be used to advantage. Stops and guides permit placement of each member quickly and accurately. Drilling templates are also employed to speed up production.

After installation and when fully loaded, the lower chord of the truss assembly will settle. To allow for this, the bottom chord is made with a slight camber, as shown in Fig. 11-58. The amount of camber depends on the span, load and other factors.

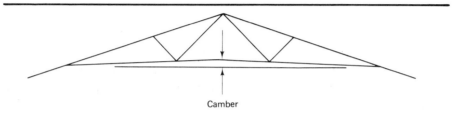

Camber

Fig. 11-58 Camber. This allowance is required in the truss to allow for sag.

Fig. 11-59 Trusses are erected by tipping them with the aid of a truss pole.

Truss assemblies are easily installed by placing them upside down on the walls so they can be rotated into position, as above in Fig. 11-59. Trusses are toenailed to the top plates. In addition to the toenailing, metal connectors are often used.

Roof Framing

Roof Openings

Roof openings for chimneys, skylights, trap doors and the like are considered major openings in a house. They usually require interruption of the roof framing and are treated in the same manner as floor openings. Headers and trimmers are doubled around the openings. For small openings, the roof is framed in the usual manner. The openings are then framed in afterwards. Large openings are constructed as the roof is being framed. As in floor framing, rafter spacing is not altered because of the openings. See Fig. 11-60 below at the left.

Dormers

Dormers are projections built in a sloping roof so as to house a vertical window. There are several types, the most common being the gable and shed. They may be used ornamentally, but they usually are installed to provide illumination in attic rooms.

The **gable dormer** is like a miniature roof. It has the usual roof framing members and construction techniques are similar. Fig. 11-61 shows a typical gable dormer. Double trimmers and headers are constructed in the usual manner. The gable framing is then installed.

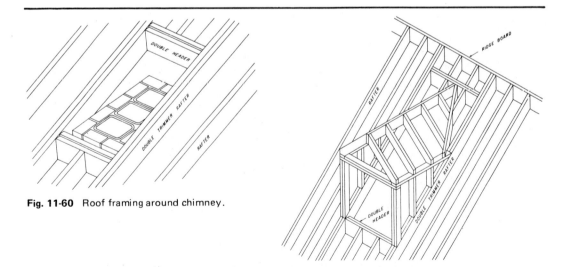

Fig. 11-60 Roof framing around chimney.

Fig. 11-61 Framing of a gable dormer; nailers must be added to the trimmers to support the sheathing.

Small ornamental gables need not be framed into the roof framing. They can be installed after the roof has been sheathed. Often they are fitted with louvers to provide ventilation. Screening should be installed behind the louvers to keep out birds and insects.

Shed dormers provide greater use of floor space in upstairs rooms. Construction is similar to the gable dormer except that full side walls are erected and the roof is usually of low pitch. Figure 11-62 shows a typical shed dormer, below at the left.

Reinforcing End Walls

When ceiling joists run perpendicular to the rafters, some way of reinforcing the wall is necessary. One method is to use stub joists with metal strapping, Fig. 11-63, below at the right.

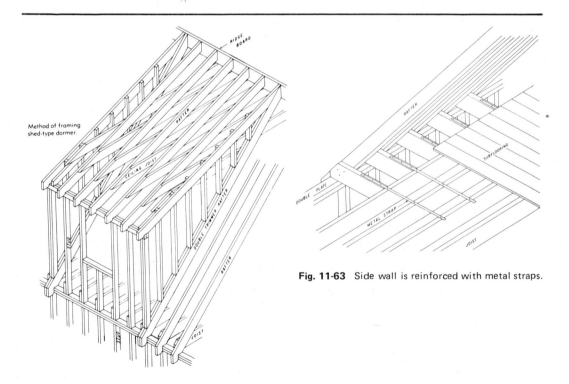

Fig. 11-63 Side wall is reinforced with metal straps.

Fig. 11-62 Shed dormer framing; nailers must be added to trimmers before roof is sheathed. *(National Forest Prod. Assoc.)*

Roof Framing

Roof Sheathing

The roof is sheathed as soon as framing has been completed. This affords protection from the elements for other trades.

Roof sheathings are many and varied. The material used depends on the final roof covering. Some of the more common roofing materials include asphalt shingles, wood shingles and shakes, slate, ceramic tile, roll roofing, asbestos-cement shingles, terne and aluminum shingles. Some may be applied on spaced boards; others require solid sheathing.

The *asbestos shingle* is perhaps the most widely used roofing material. They are applied over a solid deck. The deck may be of solid boards or plywood panels. The solid boards may be common, tongue-and-groove, or shiplap, and should be no wider than six inches. Plywood panels measure 4 × 8 feet, and their thickness varies according to the live load, roofing material and rafter spacing involved. The panels are applied with the face grain perpendicular to the rafters and with joints staggered as shown in Fig. 11-64. Exposed plywood sheathing as in open soffits must be exterior grade. The balance can be interior grade.

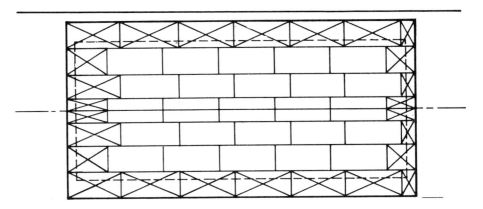

Fig. 11-64 Plan view of roof sheathing: panels marked with "X" must be made of exterior grade plywood if the soffits are the exposed open type.

Exposed soffit panels should be at least 1/2" thick to keep the roofing nails from protruding on the underside. To cut costs, some carpenters combine both 3/8" and 1/2" panels on the same roof, using the 1/2" panels at the edges only. Shims must be used to provide a flush joint at areas where the plywood thickness changes. See Fig. 11-65.

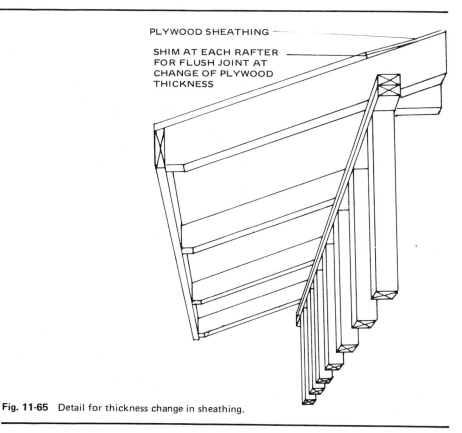

PLYWOOD SHEATHING

SHIM AT EACH RAFTER
FOR FLUSH JOINT AT
CHANGE OF PLYWOOD
THICKNESS

Fig. 11-65 Detail for thickness change in sheathing.

Special soffit panels are available to use as exposed soffits. Install them with the good face down.

Plywood is installed with a 1/16" space at the panel ends and a 1/8" space at the edges. In high humidity areas double these dimensions. Joints at hips and valleys, however, should be tight. Fasten the panels with 6d common, ring shank or spiral thread nails.

Solid wood sheathing is used for asphalt or other *composition* roofs. *Shingles* may be applied on open or solid sheathing. Spaced sheathing is used for shakes or shingles in damp areas. The spacing provides the necessary ventilation. See Fig. 11-66. The center-to-center spacing for the boards is equal to the exposure dimension of the shingles. Solid sheathing which extends beyond the end walls of a gable roof should span at least three rafters for proper support. A 3/4" clearance in the sheathing must be used around chimney openings.

Roof decking should be stacked so the carpenter can easily reach the panels. The panels are not cut to size before installation. Instead,

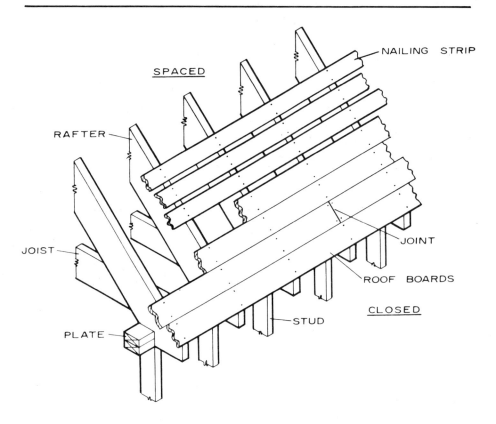

Fig. 11-66 Method for applying open and closed sheathing. *(Forest Products Lab.)*

they are made to project past the ends. They are then trimmed flush with the end rafter. Use care when working near the edge of the roof. The joints at the ends of all panels must occur at the centers of the rafters. Occasionally a panel may require trimming. Tack down all the panel edges except the edge that is to be trimmed. Mark the cutting line on the edge, then cut. (Do not use nails on the edge that is to be trimmed with the saw.) The saw blade should project a trifle more than the thickness of the decking. This procedure saves time and is safe as the panel is solidly supported while being cut.

Sheathing Clips: Plywood sheathing clips are H shaped fasteners used for stiffening plywood edges in roof construction. The clips are made of aluminum and are available for 5/16, 3/8, 1/2, 5/8, 3/4, and

13/16 inch plywood. The flanges are slightly tapered so they can be installed easily, Fig. 11-67. A slight tap with a hammer seats the clip fully. See Fig. 11-68.

Fig. 11-67 Tapered flanges slip over plywood edge.

Fig. 11-68 Decking being installed on warehouse roof.

GLOSSARY

camber: a slight arch given to a structural member; designed to straighten out under a load.

collar beam: a connecting member between rafters, used to stiffen the roof.

dormer: a projection built out from a sloping roof to house a vertical window.

facia (fascia): a board used as a facing across the ends of rafters or box cornice.

gable: the triangular end of a double sloped roof.

gable roof: a double-sloped roof.

gambrel roof: a two-sided roof having two slopes on each side, with the lower slope being steeper than the upper slope.

hip roof: a roof sloped on four sides.

lookout: a wood framing that overhangs part of a roof.

mansard roof: a roof having two slopes on all sides, the upper slope being very shallow and the lower slope being very steep.

purlin: horizontal members supporting common rafters.

rake: the trim at the end of a gable roof; it runs parallel to the slope of the roof and forms a finish between the wall and gable extension.

Roof Framing

ridge: the horizontal line formed at the top of the roof where the sloping surfaces meet.

ridge board: the board placed at the top edge of the roof into which the rafters are nailed.

shed roof: a roof with a single slope.

toenailing: the driving of a nail at an angle to join two pieces.

truss: a rigid framework consisting of beams, bars and ties; they can be used on very wide spans.

valley: an inside angle formed where two sloping sides meet.

12

Finish Roofing

The finish roof of a structure is both its crowning glory and its functional climax. Properly chosen and applied, the roofing adds beauty and value to a house while it also serves as the primary covering unit of a structure meant to protect its inhabitants from water, wind, dust, sun and/or snow. Many roofing materials are available, and new and better ones are being developed constantly. The roofing is installed only after all other trades need no longer work on the roof.

The type of material used depends on several factors. Cost, local codes, location and roof design must all be considered when choosing the roofing material. Shingles are ordinarily not practical on a flat or low-pitched roof. Likewise, standard built-up roofing should not be used on a high-pitched roof. Generally, when the pitch of a roof is less than 4" per foot, roll or built-up roofing is recommended. The lower the slope of the roof, the more susceptible it is to water penetration. With special treatment, however, shingles may be applied to roofs with less than 4" slopes. Also, by using special methods and materials, built-up roofing may be applied to steeper slopes.

Generally, the materials used for pitched roofs include asphalt, wood, asbestos, tile and slate. Sheet materials used also include roll roofing and various metals. For flat or low sloping roofs, the built-up roof with gravel or slag topping is common.

Deck Preparation

Regardless of the type of roofing used, the deck must meet certain requirements. Its surface must be smooth and free of rough spots, projections and depressions; and it must be securely nailed. The planks of wood decks must be well-seasoned and at least 3/4" thick because any covering would buckle if the sheathing boards were to shrink after any roofing was laid.

Attic and soffit ventilation play an important role in the success or failure of a roof. The space under the deck or attic and the soffits must be well ventilated to prevent condensation, Fig. 12-1. Louvered vents, cupolas and exhaust fans may be used to provide the necessary ventilation. FHA regulations require one square foot of ventilating area per 150 square feet of ceiling. If the ceiling has a vapor barrier, the mandatory ratio is one per 300 square feet.

Fig. 12-1 Several types of attic ventilators.

Roofing Underlayment

Underlayment is used to ensure a dry roof deck when roofing is being applied. It also keeps wind-driven rain from wetting the deck if any shingles are lifted or damaged during a storm. Direct contact of resinous areas of the wood deck with the shingles is also prevented.

Underlayment is made of dry felt which has been saturated with asphalt or coal tar. It is available in roll form and in various weights. The weights most commonly used for roofing are No. 15 and No. 30. The No. 15 weighs about 15 pounds per square and the No. 30 weighs approximately 30 pounds per square. A square equals 100 square feet of roof surface.

For roofs with a pitch of 4 inches or more per foot, the No. 15 asphalt-saturated felt is used. It is applied horizontally with a 2" top and 4" side lap, as shown in Fig. 12-2. A 2" overlap is being installed in Fig. 12-3.

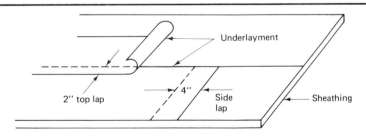

Fig. 12-2 Top and side laps in underlayment.

Fig. 12-3 Asphalt underlayment being applied with 2-inch overlap.

Low slope: Roofs with less than a 4" per foot rise use a double layer of No. 15 asphalt-saturated felt for underlayment, Fig. 12-4. The first course is a 19-inch wide strip laid along the eaves. This is followed with a 36" wide piece which overlaps the starter strip. Succeeding strips are laid with a 19" overlap. A metal drip edge is installed along the rakes and eaves to keep rain water from entering the edges of the sheathing. For additional protection from the elements, a strip of roll roofing felt is placed at the eaves over the drip edge and underlayment. This strip should extend from the eaves to a point at least 12 inches beyond the wall line. This prevents water, snow, melting ice and slush from backing up under the shingles, Fig. 12-5.

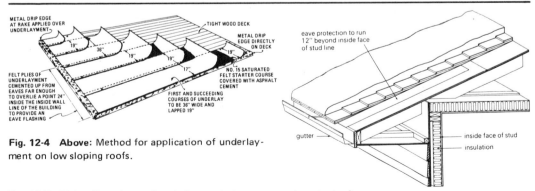

Fig. 12-4 Above: Method for application of underlayment on low sloping roofs.

Fig. 12-5 Right: Extra layer of underlayment at eave prevents water backup.

Finish Roofing

Flashing

Flashing is necessary to protect and waterproof all joints and interior angles on the roof surface, such as the intersections of roof slopes, areas around chimneys and vent pipes and wherever there is a projection through the roof surface. Since the function of these flashing strips is to provide watertight joints, the material used must be durable and impervious to water. Copper, aluminum, lead, galvanized metal and tin plate are the metals widely used for flashing. Asphaltic roll roofing is also used, as well as plastic and rubber materials. Regardless of the type of flashing used, it must be applied properly to be effective. The flashing must carry water over the joint and not into it. Various methods of applying flashing are described below:

Chimney flashing: Chimneys and the roof structures surrounding them are subject to varying stresses and certain opposing structural movements due to winds, temperature changes, settling, etc. Therefore the flashing must be applied so as to permit such necessary movement without affecting the water seal. This is assured by putting the flashing on in two sections: first, the base consisting of roll roofing is applied to the deck; then, the cap or counterflashing is secured to the chimney.

To cut the flashing pieces to size, follow the pattern layouts shown in Fig. 12-6. (The roof shingles have already been installed up to the bottom edge of the chimney.) Install the lower base flashing on top of the roof shingles; next install the side pieces. Then add the rear part of the base flashing. Finally, cement all the pieces to each other and to the roof with mastic.

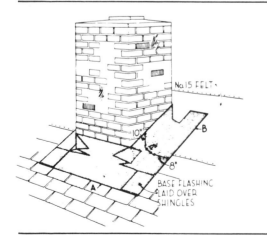

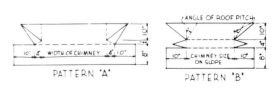

Fig. 12-6 Application of chimney base flashings; pattern layouts show measuring and cutting technique. *(Asphalt Roofing Manufacturers Assoc.)*

When a chimney is large, a cricket or saddle is sometimes used to divert the water. It goes on the deck before the underlayment is applied and is a ridge-shaped piece at the chimney back. It is constructed of wood and installed as shown in Fig. 12-7. The rear flashing is then cut and fitted, Fig. 12-8.

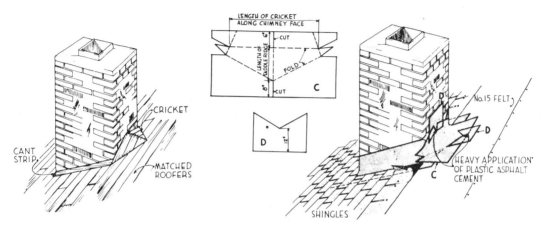

Fig. 12-7 Cricket construction on roof deck at back of large chimney.

Fig. 12-8 Pattern for chimney flashing which fits over cricket.

Cap flashing is usually installed by the brick mason when the chimney is being erected. The caps are made of sheet copper or galvanized steel and are imbedded in the mortar, Fig. 12-9. The strips are left to project upward and then bent down after the base flashing is completed. Because of the slope of the roof, the side caps are stepped as shown.

If cap flashing is installed after the chimney has been completed, the mortar should be removed to a depth of three-quarters of an inch. Then flashing strips are inserted and the joint is filled with bituminous mastic.

Wall flashing around the dormer extends under the siding and over the shingles. As for the chimney, the front flashing piece is installed after the shingle course at the bottom of the dormer is in place.

Individual metal shingles are used as flashing at the side of a dormer or between a sloping roof and vertical wall. The metal shingles are usually about 6" long and slightly wider than the finish roof shingles. They are bent at right angles and inserted at the joint between the two

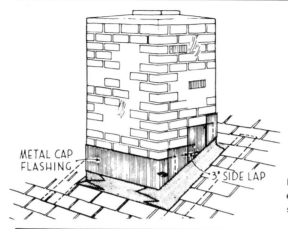

METAL CAP
FLASHING

3" SIDE LAP

Fig. 12-9 Metal cap flashing, made of either sheet copper or galvanized steel, is applied over the base flashing.

surfaces at the end of each course of roof shingling. They are inserted with a 2" lap along the sloping roof and the other lap, either 4" or more, extending upward along the vertical wall or dormer wall. One nail is used at the top of each metal shingle. Each metal shingle is then covered with a regular roof shingle as that course is laid. On the vertical wall or dormer wall, the metal flashing will be covered later with siding.

Valley flashing must be done with care since the valley serves as the waterway for two downward sloping roof surfaces and is subject to leakage problems. Valley flashing may be one of three types—open, closed or woven.

Open valley construction is widely used in roofing. The flashing consists of 90 lb. mineral-surfaced asphalt applied in two layers. The first layer consists of 18" wide strips applied face down. The second of 36" wide strips is installed face up, Fig. 12-10. If pieces must be joined, they are lapped at least 12 inches and bonded with mastic.

The valley width between shingles increases toward the bottom of the slope, Fig. 12-11. The minimum 5" width at the top of the valley should increase downward at the rate of 1/8" per foot. Chalk lines are struck to indicate the shingle edges.

When the slope of two intersecting roofs varies greatly, rain falling on the steeper slope has a tendency to overrun the shallower one and penetrate under the flashing of the lower. To prevent that, a metal flashing with a standing seam is utilized, Fig. 12-12. Be sure that the height of the seam is at least one inch.

Closed and woven valleys are sometimes used for the sake of

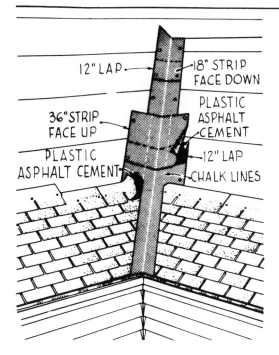

Fig. 12-10 Open valley flashed with roll roofing—in two layers. *(Asphalt Roofing Mfgrs. Assoc.)*

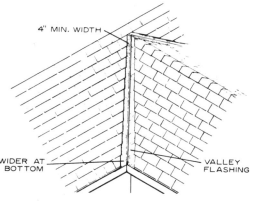

Fig. 12-11 Above: Exposed (open) flashing is wider at the bottom than at the top.

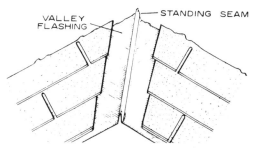

Fig. 12-12 Flashing has standing seam on intersecting roofs with different slopes. *(Forest Products Lab.)*

appearance. Shingles are cut and butted at the center of the valley thus eliminating the open waterway. Both the materials for and the application of closed and woven valley flashing differ from that of the open valley.

For the closed valley, a 36" wide 50 lb. piece of roll roofing is centered over the valley. It is installed on top of the underlayment. See Fig. 12-13. The shingles from the first roof slope covered are extended at least 12 inches past the valley centerline. The shingles applied to the second roof slope are cut at the valley centerline. A 3" wide strip of plastic asphalt cement holds the ends of the shingles fast.

In the woven valley, the shingles are laid alternately and interlocked as in Fig. 12-14. The flashing itself is the same as for the closed valley.

Fig. 12-13 Closed Valley flashing.　　　**Fig. 12-14** Woven valley flashing.

Vent flashing for stack vents is available ready-made in the form of a collar or boot. Made of metal, plastic or rubber, they are formed in one piece and are simply slipped over the vent pipe, Fig. 12-15. The flange at the upper part of the slope is placed under the shingles. The lower part of the flange lays on top of the shingles. All joints are caulked with roofing mastic. How to layout and install flashing for a soil stack can be seen in Fig. 12-16. The materials used are 50 lb. roll roofing and asphalt plastic cement.

Roofing Terms

Coverage is the measure of weather protection provided to a roof and is based upon the amount that its covering materials overlap. If the covering is roll roofing, the coverage is considered single. Shingles usually give double coverage, but triple coverage is not uncommon.

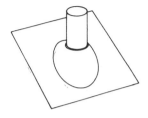

Fig. 12-15   One-piece collar made to slip over vent pipe.

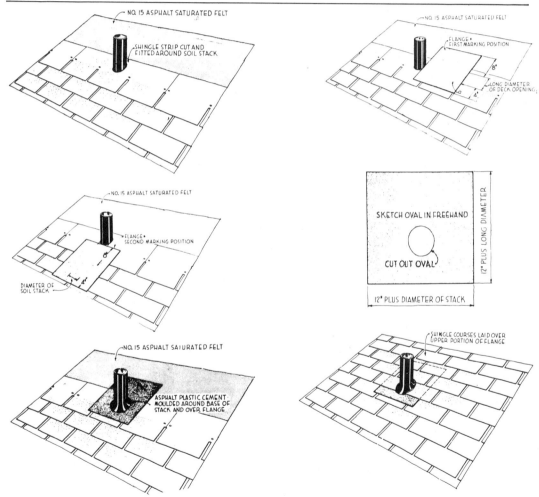

Fig. 12-16   Flashing installation around soil stack: **top left,** roofing is fitted around pipe; **top right,** first step in marking opening; **center left,** second step in marking opening; **center right,** cut oval in flange; **bottom left,** cement collar is applied around pipe; **bottom right,** shingling is completed.   *(Asphalt Roofing Manufacturers Assoc.)*

Finish Roofing

Exposure is the distance in inches from the edge of one course of shingles to the next. Thus a fairly steep roof may have a 5″ exposure while a flatter slope would probably have a 3-1/2″ exposure.

Square is a roofing unit of measure. One square of roofing will cover a 100 square foot (10' × 10') area. This applies to the finished roof surface and is not to be confused with the actual square footage of material used.

Asphalt Shingles

Shingles are available in a number of forms and sizes. Made with a base mat of cellulose, glass or asbestos fibers saturated and coated with asphalt, they are topped with mineral granules. The asphalt provides the necessary waterproofing and the granules protect the shingles from the sun's rays. The mineral granules also give the shingles color and fire protection.

The square-butt strip shingle is the most common of the asphalt shingles in use. They have one, two or three tabs. The strips measure 12" × 36" and are available with self-adhesive spots. The spots are activated by the sun's rays thus eliminating the need for tedious hand sealing of each tab. A space between each seal helps the roof to breathe.

Square-butt strip shingles are usually laid with a 5" exposure. This is considered double coverage. A 4" exposure gives triple coverage.

Starter course: Applied to the eave edge before the shingles are installed, it may be a 9″ wide strip of roll roofing or a row of the asphalt shingles with the tabs cut off, as shown in Fig. 12-17. The starter strip provides a backing for the first regular course. It also fills in the space between tabs.

Applying asphalt shingles: Chalk lines struck parallel to the ridge and rake are essential to ensure that the shingles are installed in a straight line. Space the chalk lines closely. Then, starting from either end, lay one full shingle directly on the starter course. This first course should overhang the eaves and rake by approximately three-eighths of an inch. This will prevent water from backing up under the shingles. Continue, laying the shingles with either a 5" exposure or, if triple exposure is preferred, with the 4" exposure explained earlier.

The nailing pattern is important. In high wind areas, six nails in each 12" × 36" shingle is the recommended practice. Otherwise, four

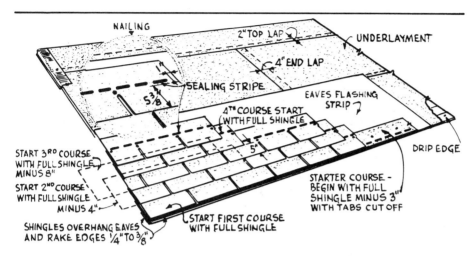

NAILING

2" TOP LAP

UNDERLAYMENT

4" END LAP

SEALING STRIPE

EAVES FLASHING STRIP

5⅜

4ᵀᴴ COURSE START (WITH FULL SHINGLE

5"

DRIP EDGE

START 3ᴿᴰ COURSE WITH FULL SHINGLE MINUS 8"

START 2ᴺᴰ COURSE WITH FULL SHINGLE MINUS 4"

STARTER COURSE - BEGIN WITH FULL SHINGLE MINUS 3" WITH TABS CUT OFF

SHINGLES OVERHANG EAVES AND RAKE EDGES ¼" TO ⅜"

START FIRST COURSE WITH FULL SHINGLE

Fig. 12-17 Method of applying shingles: starter row has tabs removed.

nails per shingle is adequate. Fig. 12-18 shows the nailing pattern for square-butt and hex-strip shingles. Nails for asphalt roofing should be barbed, have large heads and sharp points. They must be galvanized and long enough to penetrate into the sheathing at least three-quarters of an inch, Fig. 12-19.

The nails must be driven carefully to prevent damage to the shingles. Drive them squarely; if they go in at an angle, the heads may cut into the shingle.

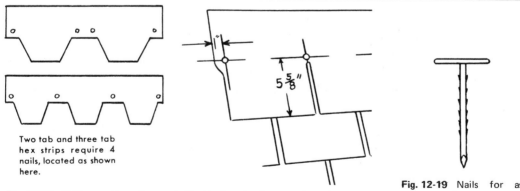

Two tab and three tab hex strips require 4 nails, located as shown here.

$5\frac{5}{8}$"

Fig. 12-18 Nailing pattern for asphalt shingles: both two-tab and three-tab strips

Fig. 12-19 Nails for asphalt roofing should be hot- galvanized, and barbed.

Finish Roofing

Shingle patterns: Square-butt shingle strips may be laid out in one of several patterns. The patterns are formed by the placement of the cutouts in succeeding courses. Fig. 12-20 shows a roof with cutouts breaking joints on halves. The second course starts with 6" removed from the first tab. The third row starts with a shingle whose entire first tab of 12 inches is removed. The fourth row is a repeat of the second. The entire roof is laid out in this manner. The result is that each row will have a cutout centered on the tab directly below it. "Random" spacing is illustrated in Fig. 12-21.

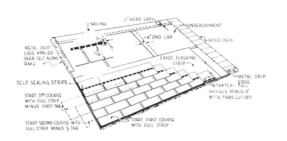

Fig. 12-20 Three-tab square butts with cutouts centered over the tabs directly below.

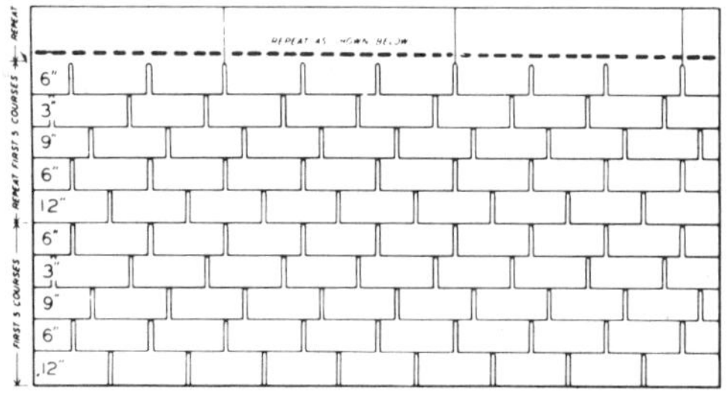

Fig. 12-21 "Random" spacing for three-tab square butt strips of roofing shingle.

Hips & ridges: To finish them, special shingles are available. They may also be made by cutting strip shingles. Their size should be 9" × 12".

Hex-strip shingles are available in two or three tab strips. Application is similar to that for the strip shingle. The starter course is inverted so that the space between tabs will be the same color as the shingles themselves. The strips are applied so that the lower edges of the tab center over the top cutouts of the preceding course.

Re-roofing: Asphalt shingles may be applied directly over old roofing provided the roof itself is strong enough to support the new roofing, any snow loads, the weight of the workers and the like. The deck must also have good nail-holding ability. Nail down all loose shingles and replace the missing ones. If the old shingles are wood, split the badly warped ones so you can nail the smaller segments flat. Use bevelled feather strips along the butts of each course. This provides a smooth surface for the new roof.

Fig. 12-22 shows a method for re-roofing in a high-wind area. The edging strips replace shingles cut away at the rake and eaves.

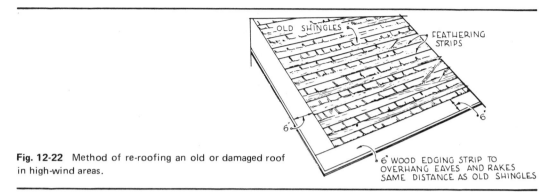

Fig. 12-22 Method of re-roofing an old or damaged roof in high-wind areas.

If the shingles to be re-covered are asphalt, replace any bad or missing pieces and cut away curled and lifted shingles. Nail as necessary. To minimize the uneven appearance of the new roofing, offset the nailing pattern by 2 inches.

Built-Up Roofing

Built-up roofing is generally used on decks with a pitch of 2 inches or less. It consists of several layers of roofing felts laid between coatings of hot tar or asphalt. The top is surfaced with gravel slag or crushed rock. For residential construction, the gravel or rock is applied in a ratio of 300 pounds to 100 square feet. Fig. 12-23 shows how the sections are built up.

On wood decks, a base layer of roofing felt is securely nailed with a 4" overlap. This is followed with applications of hot tar or hot asphalt between layers of 40 lb. felt. The crushed rock or gravel is embedded in the final coating of tar or asphalt. Finally a gravel stop, as in Fig. 12-24,

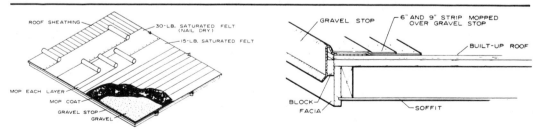

Fig. 12-23 Detail showing layers of built-up roof.

Fig. 12-24 Gravel stop keeps gravel from working off edge of built-up roof.

is installed to keep the gravel from working off the roof. Note in Fig. 12-25 how this built-up roof flashing is applied at the building line.

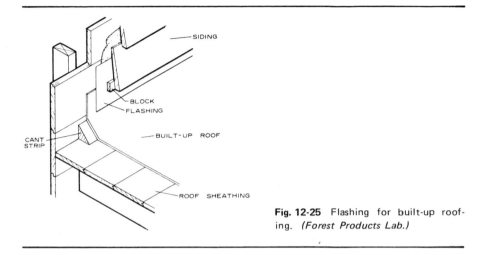

SIDING

BLOCK
FLASHING

CANT
STRIP

BUILT-UP ROOF

ROOF SHEATHING

Fig. 12-25 Flashing for built-up roofing. *(Forest Products Lab.)*

Roll Roofing

Roll roofing is generally used when economy is a dominant factor. It consists of asphalt-impregnated felt made in varying widths and lengths and surfaced with mineral granules in a wide variety of colors. The edges may be straight or patterned.

Roll roofing may be applied with either exposed or concealed nailing. The concealed method makes for greater durability, however. Nails used must be 11 or 12 gauge, hot dipped, galvanized and with 3/8" heads. Shanks should be at least 7/8" long. The lap cement should be stored in a warm place until ready for use.

When using the concealed-nail method, apply the edge strips along the eaves and rakes with a 1/4" to 3/8" overhang.

Double-coverage roll roofing: This type of roofing is also called 19" selvage roofing. It provides double coverage over the entire roof area. It has the same effect as a built-up roof. Roofs with a rise of only one inch per foot may be covered by this method. The material comes in 36" widths and is applied with a 19" lap and 17" exposure. Because specifications differ among manufacturers, their directions for application should be followed carefully.

Wood Shingles

Wood shingles provide a durable, rugged and decorative covering for residential and commercial buildings. Because of the way they are applied, wood shingles actually add strength to the structure since the interlacing and overlapping of the individual shingles has a bridging effect on the roof. One drawback is the fire hazard such roofs present, however. The shingles can be pressure treated with a fire-retardant compound, but some building codes still prohibit their use in spite of that.

Cedar, cypress and redwood are the woods most widely used for shingles. Their most important functional quality is their high resistance to decay.

Shingler's Tools: The installation of wood shingles requires the use of a special shingler's hatchet. The hatchet should be lightweight, have a sharp blade and heel and a gauge for checking shingle exposure. See Fig. 12-26 below. A small portable trimming saw is also recommended for installing wood shingles. A straight edge is also needed to keep the rows straight and true. For safety's sake, wear sneakers, crepe soles or other good gripping shoes when working on a roof.

Fig. 12-26 A lightweight shingler's hatchet.

Roof application: Wood shingles are normally applied in a straight single course, but variations are possible. Some common styles are thatch, ocean wave and serrated. Regardless of the style chosen, double the shingles at the eaves. Butts of the first course should project 1-1/2" beyond the sheathing at the eave and 3/4" at the rake. Joints must be 1/4" wide between shingles to allow for expansion when wet. Maintain at least 1-1/2" spacing between joints in adjacent courses. The joints in succeeding courses must be made so they do not line up with the joints in the two courses below. See Fig. 12-27. The first course at the eave is always doubled, Fig. 12-28.

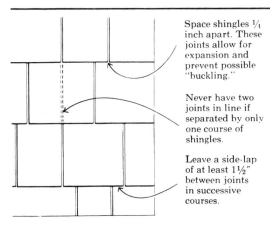

Space shingles ¼ inch apart. These joints allow for expansion and prevent possible "buckling."

Never have two joints in line if separated by only one course of shingles.

Leave a side-lap of at least 1½" between joints in successive courses.

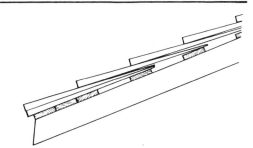

Fig. 12-28 Above, Double-coursing at the eave.

Fig. 12-27 Left, Shingle spacing on building roofs.

To prevent dripping at the gable, a length of beveled siding is installed, Fig. 12-29. This directs the water toward the eaves.

Roof sheathing: Wood shingles may be applied over open or solid sheathing. Underlayment is not required but if it is used it should be a "breather" type. That means an underlayment which has not been saturated with a waterproofing compound. Saturated felts may cause condensation problems.

If the climate is snow-free, open sheathing is preferred. Boards 1 × 4 or wider are used with spacing equal to the shingle exposure to be used, Fig. 12-30. In areas where wind-driven snow is encountered, solid sheathing is recommended.

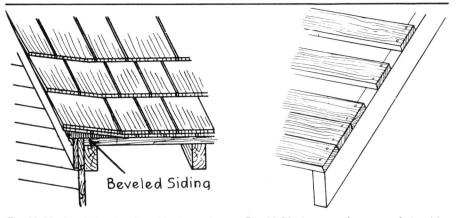

Beveled Siding

Fig. 12-29 Beveled strip tilts shingles and directs water away from gable end.

Fig. 12-30 Layout of open roof sheathing for wood shingles.

Roof, pitch and exposure: Weather exposure depends on the roof pitch. On roofs with slopes of 4 in 12 or steeper, the exposures are 5" for 16" shingles, 5-1/2" for 18" shingles and 7-1/2" for 24" shingles. On slopes less than 4 in 12 and no less than 3 in 12, the exposures are 3-3/4", 4-1/4" and 5-3/4" respectively.

Shingle nailing must be done with rust-resistant shingle nails made from either aluminum or hot-dipped zinc. Two nails are used per shingle, regardless of the shingle width. The nails are placed 3/4" from the edge of the shingle and no more than one inch above the exposure line. For 16" and 18" shingles be sure to use 3d nails (1-1/4" long) and for the 24" shingles 4d nails (1-1/2" long). See Fig. 12-31. Drive the nails until the head meets the shingle but no further. A typical application with exposure recommendations is shown in Fig. 12-32.

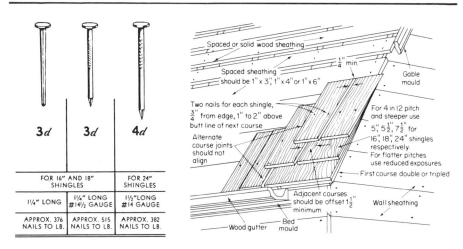

Fig. 12-31 Nails for wood shingles.

Fig. 12-32 A typical application of roof shingles with exposure recommendations.

Hips and ridges: Proper installation of the hips and ridges is important. They must be tight and without exposed nails. Factory-made hip and ridge units are available. When assembled on-site, uniform shingles should be selected.

The width of the shingles should match the exposure. To serve as a guide, tack two wood straight edges on the roof, one on each side of the hip. Then double the starting course. Nail the first shingle in place with its edge aligned with the guide strip. Then bevel the edge at the ridge. Next, install the opposite shingle and bevel its edge to fit also,

Finish Roofing

Fig. 12-33. Install the succeeding courses, alternating the procedure in each course. This is usually referred to as the "Boston Ridge." Be sure to use nails long enough to penetrate into the sheathing.

Valleys: For roofs with one-half pitch or steeper, the valley flashing should extend at least 7 inches on each side of the valley center. For pitches under one-half, extend the flashing at least 10 inches on each side. See Fig. 12-34. The metal must be of good quality and should last as long as the shingles. Center-crimped metal is recommended.

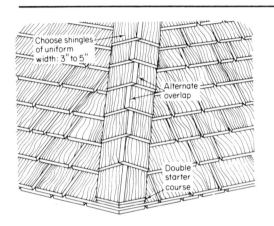

Fig. 12-33 Application details for hips and ridges. *(Red Cedar Shingle and Handsplit Shake Bureau)*

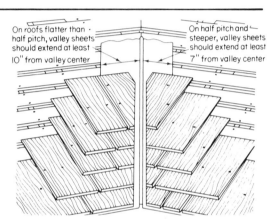

Fig. 12-34 On half-pitch and steeper roofs, flashing must extend at least 7 inches on each side of valley center.

Wood Shakes

Few roofing materials can match the rugged beauty of wood shakes. The textured surface of shakes together with their various colors and thicknesses give a distinctive character unlike that of any other roof surfacings. Three types of shakes are common: handsplit and resawn, tapersplit and straight-split.

Shakes are manufactured in several lengths and thicknesses and varying widths. Because of their length and thickness, they have greater exposure than wood shingles. Typical exposures are 7-1/2" for 18" shakes, 10" for 24" shakes, and 13" for 32" shakes.

Application is similar to that for wood shingles. Pitch must not be less than 4 in 12, however. Fig. 12-35 shows how roofing felt is applied in strips over the top portion of each course. The 30 lb. 18" wide felt is positioned a distance above the butt equal to twice the exposure. For example, 24" shakes with a 10" exposure would require the felt strip to be 20" from the butt end. The felt will then cover 4" of the top end of the shake, the remaining 14" will rest on the sheathing.

In snow-free areas, roofing felt overlay need not be used with straight-split or taper-split shakes.

Fig. 12-35 Application of roofing felt strips over top parts between wood shakes.

Valleys may be open or closed. The open valley is the most common and in the opinion of most builders, the most practical. The open valley should be underlaid with 30 lb. felt and a metal sheet at least 20" wide. The metal should be 26 gauge galvanized iron—or heavier.

Proper nailing is of great importance. Ordinary galvanized nails are not suitable for cedar shakes. Only hot-dipped zinc coated nails are recommended. Drive nails in one inch from each edge and about one and one-half inches above the butt of the following course.

Various patterns produce striking effects with shakes. A perspective effect can be achieved by graduating the exposures from eaves to ridge.

Finish Roofing

Shakes of several lengths are required. Starting with 24" shakes laid 10" to the weather, reduce the exposure gradually to about 8-1/2" at the middle of the roof. Now switch to 18" shakes and continue reducing the exposure toward the ridge, Fig. 12-36. The result is an exaggerated perspective from eaves to ridge.

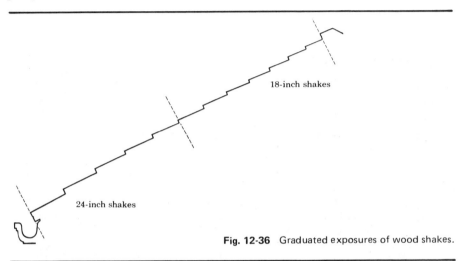

18-inch shakes

24-inch shakes

Fig. 12-36 Graduated exposures of wood shakes.

Another interesting effect is obtained by staggering the butts. An even more dramatic effect is possible by occasionally interspersing longer shakes among the regular shakes. The butts are made to extend several inches below the course line.

Shingle and Shake Panels

Shingles and shakes are available in panel form. They are bonded to 1/2" sheathing-grade plywood to form 8-foot panels. The panels are designed for roofs with a 4 in 12 or steeper pitch. Special starter strips are available. These starter strip ends must break on rafter centers. If panels other than starter panels meet between rafters, use plywood clips at the panel joints. Eight-foot panels being applied to a truss roof are illustrated in Fig. 12-37.

Flashing around vents and chimneys is applied as it is for regular shakes and shingles. Make the cutouts in shakes with a saber saw. Apply hips and valleys in the regular manner.

Asbestos-Cement Shingles

Fig. 12-37 Installation of 8-foot cedar shake panels. *(Shakertown Corp.)*

Asbestos-Cement Shingles

These shingles are made by combining asbestos fibers and portland cement under high pressure. They are highly fire-resistant and wear very well. They are available in many sizes, styles and colors, and their cost is higher than either asphalt or wood shingles. Since they are extremely long wearing, the higher cost may be justified where that extra durability is desired. Unlike wood or asphalt, asbestos-cement shingles are not easily cut with ordinary tools. The material is very hard and brittle. Cut either with a shear-type cutter or by hand. Since nails cannot be driven through the shingles, they are furnished with prepunched holes. If other holes are required, they are usually made with a special punch fitted to the cutter.

Finish Roofing

If a cutter is not available, you can punch and cut them by hand. For straight cuts score the surface with a chisel; then snap the shingle over a straight edge. You use a drift punch to make the nail holes. While punching, back up the work with a piece of end-grain wood.

Make cutouts by piercing the outline of the cut with the punch. Then use a well-placed hammer blow to knock out the waste, Fig. 12-38.

Installing asbestos-cement shingles: The roof deck must be of sound construction and strong enough to support the weight of the shingles. Cover the deck with one layer of waterproof felt and lay the shingles with a 2" head lap. Six-inch end laps are a must; and remember that hips, ridges and valleys require a double thickness of felt.

Nailing strips must be used along hips and ridges, Fig. 12-39. Butt the roof shingles against these strips.

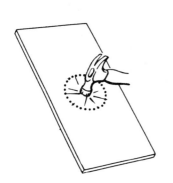

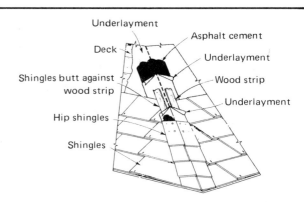

Underlayment

Deck

Asphalt cement

Underlayment

Shingles butt against
wood strip

Wood strip

Underlayment

Hip shingles

Shingles

Fig. 12-38 Knocking out holes outlined first with a punch in asbestos shingle.

Fig. 12-39 Method of applying hips and ridges with asbestos cement shingles.

Copper or *stainless steel flashing* may be used successfully with asbestos-cement shingles. Aluminum flashing cannot be used because these shingles corrode it. The flashing must be of sufficient width to prevent water penetration. To further protect the roof against leaks, bed the shingles in contact with the flashing in roofing cement.

Proper storage of the shingles before use is important. Keep the packages flat and dry. If they are left outdoors in bundles, they must be covered and kept dry. Moisture which accumulates on stored shingles can cause discoloration. But that is not a danger once they have been installed.

Slate Roofing

Slate roofs are among the most expensive of all roof coverings. Their beauty is unexcelled, however; in addition they are fireproof and will last indefinitely. Slate shingles are made by splitting natural rock and are available in several colors and sizes. Like asbestos-cement shingles they are somewhat brittle and must be cut and handled in a similar manner.

Slate is usually laid with a 3" head lap on an underlayment of saturated felt. Copper or slater's nails are used for fastening.

Tile Roofing

Tiles used for roofing are manufactured by shaping a mixture of shale and clay in molds and then firing it. Various patterns and colors are available; the most commonly used styles are mission and Spanish. Since the tiles are made in molds, their uniformity is assured. They are put down on felt underlayment and fastened with nails. Ridge and hip tiles are also available in matching styles and colors. The valleys are usually of sheet copper.

Metal Roofing

Various metal roof coverings are available: they include galvanized iron, tin, copper, aluminum, lead and zinc. A few of these are used on residential dwellings. One of them, terne metal roofing, is a combination of lead and tin coated on steel. It has excellent weather resistance and is available in both sheets and rolls. It is also used for termite shields, flashing, gutters and downspouts.

Roof Drainage

The drainage system of a roof is very important. Rain and snow falling on a roof must be directed away from the house or into a storm sewage system. This is accomplished with gutters and downspouts, also called conductors. The gutters collect the water at the edge of the roof

Finish Roofing

and direct it toward downspouts; the downspouts carry it downward to its final destination.

Gutters may be made of wood, metal or plastic. Some woods commonly used for gutters include redwood, cedar, cypress, Douglas fir and pine. Unlike metals, these woods are not affected by corrosive fumes in the atmosphere and will give many years of service.

Wood gutters are fastened directly to the facia board with either galvanized or brass screws. The brass are preferable for these gutters which are usually installed before the roof covering is applied. Blocks set on 24" centers provide the necessary air space between gutter and facia. Wood gutters are usually treated by the manufacturer. If not, give them several coats of preservative.

Metal gutters are generally made of aluminum, copper or steel. Steel gutters may be either galvanized or enameled. Aluminum gutters are available either finished or unfinished, but they are not as strong as steel and will dent easily. Copper gutters are the most expensive of the metal gutters and the most difficult to install. They require soldered joints whereas steel and aluminum gutters are simply joined with rivets and caulked. But copper is corrosion-resistant and gives good service.

Copper gutters are usually left unpainted because the greenish oxidation has a pleasing appearance. The oxidized coating also serves as a protective coating for the metal.

Gutters must be installed with a slope toward the downspout. Standard practice is to drop the gutter about one inch for every 20 feet of gutter. When the gutter run is more than 35 feet, use two downspouts, one at each end of the run.

Gutters are available in various widths. The size used depends on the area of the roof. For roofs with an area of up to 750 square feet, a 4" gutter will do. Over 750 square feet, a 5" gutter is needed; and anything over 1500 square feet will require a 6" gutter.

Built-in gutters allow for many design variations. Architectural lines may be preserved by concealing the gutters altogether or they can be designed so that they appear less conspicuous. One method is to conceal the gutter behind a facia board.

Downspouts must be large enough to quickly remove the water collected by the gutters. The number and size of downspouts used depends on the area of the roof. Generally one 3" downspout is used for each 1000 square feet of roof served.

The component parts of metal gutters and downspouts are shown in Fig. 12-40. The installation procedure for a typical system is shown step-by-step in Figs. 12-41 through 12-48. Plastic gutters are basically the same except that their only metallic parts are the hangers—used instead of spikes.

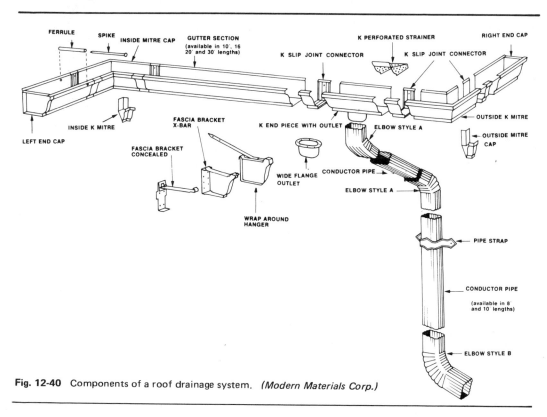

Fig. 12-40 Components of a roof drainage system. *(Modern Materials Corp.)*

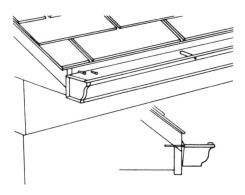

Fig. 12-41 First, tack a string on the facia as a guideline for the top edge of each gutter; then start the installation at the end opposite the downspout. Position each gutter length so its top edge lines up with the string; then insert a ferrule inside the gutter bead and spanning the trough of the gutter. Drive a spike through gutter edge, hollow ferrule, and facia into the first rafter. Locate the succeeding spikes similarly on every other rafter.

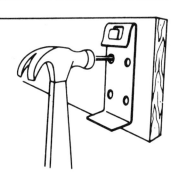

Fig. 12-42 Alternatively, you may use concealed facia brackets instead of spikes. If you do, nail the brackets to the facia with their top edges aligned with the line-up string (string not shown here).

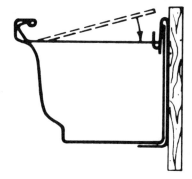

Fig. 12-43 Now insert the strap hangers under the front lip of each gutter. Then slip the rear end of each strap over a bracket hook which when depressed holds the strap securely.

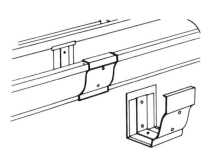

Fig. 12-44 Use slip-joint connectors to unite the several sections in the run of the gutter. Before you actually join them, fill all joints with gutter mastic.

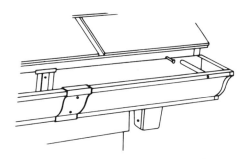

Fig. 12-45 Install end pieces with an outlet at each corner of the building; join with connectors, then add the end caps.

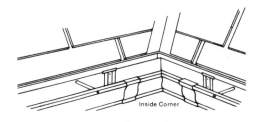

Inside Corner

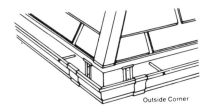

Outside Corner

Fig. 12-46 Use either inside or outside miter corners where the gutters make a turn; and support the corners with wrap-around hangers nailed to the facia.

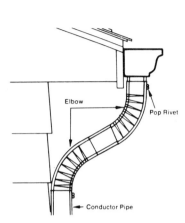

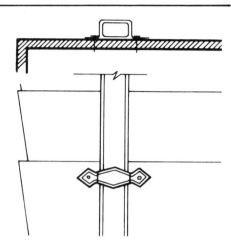

Fig. 12-47 Next, connect the gutters to the conductor pipes (downspouts) with elbows. Join them by force-fitting the large ends over the small ends.

Fig. 12-48 Finally, fasten the conductor pipes against the walls with pipe straps; attach each strap with two nails.

GLOSSARY

Boston Ridge: a method of applying shingles to a ridge to make it watertight.

butt: the lower edge of a shingle.

conductor: a pipe or connector which carries rain water from a roof to the ground.

cricket: a double-sloped structure for diverting rain water away from a chimney.

drip edge: in roofing, a metal strip used to throw off water; it projects beyond other parts.

eave: the projection of a roof over the sidewall.

exposure: the amount that a shingle is exposed or the distance from the butt of one shingle to the butt of the one directly above.

flashing: sheet metal or other material used in roof and wall construction to prevent water penetration.

gravel stop: a formed metal projection turned upward at the edge of a built-up roof to keep gravel from washing off.

gutter: a channel of wood, metal, or plastic placed at the edge of a roof below the eaves to carry off rain water; also called eave trough.

sheathing: a covering, usually of wood boards or plywood, applied to studs and rafters; building board is also used as sheathing for walls.

soffit: the underside of a cornice.

square: a unit of measure in roofing materials, including necessary overlap. One square covers an area of 100 square feet.

13

Windows

The primary function of windows is to provide light and ventilation for the interior of a building. Until recently, if the windows in ordinary small dwellings served these purposes, most aesthetic considerations were usually subordinate in American building design. Today however, both modern production techniques and contemporary architectural trends offer a wide range of distinctive styling for windows. They are made in many materials and designs to satisfy a variety of tastes as well as different needs. A contemporary bow window made of ponderosa pine is illustrated in Fig. 13-1.

Fig. 13-1 Factory-made bow window of ponderosa pine adapted to contemporary use.

Windows

Factory-made Windows

At one time, windows and doors were made by the carpenter on the job. Later on, they were made in local millwork shops, mostly by hand, and delivered to the job site. Now, windows are mass-produced in large factories under strict quality control methods and delivered completely mounted in the frame. Today's carpenter usually only has to install them in the rough opening (RO), Fig. 13-2.

Materials: Windows are generally made of wood, steel or aluminum. Each material has advantages: wood has greater insulating qualities than either aluminum or steel; steel is much stronger than wood; and aluminum windows do not require painting. So these factors as well as cost and climate influence the choice of a particular material. For instance, wood windows cost more than aluminum or steel, but their use may well result in lower fuel bills since aluminum loses heat more than 1700 times faster than wood. In warm climates that is not a factor but it must definitely be considered in cold climates. Wood windows also result in lower air conditioning bills in summer as less heat is conducted into the room from outside than with metal window frames. The disadvantage of wood windows is that they will warp and swell if subjected to excessive moisture, and therefore they should be always painted. Lumber used for windows should always be preservative treated against decay and insects, and to reduce moisture absorption qualities.

Steel and aluminum windows are made with frames much narrower than required for wood, since metal frames are stronger and more rigid than wood frames. See Fig. 13-3 which shows a steel casement window.

Fig. 13-2 Window units delivered to the job site ready for installation.

Fig. 13-3 A steel casement window with narrow frame installed in a house.

Types of Windows

New developments in aluminum windows are being made to cut down on the heat-loss factor, such as the insertion of insulation between the walls of the window frame.

Plastic is also coming into use for windows. At present, it is combined with wood or metal, but all-plastic windows are a possibility in the near future. Plastic windows would be lightweight and easy to handle, easily cleaned, never have to be painted, and be corrosion free.

Types of Windows

While there are many stylistic variations, windows can be classified mechanically as either stationary, sliding or swinging, Fig. 13-4. Various types include double-hung, casement, awning, bow, bay, jalousie, and fixed windows.

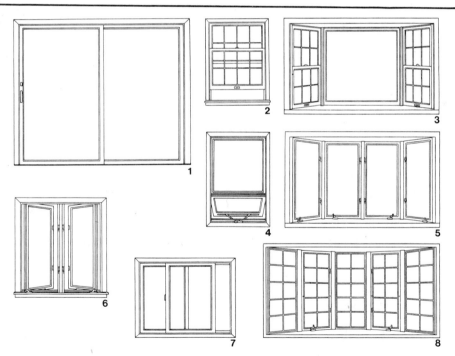

Fig. 13-4 *WINDOW STYLES:* 1) Gliding door; 2) Double-hung window; 3) Angle-bay with center-fixed picture window flanked by operating double-hung windows—available in 30- and 45-degree styles; 4) Awning window with fixed sash above; 5) Angle-bay with operating casement windows, available in 30- and 45-degree styles; 6) Double casement window unit; 7) Gliding window; 8) Bow window with both fixed and operating casements. *(Credit: Andersen Corp.)*

Windows

Double-hung windows consist of two vertical sliding sash members—each covering half of the window opening. Double channels allow the sashes to by-pass each other. Different counterbalancing methods are used to hold the window in either open or closed positions. Sash weights were common in the past, but they have been largely replaced by spring type devices.

Double-hung windows provide 50 percent ventilation when fully opened. A disadvantage of this type of window is that it is difficult to clean from inside the room. To overcome this problem, some manufacturers now make removable sashes.

Casement windows: In these, the sash is hinged at the side. A crank mechanism or lever controls the opening and closing; and the swing is outward. The casement is ideally suited to areas where access is limited, such as over a kitchen sink. Because of their design, casement windows can be sealed very tightly against the weather. Their ventilation area is 100 percent.

Sliding windows consist of two or more sashes which slide horizontally in a common frame. In two-sash units, both members are movable. In three-sash windows, the center unit is usually fixed. The ventilation area is generally 50 percent. Fig. 13-5 illustrates the construction details of the sliding unit in this type of window.

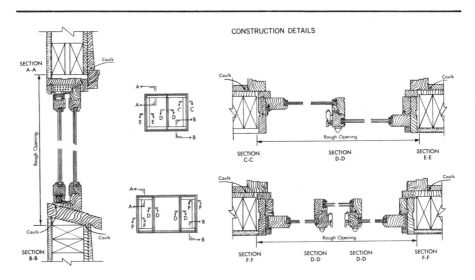

Fig. 13-5 Construction details of the sections of a sliding window.

Types of Windows

Awning windows swing outward and have hinges at the top. They are used individually in stacks, in rows or in combination with fixed windows. A similar unit called a hopper works like the awning but hinges at the bottom. Its top swings inward.

Bow and bay windows: Bow windows have a graceful outward curve. Bay windows are straight at the center and angled at each end. Both were widely used in Georgian and colonial architecture and are still popular for suitable structures. Bow windows usually have fixed sections; however, the one shown in Fig. 13-6 has casement sections at each end. Bay windows commonly have a fixed center and operating end units.

Knee brackets are made as reinforcements for bow windows. In addition to being decorative, they offer important extra support, Fig. 13-7. They are recommended for use on large units and those glazed with double pane insulating glass.

Fig. 13-6 Bow window with fixed center and operating casement end units.

Fig. 13-7 Bow window with supporting knee bracket. *(Fred Reuten, Inc.)*

If they are installed without overhang protection, either bow or bay windows require a roof. Use wedge-block rafters to form the rough roof and then cover with wood decking. Complete them by adding flashing and finish roof. See Fig. 13-8 for details of a bow window roof.

Jalousies consist of slats mounted horizontally between a frame and sloping upwards from without. Made of glass, metal or other materials the slats are operated by a mechanism which links them to a single

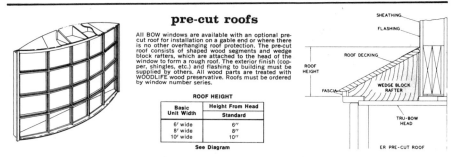

pre-cut roofs

All BOW windows are available with an optional pre-cut roof for installation on a gable end or where there is no other overhanging roof protection. The pre-cut roof consists of shaped wood segments and wedge block rafters, which are attached to the head of the window to form a rough roof. The exterior finish (copper, shingles, etc.) and flashing to building must be supplied by others. All wood parts are treated with WOODLIFE wood preservative. Roofs must be ordered by window number series.

ROOF HEIGHT

Basic Unit Width	Height From Head
	Standard
6' wide	6"
8' wide	8"
10' wide	10"

See Diagram

Fig. 13-8 Details of a pre-cut bow window roof. *(Fred Reuten, Inc.)*

control. When the slats are partly or completely open (tilted downward or nearly horizontal), they admit air and light. When they are closed, each slat overlaps the one below it slightly, thus keeping out wind and rain to some degree. Because of this movable design feature, however, they cannot be weatherstripped effectively. Their use is therefore limited to warm climates—climates as mild as those around the Mediterranean where they originated.

Fixed windows are used mostly for effect. They can be used alone or in combination with other types. Usually quite large, they may be set in sash or mounted directly into a rabbeted frame. They do not open for ventilation but can be excellent sources of light if well placed.

Skylights are very useful for admitting light to attics, interior rooms and windowless hallways. Some are made with mechanisms which permit them to be opened to admit air. Various sizes and shapes are available. Those made of formed plastic and factory-assembled are easily installed on flat or sloping roofs. Fig. 13-9 shows a unit installed in a dormer. As a safety measure, skylights should be made of reinforced glass or plastic.

Storm Windows

Storm windows consist of a second pane of glass placed close to the main window. The air space formed between the two glass surfaces greatly reduces heat loss from the interior of rooms to the outside.

Various methods are employed to hang storm windows. One utilizes a storm sash fitted with brackets at the top. The storm windows themselves then hang on hooks fitted to the window frame. Other designs include those with center hinges which facilitate ventilation control from inside the house.

Fig. 13-9 Skylight installation in the dormer of an attic room. (*Ventarama Skylight Corp.*)

Still another type includes combination self-storing sashes and screens. These storm windows have an aluminum frame with vertical channels that accept both upper and lower storm sashes and screens. The units are removable for cleaning but are not removed for storage.

Some storm units are made with expander strips. These strips permit adjustments for variations and/or faults in non-standard openings, Fig. 13-10.

Windows

Double-pane windows have excellent insulating qualities. The double pane is made with two sheets of permanently-sealed glass with a dead air space between, Fig. 13-11. Double pane windows are used instead of conventional storm windows when the buildings have large expanses of fixed glass. They are better insulators than single pane windows; they therefore cut down on fuel bills in the winter and save on air conditioning costs in the summer. The double pane windows also reduce body radiation of heat to colder surfaces by keeping the glass surface warmer in the interior of the room. They eliminate the need for storm windows and minimize condensation on the glass panes.

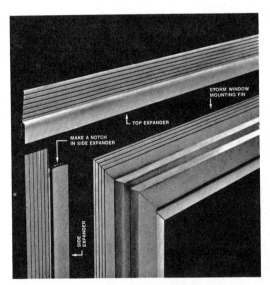

Fig. 13-10 Details of expander strips used in one type of storm window frame. *(Louisiana-Pacific)*

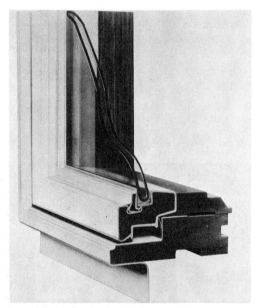

Fig. 13-11 Cutaway section of insulated sash in double-pane window. *(Andersen Corp.)*

Special Windows

Some windows are made for special applications, such as those installed in gable ends, hallways, garages and stairways. They provide some light and are decorative as well. Two types are shown in Fig. 13-12; both have fixed sashes. They are also made with a movable sash.

Fig. 13-12 Decorative windows: **left,** octagon; **above,** half-round.

Removable Grilles

Removable grilles are laid against the glass of modern windows and doors to give the effect of muntins. Muntins developed as internal supports that held together the small pieces of glass which were available for the first windows. Today they are ready-made in numerous sizes and patterns. They snap into place inside the frame and are easily removed for cleaning and painting.

Typical Window Parts

The components of a double-hung window are shown in Fig. 13-13. The three basic parts are head jamb, side jambs and sill. The head jamb is the top horizontal part of the frame; the two side jambs are the vertical frame sides; and the sill is the lower horizontal part. Fig. 13-14 shows the cross section of a double-hung window. The mullion is a vertical member between two adjacent windows.

In frame construction, the window head may contain a drip cap to protect the casing from rain water. The drip cap is not necessary when the window has a deep overhang. Likewise, brick structures do not need a drip cap since the head bricks serve the same purpose.

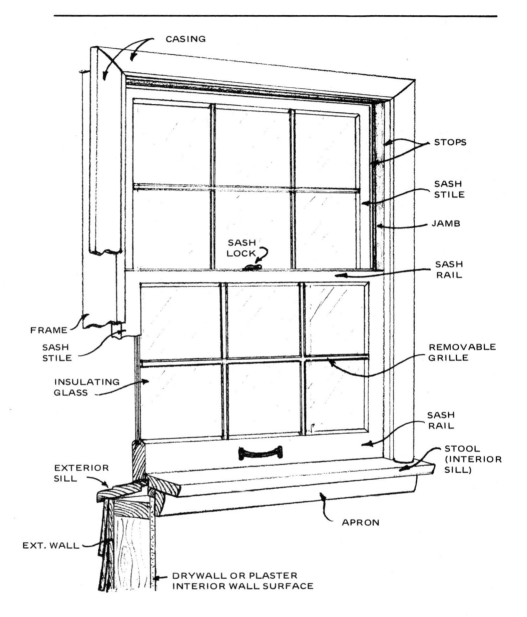

CASING

STOPS

SASH STILE

JAMB

SASH LOCK

SASH RAIL

FRAME

SASH STILE

INSULATING GLASS

REMOVABLE GRILLE

SASH RAIL

STOOL (INTERIOR SILL)

EXTERIOR SILL

APRON

EXT. WALL

DRYWALL OR PLASTER INTERIOR WALL SURFACE

Fig. 13-13 Components of a double-hung window—including the framing parts, casement, and exterior wall cross-section.

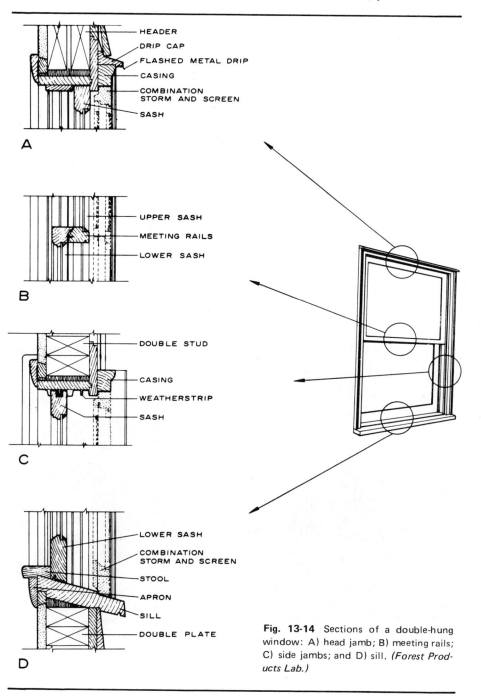

A

- HEADER
- DRIP CAP
- FLASHED METAL DRIP
- CASING
- COMBINATION STORM AND SCREEN
- SASH

B

- UPPER SASH
- MEETING RAILS
- LOWER SASH

C

- DOUBLE STUD
- CASING
- WEATHERSTRIP
- SASH

D

- LOWER SASH
- COMBINATION STORM AND SCREEN
- STOOL
- APRON
- SILL
- DOUBLE PLATE

Fig. 13-14 Sections of a double-hung window: A) head jamb; B) meeting rails; C) side jambs; and D) sill. *(Forest Products Lab.)*

Windows

Installation of Windows

Stock windows are generally delivered to the site, primed, braced and ready to install. Rough handling may have put the unit out of square, however, so check it very carefully before installation. Leave the braces on the windows during installation. Remove them only when the trim is ready to be added.

Usually the rough-opening allowance is 1/2" at the sides and 3/4" at the top. This RO allowance permits any necessary adjustments when plumbing and leveling the unit. Install wedges or shims as needed and carefully in order to prevent distorting the frame. Then fill the spaces between the rough opening and frame with insulation. Install double-hung frames from the outside, Fig. 13-15. Be sure to use strips of building paper to line the perimeter of the opening as protection against air infiltration.

If panel siding is used in place of sheathing, install the siding first, then the window frame. Place a bead of caulking under the side and head casing.

Fig. 13-15 Double-hung window frames installed in the rough openings from the outside.

Nailing: Use galvanized nails spaced about 12 inches apart. Before driving the nails all the way, check the windows to be sure that they slide freely. These nails should be long enough to penetrate well into both studs and header.

Glass Blocks

Glass blocks are often used in window construction. They provide light, privacy and, since they are not movable, a degree of burglar-proofing. They are also used as room dividers and partitions.

The blocks come in varying sizes, the most common widths and heights are 6" × 6", 8" × 8", and 12" × 12"—and all 4 inches thick. Attractive patterns and designs are obtainable. Some blocks diffuse light, while others direct it. Because of their double-wall construction they offer a considerable degree of insulation.

Installation of glass blocks: The blocks are installed in a manner similar to bricklaying. They are set in a bed of mortar if installed in a masonry wall. In frame construction, the blocks are supported by wood members. In frame construction, always prime the wood next to the mortar with asphalt emulsion.

If extra security is desired, steel reinforcing rods may be used between the horizontal rows of blocks. Bed them directly in the mortar. Then, after the blocks are all laid, apply caulking around the perimeter of the window.

GLOSSARY

apron: the inside trim of a window placed against the wall directly below the stool.

bay window: a rectangular or polygonal window projecting outward from the wall of a structure.

bow window: a curved window projecting outward from the wall of a structure.

casement window: a wood or metal window hinged vertically, usually at one side and swinging fully outward.

casing: the finish trim around door and window openings.

double-hung: a type of window with upper and lower sashes; the sashes are both movable in a vertical plane.

Windows

drip cap: a molding used over doors and windows to direct water away from the frame.

jalousie: a window with louvered panes.

jamb: the inner surface lining the sides and top of door and window frames.

mullion: a vertical divider between two adjacent windows.

muntin: the molding separating the glass panes or lights of a window.

rail: the horizontal crosspiece of a door or window; it is also the top piece of a balustrade.

rough opening (R O): the opening in framing members for windows, doors and the like.

sash: either the framework which holds the glass in a window or the movable part (or parts) of the window itself.

sill: in a door or window the lowest horizontal member.

stile: the vertical outside framing piece of a door or window.

stops: narrow wood strips which keep double-hung windows in place.

stool: a flat molding fitted over the window still between the jambs.

14

Exterior Doors & Frames

Main entry doors, Fig. 14-1, are generally 3' 0" wide and at least 6' 6" high. The usual height however is 6' 8". Secondary doors for side and rear entrances are usually 2' 8" wide. Except for storm doors, exterior doors are 1-3/4" thick. They may be of flush or panel construction,

Fig. 14-1 Main entry door of house with sidelights and pediment.

Fig. 14-2. The frames for exterior doors are heavier than those for interior use; and framing lumber is a minimum of 1-1/8" thick. Exterior

Exterior Doors and Frames

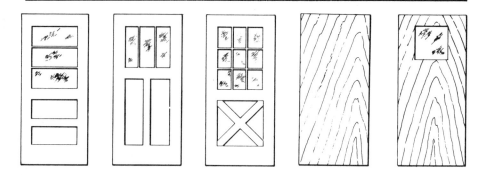

Fig. 14-2 Commonly-used types of exterior doors: three panel styles and two flush styles.

door parts are shown in Fig. 14-3: cross sections A, B, and C shows the layout details for head, side, and sill construction.

Wood Doors

Exterior wood doors are available prehung, complete with all trim. Most manufacturers furnish them preservative-treated and primed, ready for installation in the rough opening. To accomodate various wall thicknesses, most door frames have reversible extension strips.

Manufacturers specifications sheets clearly indicate RO widths and heights, see Fig. 14-4 on page 254.

Exterior door frame installation: Install exterior door frames with great care. To maintain a level surface between the finish floor and door sill, cut and trim off a portion of the header joist or stringer. Usually this is done with saw, ax and chisel. When the joists run parallel to the door, a support member should also be added, as the dotted lines shown in Fig. 14-5 (page 254). Note that the joint between finish floor and sill is covered by the threshold. In the same Fig. 14-5 study the details of door frame construction at the sill.

The frame is secured to the header and studs of the rough opening after it has been carefully leveled and plumbed. Use wedges to ensure a snug fit between the casing and studs. If necessary, the sill can be shimmed so that it will bear solidly against the rough framing. Then fasten the frame to the studs and headers by nailing through the outer casing. Add the threshold later after the finish floor is installed. A 1/2" clearance is required between the bottom of the door and the top of

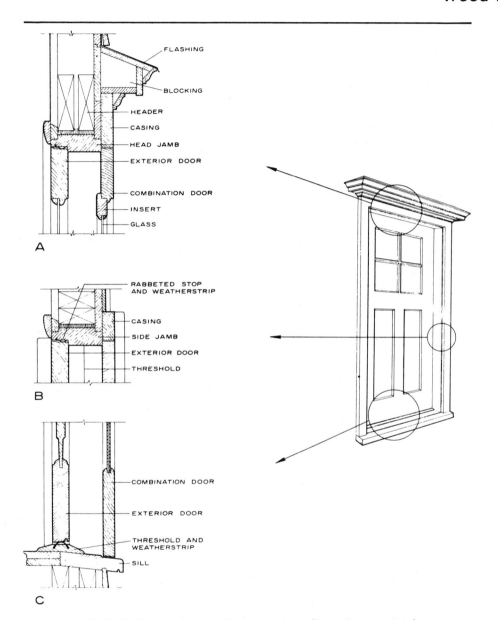

FLASHING

BLOCKING

HEADER

CASING

HEAD JAMB

EXTERIOR DOOR

COMBINATION DOOR

INSERT

GLASS

A

RABBETED STOP
AND WEATHERSTRIP

CASING

SIDE JAMB

EXTERIOR DOOR

THRESHOLD

B

COMBINATION DOOR

EXTERIOR DOOR

THRESHOLD AND
WEATHERSTRIP

SILL

C

Fig. 14-3 Exterior door parts in cross section. *(Forest Products Lab.)*

the finish flooring. If a threshold is used, the clearance between the bottom of the door and the top of the threshold should be at least 1/4". These are industry standards set up by the National Woodwork

Exterior Doors and Frames

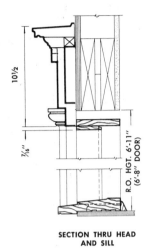

SECTION THRU HEAD
AND SILL

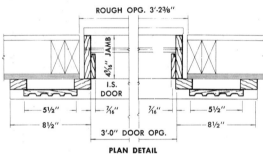

ROUGH OPG. 3'-2⅜"

4⁹⁄₁₆" JAMB

I.S. DOOR

— 5½" — ⁷⁄₁₆" — ⁷⁄₁₆" — 5½" —

— 8½" — — 8½" —

3'-0" DOOR OPG.

PLAN DETAIL

Fig. 14-4 Rough opening sizes for entrance door like that shown **(upper left)**: the cross section through of head and sill is above; the plan detail left. *(Morgan Co.)*

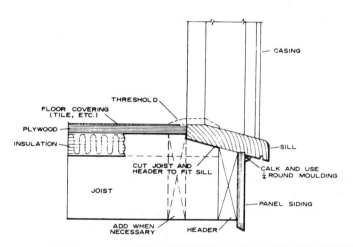

CASING

THRESHOLD

FLOOR COVERING
(TILE, ETC.)

PLYWOOD

INSULATION

SILL

CUT JOIST AND
HEADER TO FIT SILL

CALK AND USE
¼ ROUND MOULDING

JOIST

PANEL SIDING

ADD WHEN
NECESSARY

HEADER

Fig. 14-5 Details of a door frame at the sill.

Manufacturers Association. Different clearances may be needed for special weatherstripping or for special purpose doors; follow the manufacturer's specifications.

Most wood doors are made of ponderosa pine, except for the sill and threshold which are usually oak. The oak is used for these parts because it is tough and durable. For further information on doors, see also Chapter 21 on interior doors and trim.

Cap pieces (head casing) are the horizontal members above the doorway. They come in various styles, Fig. 14-6. Matching pilasters are also made. Both cap pieces and pilasters are available as a complete door unit, or they may be purchased separately for use in remodeling work.

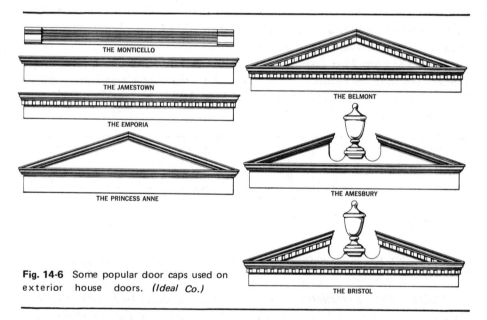

Fig. 14-6 Some popular door caps used on exterior house doors. *(Ideal Co.)*

Sliding Glass Doors

Sliding glass doors consist of two or more framed glass panels which move horizontally in a wood or metal framework. The most common type contains two panels, one movable, one fixed. Some wood frameworks have a rigid vinyl sheathing on all exposed parts. The vinyl protects the wood surface and eliminates the need for paint.

Exterior Doors and Frames

Manufacturers specifications sheets list exact sizes of sliding units available and the opening size requirements. To facilitate ordering these units and to make for uniformity in the industry, certain standard designations have been adopted by the manufacturers. These designations appear on specifications sheets and on plans as combinations of the letters X and O. These letters indicate how many panels make up the sliding door unit, which panels move, and the direction of movement. The letter X always indicates a movable panel; the letter O always indicates a fixed panel; and an arrow indicates the direction of movement. This symbol is always read as if the sliding door is being viewed from the outside of the building or room. When ordering a door, the carpenter specifies not only the model or stock number of the unit, but also the letters to show desired direction of movement of the panels. For example, a sliding door marked XOX has three panels, the middle one is stationary, two outer ones are movable; while a door marked OXO has two fixed panels with a movable middle panel only. These designations facilitate the ordering of door units and are used in the entire industry, as set up by the National Woodwork Manufacturers Association; see Fig. 14-7.

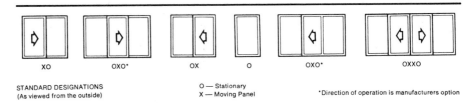

XO   OXO*   OX   O   OXO*   OXXO

STANDARD DESIGNATIONS   O — Stationary
(As viewed from the outside)   X — Moving Panel   *Direction of operation is manufacturers option

Fig. 14-7 Stock sliding doors with standard designations for operational units.

Sliding door installation: The installation described here is for a door with an aluminum frame. Check the rough opening to make sure that it is correct in both height and width. Seal all joints with a good quality sealant. Apply heavy beads of caulking at the sill area. Install the frame and tack temporarily through holes in the jamb, Fig. 14-8. Level the frame carefully and shim if necessary, Fig. 14-9. Be sure to locate shims under fasteners. Secure the frame with wood screws. Fill all spaces between shims with insulation.

Use a piece of wood to lock the door until hardware is installed. Be sure to place a protective block at the jamb, Fig. 14-10.

Note: As a safety precaution, all glazing, especially in sliding doors, should be taped with a large "X". This reminds workers that glass is in place.

Sliding Glass Doors

Fig. 14-8 First, door frame is installed temporarily in the rough opening.

Fig. 14-9 Then, head and sill are leveled, sides plumbed and frame shimmed as necessary.

Fig. 14-10 Method of temporarily locking sliding doors by wedging a block at the jamb.

Exterior Doors and Frames

Garage Doors and Frames

Garage doors for American cars should be a minimum of 8 feet wide and 6 feet 6 inches high for single-car garages. Two-car garage doors must measure at least 16 feet wide. These large openings require substantial framing to provide sufficient rigidity.

The most common garage doors are the roll-up and the swinging type, Fig. 14-11. Both are used extensively and have replaced the older hinged doors. Installation procedures differ among manufacturers, but there is little difference in the basic designs. Counterbalance springs permit the doors to be raised or lowered effortlessly. Electrically operated doors actuated by sound, light or radio waves are also common.

Fig. 14-11 Typical overhead garage door equipped with a torsion spring mechanism.

Some of the materials used for garage doors include fiberglass, metal and wood. Garage doors are generally listed by the finished opening size.

Garage door frames: In frame construction, the RO size for garage doors is normally 3 inches wider and 1-1/2 inches higher than the finish size. The lumber used for the jambs and header depends on the design. The details shown in Fig. 14-12 are for designs widely used in brick veneer and shingle construction.

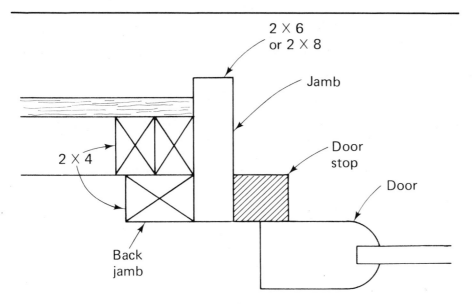

Fig. 14-12 Typical framing details for a garage door installation.

Be sure to consider the regular head room allowance when ordering garage doors. Special hardware is available and must be used for low headroom clearances. Installation procedures vary among manufacturers, therefore it is necessary to refer to instruction sheets when installing the doors.

Most electric automatic garage door openers are made with a safety slip-clutch which stops door travel instantly when any obstruction is encountered, such as a car, a person entering on foot, or the like.

Exterior Doors and Frames

GLOSSARY

blocking: in door construction, the wood pieces used to frame a door cornice.

cap: the cornice over a door.

counterbalance: a device, usually either a spring or weight used to offset the moving weight of doors, windows, and disappearing stairs that either slide, by-pass one another, or rise when not locked in place.

finial: a crowning ornament, as over a door.

header: the horizontal structural member over doors and windows; also called a lintel.

pediment: a triangular decoration often used above exterior doors; in classical architecture, the triangular space forming the gable of a two-pitched roof.

pilaster: a decorative column used at the sides of exterior doors; also a structural pier used for reinforcement.

threshold: a strip of wood or metal with beveled edges covering the joint between the finished floor and sill of a doorway.

weatherstrip: a narrow piece of wood, rubber, plastic or other material used around doors and windows to prevent air and moisture infiltration.

15

Roof Cornices

A cornice is generally defined as a decorative projecting horizontal trim that surmounts and crowns a wall. Cornices are often used in room interiors to conceal drapery and curtain hardware. In house and roof construction, the cornice is the exterior finish and trim that connects the side walls to the overhanging horizontal projections of the roof. In a hip roof the cornice extends completely around the perimeter of the structure. In a gabled roof the cornice is formed only at the sides of the house below the eaves or roof rafter ends. The section of a typical cornice can be studied in Fig. 15-1.

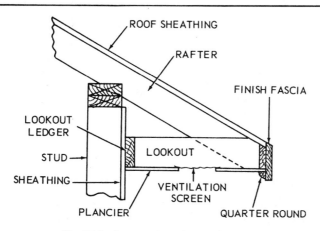

Fig. 15-1 Cross section of a cornice.

Types of Cornices

Figure 15-2 shows both a box and an open cornice. The box type is the most commonly used. It offers ample protection to the side walls and gives the structure a finished appearance. The open cornice is similar, except that it is open on the underside. That is, lacking a soffit or plancier, the projecting roof rafters are exposed. The close cornice

Roof Cornices

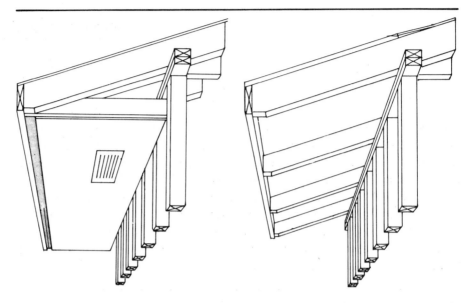

Fig. 15-2 The commonly used box cornice (**left**), and the open cornice (**right**) without soffit.

consists only of facia board and molding trim since the roof rafters do not project beyond the side walls of the house. The simplest and least expensive to install, the close or simple cornice offers little if any protection to the side walls. Even so, it still enhances the architectural beauty of the house.

The cornice projection depends on the style of the building. A 4-foot overhang is common on ranch-style houses; while colonial-style houses may have an extension of only 18 to 24 inches.

Box Cornice: It may be either wide or narrow and conceals the roof rafters with soffit, facia board, and molding trim. In the narrow cornice, the rafter serves as the nailing surface for both the soffit and the facia board, Fig. 15-3. For the wide cornice, additional horizontal supports called lookouts and usually made of 2 × 4's are required. Nail them to each rafter and toenail them to a ledger strip which has been nailed to the wall. Sometimes a 2" × 2" ledger is used in conjunction with notched lookouts. Join siding and soffit with frieze board.

Sloping Soffit Cornice: In some methods of construction a boxed cornice is made without lookouts. The soffit is then fastened to the underside of the rafters so as to form a sloping surface. See Fig. 15-4.

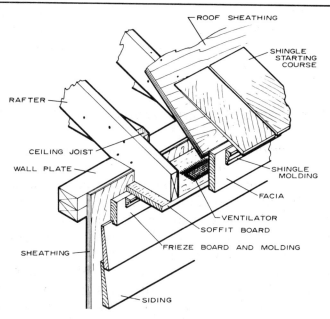

ROOF SHEATHING

SHINGLE STARTING COURSE

RAFTER

CEILING JOIST

WALL PLATE

SHINGLE MOLDING

FACIA

VENTILATOR

SOFFIT BOARD

FRIEZE BOARD AND MOLDING

SHEATHING

SIDING

Fig. 15-3 Construction layout for a narrow box cornice. *(Forest Products Lab.)*

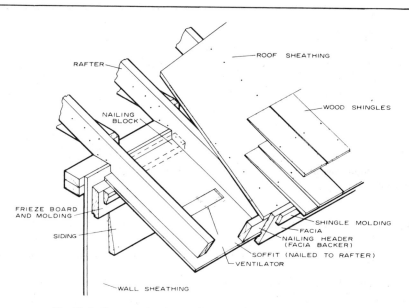

RAFTER

ROOF SHEATHING

NAILING BLOCK

WOOD SHINGLES

FRIEZE BOARD AND MOLDING

SIDING

SHINGLE MOLDING

FACIA

NAILING HEADER (FACIA BACKER)

SOFFIT (NAILED TO RAFTER)

VENTILATOR

WALL SHEATHING

Fig. 15-4 Construction layout for a cornice with a slanting soffit.

Roof Cornices

Open Cornice: This cornice is similar in construction to the sloping soffit cornice. The basic difference is that the soffit itself is eliminated. In this case, use exterior plywood since the underside of the roof deck is exposed to the weather.

Finish Trim for Gable End Walls

Rake Construction: The rake is the exterior finish and trim applied parallel to the sloping end walls of a gabled roof. Either the rake may be close with small or no projection or it may be wide with projections up to 2 feet or more. The close rake may consist only of facia board nailed directly to the frieze board. See Fig. 15-5. However, the greater the projection of the rake section, the better the protection of the end walls. For narrow projections up to 16 inches, a rake or fly rafter, supported by the ridge beam and rafter header, supports lookouts which in turn help support the soffit. See Fig. 15-6. Spacing of the lookout blocking is determined by the soffit material used. For wider overhangs beyond 16 inches, ladder framing with lookouts is necessary.

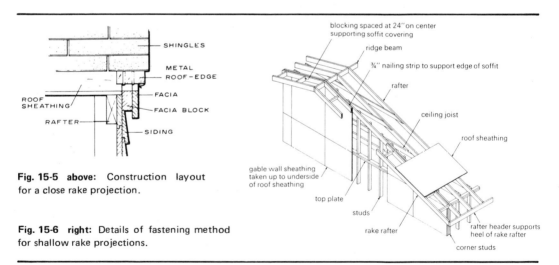

SHINGLES

METAL
ROOF-EDGE

FACIA

ROOF
SHEATHING

FACIA BLOCK

RAFTER

SIDING

Fig. 15-5 above: Construction layout for a close rake projection.

Fig. 15-6 right: Details of fastening method for shallow rake projections.

blocking spaced at 24" on center supporting soffit covering

ridge beam

¾" nailing strip to support edge of soffit

rafter

ceiling joist

roof sheathing

gable wall sheathing taken up to underside of roof sheathing

top plate

studs

rake rafter

rafter header supports heel of rake rafter

corner studs

Cornice Return: On a hip roof, the cornice is continuous around the entire roof. On a gabled roof, the cornice must be joined to the rake projections. This is called the cornice return. There are several ways in which the return may be constructed. In one method, the sloping soffit is continued on through the gable end projection. In another method, the boxed end is used when the eave soffit lies in a horizontal plane. The facia board must be widened at the end to close off the soffit, as shown in Fig. 15-7.

Soffits: Soffits may be made of various materials such as metal, plywood, and hardboard. Unless a ready-made soffit system is employed, the carpenter generally uses plywood. The face grain of the plywood must run at right angles to or across the supports. The plywood used is usually 3/8" thick when the support spacing is 24 inches or less. If the supports are spaced 48" OC (48 inches on center), then 5/8" plywood must be used. Nails must be a noncorrosive type; either box or casing nails work well. Venting must be provided either by continuous screening or by louvers. See Fig. 15-8.

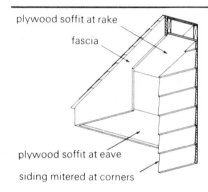

plywood soffit at rake

fascia

plywood soffit at eave

siding mitered at corners

Fig. 15-7 above: Boxed cornice return. *(Central Mortgage & Housing Corp.)*

Fig. 15-8 right: Louvers in end wall of gable provide ventilation for attic.

Contemporary soffit systems have been devised to greatly reduce installation time. One such labor-saving system utilizes aluminum soffits complete with trim molding, wall supports and facia caps. A typical installation begins when the carpenter strikes a chalk line level with the bottom edge of the subfacia. Then he attaches an appropriate molding

Roof Cornices

or channel to the wall on the chalk line. Next he cuts the soffit panels to size and nails them to the subfacia. And in the final step, the carpenter installs the facia trim, Fig. 15-9.

Strike a chalk line along the wall that is parallel and level with the bottom edge of the sub-fascia.

When using Alcoa Quarter-round molding or J-channel as a·wall support piece, cut tabs and bend as shown, at stud locations. When using F-channel as a wall support, attach directly to wall on chalk line.

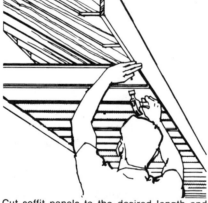

Cut soffit panels to the desired length and insert into wall support piece. Nail panel to sub-fascia. (Intermediate nailing supports are not required on soffits up to three feet in width.)

Fig. 15-9 Installation of an aluminum soffit system: **upper left,** striking a chalk line; **upper right,** installing a wall support piece; **left,** nailing the soffit panels to the sub-facia. *(Alcoa)*

Another soffit system utilizes wood fiber panels stiffened with metal channels. This construction method eliminates the need for lookouts; and only the panel edges are supported with joint molding.

To use, first insert the leading edge into a rabbeted facia board then tilt it into place. Next fasten the rear edge to a 1 × 4 nailer attached to the house. Finally, add a wood molding to conceal the nail heads.

Use stiffeners for soffit systems only if lookouts are not used. Install stiffener channels by placing the soffit panel face down on a flat surface. Next locate the channel properly on the panel, prongs down. Then insert a piece of 2 × 4 in the channel and hammer it several times. That drives the prongs into the panel.

Facia: Factory-made grooved wood facia are available in many sizes. Some are made with a double groove which gives the builder a choice of either 1/4" or 3/8" thick soffit panels, also called planciers. The use of grooved facia, whether or not a soffit system is employed, cuts installation time and cost. Fig. 15-10 shows a plowed facia used on a flat roof with narrow overhang.

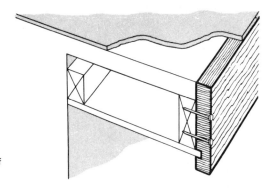

Fig. 15-10 Plowed facia on a flat roof with a narrow overhang.

GLOSSARY

cornice: the horizontal overhang of a roof, consisting of a facia, soffit, and moldings.

cornice return: the section of a cornice which returns on the gable end of the house.

facia (or fascia): the outer face of a cornice.

facia block: backing for facia when rake projection is shallow.

fly rafter: end rafter of the gable overhang supported by roof sheathing and lookouts.

frieze: a horizontal trim member connecting the soffit and wall.

lookout: the structural support of an overhang such as in a roof.

Roof Cornices

plancier: the underside of a cornice, especially one made of wood.

soffit: the underside of an architectural member, especially an overhanging cornice; a plancier.

16

Exterior Wall Covering

The exterior wall covering of a house is added after the roof and cornices are completed. The covering is usually installed over sheathing, but some wall coverings are applied directly to the frame. In such applications, the covering serves as both sheathing and siding. A wide range of materials is available for exterior walls, including wood, plastic, metal, glass, wood composition, hardboard, masonry, asphalt and asbestos. Fig. 16-1 shows a house finished with western red cedar handsplit shakes.

Fig. 16-1 House finished with sidewalls of handsplit cedar shakes.

Methods of application depend on the kind of material used. Exterior walls when properly installed with quality materials should give years of trouble-free service. In any remodeling work, loose boards

269

should also be repaired or replaced as well. The sheathing papers must always be free of tears or holes; and wood coverings must be kept at least 8 inches above the ground to prevent decay and termite damage.

Horizontal Siding

Wood siding is widely used in the United States. Some of the many types available are shown in Fig. 16-2. The most popular of these is bevel or tapered siding. Bevel siding is made by sawing boards diagonally so that one edge is thicker than the other. Sizes range upward from 1/2" to 3/4" at the thick edge to 3/16" at the thin edge or top. One surface of bevel siding is smooth, the other rough sawn.

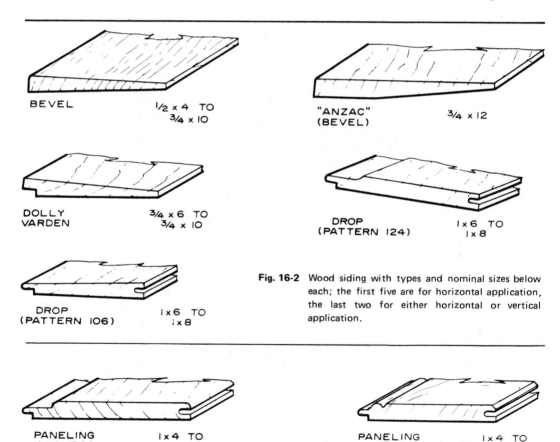

BEVEL   1/2 x 4 TO 3/4 x 10

"ANZAC" (BEVEL)   3/4 x 12

DOLLY VARDEN   3/4 x 6 TO 3/4 x 10

DROP (PATTERN 124)   1 x 6 TO 1 x 8

DROP (PATTERN 106)   1 x 6 TO 1 x 8

Fig. 16-2 Wood siding with types and nominal sizes below each; the first five are for horizontal application, the last two for either horizontal or vertical application.

PANELING (WC 130)   1 x 4 TO 1 x 12

PANELING (WC 140)   1 x 4 TO 1 x 12

Which surface is placed outward depends on the style of the house and the finish desired. For a painted finish the smooth side is normally exposed; while stained finishes are best done on the rough-sawn side.

Drop siding has tongue and groove or shiplap edges and is usually 3/4" thick. It is recommended for walls that are not sheathed, such as garages, tool sheds and other secondary buildings.

It is important that siding have the correct moisture content. Otherwise, excessive shrinkage after installation can cause problems, especially in wide boards. The recommended moisture content for siding is about 9 percent in dry southern states, and about 12 percent in the rest of the country.

Wall Preparation: As explained previously, siding may be applied to sheathed or unsheathed walls. If the siding is to cover an unsheathed wall, place an underlayer of waterproof paper over the studs first. Use it also over lumber sheathing, Fig. 16-3. Such waterproof paper is not

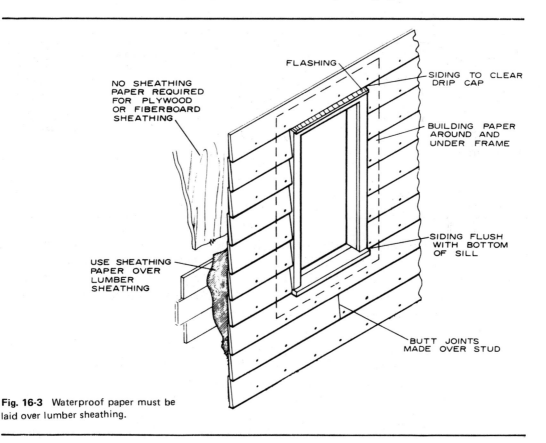

NO SHEATHING PAPER REQUIRED FOR PLYWOOD OR FIBERBOARD SHEATHING

USE SHEATHING PAPER OVER LUMBER SHEATHING

FLASHING

SIDING TO CLEAR DRIP CAP

BUILDING PAPER AROUND AND UNDER FRAME

SIDING FLUSH WITH BOTTOM OF SILL

BUTT JOINTS MADE OVER STUD

Fig. 16-3 Waterproof paper must be laid over lumber sheathing.

needed over plywood or fiberboard sheathing. Drip caps over doors and windows should be flashed, however, unless they are protected by wide overhangs.

Measuring Techniques: Use a story pole to ensure accuracy when installing horizontal siding. It helps keep the siding level and aligned around the perimeter of the building. To make the pole, select a flat straight piece of 1 × 2 stock. Then carefully mark on it the location of the upper and lower limits to which the siding will reach on the wall, Fig. 16-4. Divide that span into equal parts so as to represent the exposure of the siding, Fig. 16-5. If possible, adjust the spacing marks to allow the siding boards above and below the windows to be placed without notching.

Now transfer the reference lines on the story pole to each inside and outside corner as well as to the windows and doorways, Fig. 16-6. Those lines represent the top edge of each siding board. Chalk lines may be struck if desired.

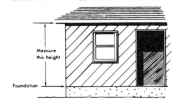
Fig. 16-4 Top and bottom markings on story pole are determined by height of dominant wall of house.

Fig. 16-5 Layout of courses is adjusted to line up with the tops and bottoms of windows.

Fig. 16-6 Story pole marks are transferred to house corner. *(Red Cedar Shingle & Handsplit Shake Bureau)*

Corners: Various corner treatments are possible with horizontal siding. The most difficult corner to install is the mitered joint. Each piece of it must be fitted painstakingly and at best is a tedious task. Corner boards and metal corners are more widely used today. If corner boards are used, you must install them before applying the siding; Fig. 16-7 illustrates two corner details. Inside corners can be made with a strip of wood, as shown in Fig. 16-8. Metal corners can also be used; they are designed to conceal nail heads.

Bevel Siding Application: Install a starter strip, nailing it to the bottom of the sheathing or sill plate. Be sure it is level. Make the strip the same thickness as the top edge of the siding, Fig. 16-9. Lay the first piece so that its butt edge is below the starter strip; Fig. 16-10 shows installation of 6" siding boards with 1" overlap. For 8" and 10" siding,

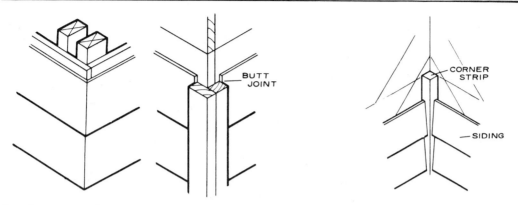

Fig. 16-7 Details of two styles of outside corners for horizontal siding.

Fig. 16-8 Inside corners made with strip of wood.

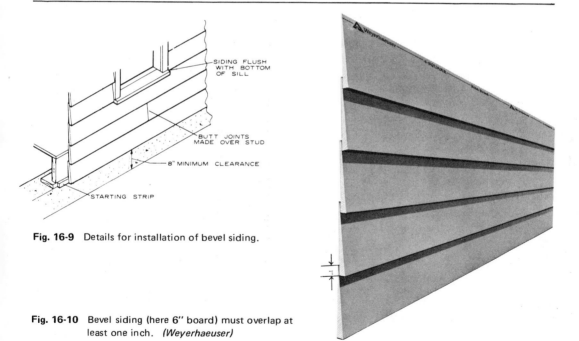

Fig. 16-9 Details for installation of bevel siding.

Fig. 16-10 Bevel siding (here 6" board) must overlap at least one inch. *(Weyerhaeuser)*

be sure to make a 1-1/2" overlap. Some siding is available with markings to indicate the lap. If not indicated on the siding, be sure to install the pieces parallel to each other.

Butt joints at windows, doors and corners must be tight to keep out moisture. Before installing cut boards, brush the ends with a preservative, Fig. 16-11. This is a very important detail. Do not overlook it. Caulk all joints for protection against rain, snow and wind; and use caulking to close the joints above doors and windows before installing flashing, Fig. 16-12.

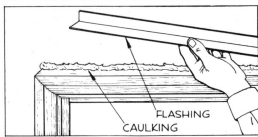

Fig. 16-11 Boards cut on the job should be treated with a water repellent.

Fig. 16-12 Joints should be caulked before flashing is installed, especially above doors and windows.

Nailing: For siding use only non-corrosive fasteners, such as aluminum, hot-dipped galvanized or stainless steel nails. Iron and copper nails should not be used since they can cause staining. Determine nail lengths by calculating the thickness of the siding as well as the type and thickness of the sheathing it is nailed to. For 1/2" siding use sixpenny nails; for thicker material, use eightpenny nails.

One way to avoid splitting boards near their ends is to use blunt-pointed nails. Another way to eliminate splitting is to predrill the nail holes. Be sure to use annular or spiral nails because of their greater holding power than smooth nails.

Nailing methods for different types of siding are shown in Fig. 16-13. Be sure to place all nails so that they clear the tip or lap of the siding below, in order to allow expansion clearance. Always penetrate each stud bearing with a nail.

For unusual effects, siding can be combined with other types of wall finish. For example, part of a wall can be stuccoed and the balance covered with either vertical or horizontal siding.

Vertical Siding

Vertical siding is popular for one-story houses. Used at entrance ways, on gable ends, and on entire walls, it tends to make a house appear taller. Interesting effects are also obtained by combining both horizontal and vertical siding in the same structure. Many different

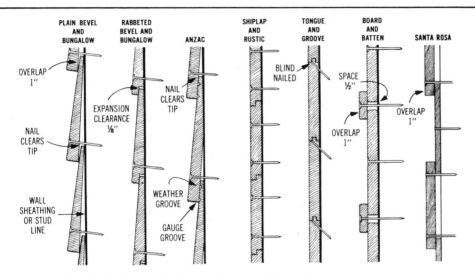

Fig. 16-13 Details of nailing methods for different kinds of siding.

styles and materials are available. Board-and-batten is a fairly common variation. Other batten combinations are "T" batten, board-on-board, and reverse batten. See Fig. 16-14. Additional variations include flush patterns and V-joint. The surfaces may be smooth or textured.

STANDARD BOARD AND BATTEN: Drive one 8d nail midway between edges of the underboard, at each bearing. Then apply batten strips and nail with one 10d nail at each bearing so that shank passes through space between underboards.

BOARD ON BOARD: Space underboards to allow 1½-inch overlap by outer boards at both edges. Use one 8d nail per bearing for underboards. Outer boards must be nailed twice per bearing to insure proper fastening; use 10d nails, driven so that the shanks clear the underboard by approximately ¼ inch.

SPECIAL BATTENS: A T-shaped batten, or standard batten nailed over a vertical nailing strip, is nailed exactly the same as the standard method; however, in this case an exceptionally good bearing is provided while driving nail through the batten.

REVERSE BATTEN: Nailing is similar to board on board. Drive one 8d nail per bearing through center of under strip, and two 10d nails per bearing through outer boards.

Fig. 16-14 Four vertical siding combinations and their nailing patterns.

Cross blocking made with horizontal 2 × 4's is a must as a nailing base for vertical siding. For siding up to 1/2" thick, 24" OC spacing is recommended. For heavier siding, 48" OC spacing is permitted.

Exterior Wall Covering

Plywood Siding

Plywood siding is often used as an exterior wall covering. Installation time is reduced considerably because large sheet sizes are available. The panels may be installed vertically or horizontally. For horizontal application, use battens to conceal the end joints. Different surface textures are made with special manufacturing processes.

Use redwood plywood when a natural wood appearance is desired. These panels can be applied directly to studs to serve as both sheathing and siding. Very interesting patterns of saw-textured redwood plywood are available in 8, 9, and 10 foot lengths. You can also install dummy battens on plywood sheets to create a board-and-batten effect.

In addition to full-size sheets, plywood is also made in 12" and 16" widths for horizontal lap application.

Proper joint treatment is important and should be done carefully. Fig. 16-15 shows different joint details. Be sure to leave a 1/16" spacing at all edges and ends. Space the nails 6" OC along the edges and 12" OC along intermediate supports. Drive the nails through the battens so that they penetrate the studs at least one inch. These nails must be a corrosion-resistant type such as hot-dipped galvanized, aluminum, or stainless steel. Use a polyurethane or silicone caulking, and give all edges a waterproofing coat to prevent moisture problems later. Installation of reverse board-and-batten cedar siding is shown in Fig. 16-16.

Wood Shingles and Shakes

Wood shingles are available for walls in a number of sizes and styles. They may be used in their natural state but they can also be painted or stained. The installation procedure is similar to that for roofs but the weather exposure is greater. Wood shingles can be applied to old or new walls.

For side wall application, use rebutted and rejointed shingles. These shingles have been machine-trimmed so that the sides and butts are square and parallel. They are used with tight-fitting joints. Made with either "sawed" or sanded face, they may be either single- or double-coursed. Grooved shakes are rejointed shingles with a machine-grooved face and must be applied double-coursed.

In single-coursing, one layer of shingles is used. In double-course work, two layers of shingles or shakes are installed.

Wood Shingles and Shakes

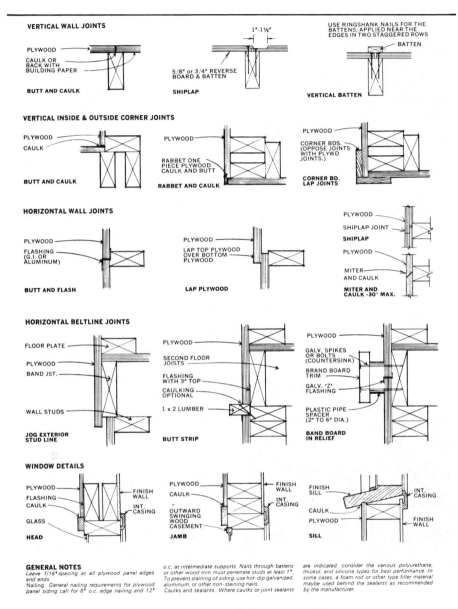

VERTICAL WALL JOINTS

PLYWOOD
CAULK OR
BACK WITH
BUILDING PAPER

BUTT AND CAULK

1"-1½"

5/8" or 3/4" REVERSE
BOARD & BATTEN

SHIPLAP

USE RINGSHANK NAILS FOR THE
BATTENS, APPLIED NEAR THE
EDGES IN TWO STAGGERED ROWS

BATTEN

VERTICAL BATTEN

VERTICAL INSIDE & OUTSIDE CORNER JOINTS

PLYWOOD
CAULK

BUTT AND CAULK

PLYWOOD

RABBET ONE
PIECE PLYWOOD
CAULK AND BUTT

RABBET AND CAULK

PLYWOOD

CORNER BDS.
(OPPOSE JOINTS
WITH PLYWD
JOINTS.)

**CORNER BD.
LAP JOINTS**

HORIZONTAL WALL JOINTS

PLYWOOD
FLASHING
(G.I. OR
ALUMINUM)

BUTT AND FLASH

PLYWOOD
LAP TOP PLYWOOD
OVER BOTTOM
PLYWOOD

LAP PLYWOOD

PLYWOOD
SHIPLAP JOINT
SHIPLAP

PLYWOOD
MITER
AND CAULK

**MITER AND
CAULK -30° MAX.**

HORIZONTAL BELTLINE JOINTS

FLOOR PLATE
PLYWOOD
BAND JST.

WALL STUDS

**JOG EXTERIOR
STUD LINE**

PLYWOOD
SECOND FLOOR
JOISTS
FLASHING
WITH 3" TOP
CAULKING
OPTIONAL
1 x 2 LUMBER

BUTT STRIP

PLYWOOD
GALV. SPIKES
OR BOLTS
(COUNTERSINK)
BRAND BOARD
TRIM
GALV. 'Z'
FLASHING
PLASTIC PIPE
SPACER
(2" TO 6" DIA.)

**BAND BOARD
IN RELIEF**

WINDOW DETAILS

PLYWOOD
FLASHING
CAULK
GLASS

HEAD

FINISH
WALL
INT.
CASING

PLYWOOD
CAULK

OUTWARD
SWINGING
WOOD
CASEMENT

JAMB

FINISH
WALL
INT.
CASING

FINISH
SILL
CAULK
PLYWOOD

SILL

INT.
CASING
FINISH
WALL

GENERAL NOTES
Leave 1/16" spacing at all plywood panel edges and ends.
Nailing: General nailing requirements for plywood panel siding call for 6" o.c. edge nailing and 12" o.c. at intermediate supports. Nails through battens or other wood trim must penetrate studs at least 1". To prevent staining of siding, use hot-dip galvanized, aluminum, or other non-staining nails.
Caulks and sealants: Where caulks or joint sealants are indicated, consider the various polyurethane, thiokol, and silicone types for best performance. In some cases, a foam rod or other type filler material maybe used behind the sealants as recommended by the manufacturer.

Fig. 16-15   Joint details for plywood siding.

The weather exposure for both shingles and shakes is determined by their length and by the application method. For single coursing, the exposure must not be less than one half the shingle length minus 1/2

Exterior Wall Covering

Fig. 16-16 An interesting use of reverse board-and-batten cedar plywood siding. *(U.S. Plywood)*

inch. This assures two layers of wood over the entire wall. Recommended exposures are 8-1/2" for 18" shakes, and 11-1/2" for the 24" shakes. For double-coursing, the exposure is much greater. It should be 14" for 18" shakes and 20" for 24" shakes. For the sake of appearance, lay out the exposures so that the butts are even with the top and bottom line of window openings.

Wood Shingles and Shakes

On double-coursed walls, low-cost undercourse shingles are used for economy since they are concealed by the upper course and do not show on the finished wall. The butt of the undercourse is set back about 1/2 inch from the exposed shingle.

Shingles are made in uniform and in random widths and in several quality grades. First grade shingles are the best and are cut from 100 per cent clear edge-grain heartwood. Second and third grade shingles are of lesser quality and some manufacturing defects are permitted. The fourth lowest grade is made for undercourse applications on double-coursed walls.

Laying Out: The layout procedure for shingles is similar to that used for siding. Your story pole should indicate the exposures as well as the soffit and base lines. Nailing strips for spaced sheathing should also be included on the S.P. While chalk lines are suitable for applying siding, straight-edge guides are better for shingles, Fig. 16-17.

Single-Course Application: The procedure is very like that used in roof construction. The main difference is that a wider exposure is allowed for wall applications. The starter course is doubled at the foundation line. Following courses are laid one over the other, as in Fig. 16-18.

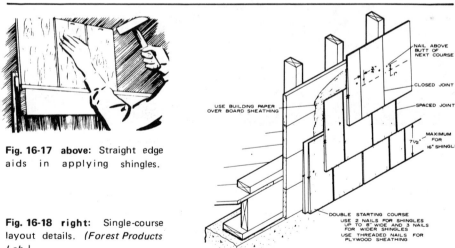

Fig. 16-17 above: Straight edge aids in applying shingles.

Fig. 16-18 right: Single-course layout details. *(Forest Products Lab.)*

For single-course side walls, concealed nailing is standard procedure. Again, be sure to use hot-dipped galvanized or aluminum nails. Drive them not more than 1" above the butt line of the following course of shingles. Use two 3d (threepenny) nails per shingle and place them 3/4"

Exterior Wall Covering

from the edge. Use a third nail if the shingle width exceeds 8 inches. Fig. 16-19 shows installation of a shake wall.

Fig. 16-19 Flashing being applied for a cedar-shake installation.

Double-Coursed Application: Double-coursed shingles permit very wide exposures. In addition to the wider exposure, double-coursing also produces deep shadow lines. Since low-grade shingles are used for the undercourse, greater coverage is obtained at a comparatively low cost.

The starter course at the foundation line requires three shingles, two of which are undercourse. This makes the bottom course as thick as the following courses. The undercourse is laid with a slight space between shingles. The top course is laid tight, with butts projecting 1/2" below the lower edge of the undercourse, Fig. 16-20.

Nails are applied in a straight line 2" above the shingle butts. Use 5d hot-dipped or aluminum nails, spaced 3/4" from the edge. Some manufacturers also fabricate colored nails to match factory stained shingles.

Corners: Outside corners are made with alternating overlaps, similar to the Boston Ridge used on roofs. The inside corners can either be butted against a square corner strip or mitered over a metal flashing. Various corner treatments are shown in Fig. 16-21. A bead of caulking at door and window joints helps keep out moisture.

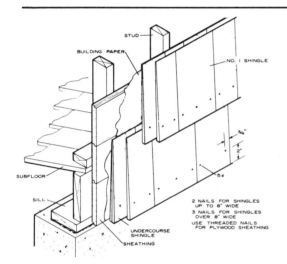

Fig. 16-20 Layout details for double-course application.

Fig. 16-21 Various corner treatments for shingle wall finishing.

Remodeling with Shingles and Shakes: Shingles and shakes can be used to modernize old walls. This overwalling may be applied to any nailable surface. On a masonry wall or other non-nailable surfaces, horizontal nailing strips must be first nailed to vertical furring. On stucco walls, eliminate the furring and nail the horizontal strips directly to the stucco. Be sure to use nails long enough to penetrate the original sheathing or studs. On wood walls, apply the shakes directly. Single or double-coursing may be used. The procedure is the same as for new walls. The additional thickness may make it necessary to add new moldings around doors and windows, however.

Shingle Coverage: Sidewall shingles and shakes are packaged in cartons and by the various grades. Each carton covers 100 square feet as compared to roofing shingles which require 4 bundles per square. When figuring wall coverage, be sure to calculate the areas to be covered by measuring the square feet involved then deduct the windows and doors and add 8 per cent for waste. Consult a coverage chart (supplied by manufacturer) to determine the number of bundles required.

Mineral-Fiber Siding

Mineral-fiber siding is also known as asbestos-cement siding and is similar to the asbestos-cement roofing shingles described in Chapter 12. Made of asbestos fibers and portland cement, the siding is available in several sizes. The most popular is a 12" × 24" shingle; thicknesses range from 5/32" to 3/16". Butts are straight or wavy-edged and surfaces may be smooth or textured. Several factory-applied colors and finishes are readily available. Because of its fine weather resistance and durability, this sort of siding is highly suitable for exterior wall finish.

As explained earlier, the impact resistance of these shingles is poor so care must be taken in their storing and handling. Keep the siding dry since dampness or moisture can cause staining.

Apply it only to a smooth, flat, dry surface. Lumber, plywood and fiberboard are all suitable wall sheathing—provided they meet the above requirements. Nailing specifications depend on the wall sheathing used. The nails should be long enough to penetrate the nailing base, and are usually furnished by the siding manufacturer. Use nailing strips when the sheathing will not support the nails, Fig. 16-22.

The flashing detail recommended for use above doors and windows is shown in Fig. 16-23. Flash inside and outside corners with 12" strips of saturated felt if corner boards are used. Otherwise, overlap the saturated felt 12 inches at the corners.

Backer boards may be used to apply shingles, Fig. 16-24. This method accents the lines of the siding and produces deep shadow lines. The backer strips should be at least 5/16" thick and 1/4" narrower than the shingle. This produces a drip edge at the butts of the shingles.

Special channels may also be employed to install asbestos shingle. Nail the channels directly to suitable sheathing or to the framing members. Application detail is shown in Fig. 16-25. Finish the corners in one of several ways. Inside corners are usually butted, but you may

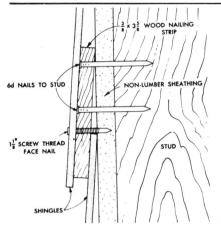

Fig. 16-22 Nailing strips are used when sheathing will not support nails.

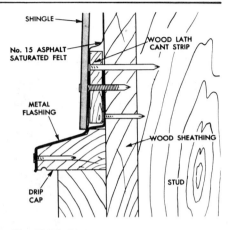

Fig. 16-23 Method of flashing the tops of doors and windows.

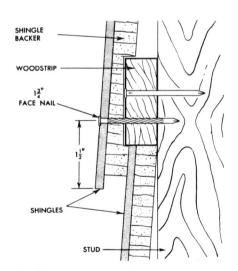

Fig. 16-24 One type of backer board used for mineral-fiber siding.

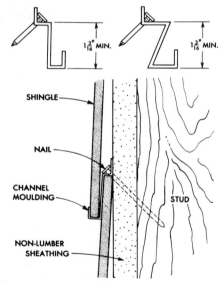

Fig. 16-25 Detail of metal channels used to install mineral-fiber siding.

also use wood or metal corners. Outside corners may be finished either with matching metal corners, Fig. 16-26, with wood corners, or with woven corners. Woven corners may only be used with gray or white shingles, however. Colored shingles are not suitable because the cut edge does not match the face color. Fig. 16-27 shows a completed installation.

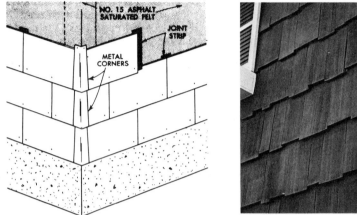

Fig. 16-26 House corners finished with matching metal corners.

Fig. 16-27 A house finished with both wood and mineral-fiber shingles.

Asphalt Sidings

Asphaltic materials for exterior siding are available in both roll and shingle form. Roll siding is generally furnished with an embossed surface, simulating brick or stone. Made of heavy felt saturated with asphalt, it usually has crushed slate embedded in its surface and provides a durable waterproof wall surface. Both roll and shingle sidings are made in many styles, colors and sizes.

Asphalt shingles are laid on sound exterior walls. Roll siding is applied like roll roofing and side wall shingles like roof shingles.

Outside corners are fitted with a metal corner bead. Inside corners are flashed with 15 lb. saturated felt. Special corner pieces must be used for roll brick siding. They are set in cement and nailed.

If the trim around doors and windows protrudes beyond the wall surface, cut the siding to fit around the edge of the trim. Before nailing it, embed it in a bead of asphalt cement.

Hardboard Siding

Hardboard sidings have a hard moisture-resistant surface which won't split, splinter or crack easily. As either panel or lap siding, they are made in many forms, sizes and textures. Special embossing techniques yield surfaces with a remarkable resemblance to wood, masonry and other materials.

Standard hardboard panels come in 4-foot widths and in 8, 9, and 10-foot lengths. Lap siding is available in up to 12 inch widths by 16 feet long. They are either primed or factory painted and can be had with matching trim. Fig. 16-28 shows a typical corner treatment with hardboard siding. Fig. 16-29 illustrates hardboard siding with a rough cedar texture. It is actually a lap siding with an embossed surface, 12 inches wide by 16 feet long. Lap siding ranges from 3/8" to 9/16" thick and the top edge is always rabbeted. Because of this 1" rabbet on the 12" width, the exposed width actually is 11 inches.

Fig. 16-28 Hardboard siding with matching metal corners. *(Celotex Corp.)*

Fig. 16-29 Hardboard siding with the appearance of wood shingles. *(Masonite Corp.)*

Underlaying wall construction: Install hardboard lap siding over either sheathed or unsheathed walls. Be sure that the stud spacing does not exceed 16" OC and that the walls are adequately braced since a solid nailing base is required.

Building paper is needed when the siding is applied to open studs or

Exterior Wall Covering

board sheathing. Fasten it with 8d galvanized nails with 3/16" heads. If the sheathing is 25/32" insulation board, use 10d nails. Hardboard siding can be cut with either a fine-toothed handsaw or a portable saw fitted with a combination blade.

Application of hardboard lap siding: Inside wood corners 1-1/8" thick are installed as shown in Fig. 16-30. They should extend from the desired height to a point 1/8" below the sheathing.

A starter strip, 3/8" thick × 1-1/2" wide, should be nailed along the bottom edge of the sill plate. See Fig. 16-31. Carefully install the first course of lap siding over the starter strip, check for levelness, then nail it so that it extends 1/8" below the starter strip. Nail into each stud, placing the nails at least 1/2" from the edges and ends. If butt joints are necessary, be sure to position them over a stud.

For outside corners use either shaped metal or wood strips, Fig. 16-32. Door and window treatment are shown in Fig. 16-33.

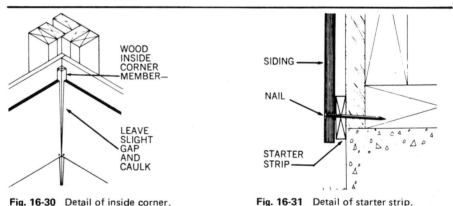

Fig. 16-30 Detail of inside corner.

Fig. 16-31 Detail of starter strip.

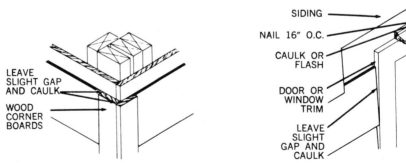

Fig. 16-32 Outside wood corner construction.

Fig. 16-33 Detail for door and window treatment.

Application of Hardboard Panel Siding: Apply panel siding over either sheathed or unsheathed walls with studs spaced 16" or 24" OC. Make sure that the butt joints fall on the studs. They must be covered with batten strips for waterproofing. Nail the panels the same way as hardboard lap siding.

Aluminum Siding

Aluminum siding comes in many styles, textures and colors. Its baked finishes prolong the life of the material and lessen maintenance requirements, and its prepunched elongated nail holes allow for normal expansion and contraction in the prefabricated panels.

For installation a hacksaw, file, tin snips, and a power saw fitted with an aluminum cutting blade are needed, as well as hammer and nails.

Aluminum siding may be installed on any sound wall, whether new or old. In remodeling work, uneven walls will have to be furred.

To increase the cutting efficiency of the portable saw lubricate its blade with wax and always saw away from you. Be sure the work is well supported.

Since aluminum is an electrical conductor, the installation must be properly grounded to prevent electrical shock hazard. The usual procedure is to run a suitable wire to an approved ground such as the cold water service or the grounded electrical system.

Vinyl Siding

New plastic processes have been developed that enable manufacturers to emboss vinyl sidings with textured surfaces. Both vertical and horizontal sidings are available in various colors and textures with matching accessories.

Vinyl siding is not affected by weather and will give years of service. Double 4" and 8" widths are available in lengths up to 12' 6". The installation method is somewhat like that for aluminum siding, and the nail holes are slotted to permit the material to expand and contract without buckling. Backer boards add rigidity.

Most extruded vinyl siding is made with a self-aligning lock. The sections snap together so that only the tops of each strip must be nailed. Note that the nails which fasten panels are centered in the slots. Check every fifth or sixth course for alignment. See Fig. 16-34.

Exterior Wall Covering

Fig. 16-34 Rows of siding must be checked for alignment, especially when windows interrupt the run.

Stucco

Stucco is an exterior wall finish similar to plaster. Actually a thin slab of concrete, it is applied much like plaster and with similar tools. It provides a strong, durable waterproof and fire-resistant surface.

Stucco is trowelled over metal lath in what is known as two-coat or three-coat work. In two-coat work, the stucco consists of a base and finish coat with a 5/8" minimum thickness. In three-coat work, the stucco must be at least 7/8" thick and is made of scratch, brown and finish coats.

Stucco can be applied to brick, masonry or wood frame structures. In wood frame construction, cover the sheathing first with backing paper, then follow with a metal lath reinforcement. Attach the lath with its long dimension crosswise to the studs and use noncorrosive nails to fasten it. Be sure that the nails penetrate the supporting members by at least 7/8 of an inch.

Veneer Walls

Wood frame walls covered with stone or brick are called veneer walls. Only decorative, they are non-loading and do not enter into calculations when wall thicknesses are considered. The veneered wall is of single thickness and is held to the frame wall with suitable metal ties, Fig. 16-35.

A 1" air space must be provided between the veneer and sheathing as well as flashing and weep holes. The holes allow moisture to escape at the bottom to the outside of the structure. Place the weep holes along the first course of bricks resting on the flashing. Insert wood plugs temporarily in the holes while laying the brick. This keeps the spaces clear of mortar. Remove them after the mortar sets, Fig. 16-36.

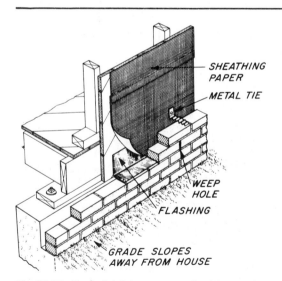

Fig. 16-35 Typical brick veneer wall with one tie for every two square-feet of wall.

Fig. 16-36 Weep hole plugs are removed from brick wall after mortar has set.

Before covering a wall with veneer, apply heavy weight paper. Install this heavyweight waterproof paper with a 4" lap to ensure that any trapped moisture is carried downward to the weep holes and outside.

Exterior Wall Covering

Lay bricks on a 1/2" bed of mortar, corners first, using the builder's square to check that their angles are true. Run a level line between corners in order to align the bricks horizontally. Then with a trowel butter the end of one brick at a time, laying it down in the 1/2" bed of mortar placed on the preceding course of bricks. Tap the brick lightly with the trowel to help position it properly. Finally, scrape off any excess mortar with the trowel. Be sure to stagger the vertical joints of the bricks to strengthen the wall. After laying a few courses, check horizontal, vertical and diagonal surfaces with your spirit level for accuracy. Also, use a story pole to check the height of each course of bricks. The alignment is best checked before the mortar hardens. Also, take care to wipe all mortar droppings off the face of the brick and tool all mortar joints before the mortar sets. Fig. 16-37 shows brick installation, and Fig. 16-38 the tooling of mortar joints. Fig. 16-39 shows how a properly finished mortar joint should look. For a completed brick veneer wall see Fig. 16-40.

When brick veneer covers a large area, movement of the wall frame can cause cracks. To minimize this hazard use vibration joints. They are made by laying an occasional dry course. That is, the mortar is allowed to harden before the next course is laid and additional ties are used above and below this joint. Stone veneer is applied in the same manner as brick. Flashing, connecting metal ties to the framework, and weep holes for water drainage to the exterior, are also required as for brick veneer installation. The individual stones may be irregular or squared (ashlar) and are set in mortar in a random or coursed pattern.

Fig. 16-37 Application of brick veneer to a house wall.

Fig. 16-38 Tooling mortar joints in a brick veneer wall.

Fig. 16-39 Tooled mortar joint in brick veneer.

Fig. 16-40 Completed brick veneer wall.

GLOSSARY

asbestos cement: a wall siding material made in the form of shingles from asbestos fibers and portland cement, also called mineral fiber.

batten: a narrow trim used to cover joints, also used decoratively with vertical siding.

bevel siding: a wall covering with a wedge-shaped cross-section.

brick veneer: a single thin facing of brick used to cover wood frame construction.

drop siding: a wall covering with a shaped face and interlocking joints.

fiberboard: sheet material made in various thicknesses and densities of various vegetable fibers.

stucco: an exterior wall-covering plaster consisting mostly of portland cement, some sand and, commonly, a small amount of lime.

tie: in brick veneer construction, a metal strap used to fasten the brick wall to the wood frame.

weep hole: a small hole at the bottom of a masonry wall which allows water to escape to the outside of the structure.

17

Thermal Insulation

Thermal insulation plays an important role in the economics and comfort of a residence. Indeed, a properly insulated house means lower costs for fuel in the winter and for electrical energy in the summer. The purpose of such insulation is to minimize the inflow of heat through building walls in the summer and its outflow during the winter. It is a matter of scientific fact that the greater the difference between inside and outside temperatures, the faster the heat flow. Insulation effectively slows down that transfer of heat and maintains a uniformly comfortable temperature throughout a house. Today many types of insulating materials are manufactured and packaged in convenient rolls and batts for the construction industry.

Heat Transfer

Heat is transferred in or out of a building through the walls and roof. It also moves through cracks and openings which are sometimes the result of poor workmanship. This exchange of heat takes place in one of or in any combination of three ways: conduction, convection or radiation.

Conduction occurs when heat travels through any solid body from one interacting molecule to another. Different materials conduct heat at different rates. For example, a piece of steel conducts heat much faster than a piece of wood. Generally, the denser the material, the faster the flow of heat.

Convection is the movement of heat through air. The density of air decreases as its temperature rises; this in turn causes heavier colder air to move in and replace the warmer air. As the cycle continues, whether upward or downward, these convection currents carry heat from one area to another.

Radiation: In this process, heat is transmitted from one body to another by electromagnetic waves which are emitted and travel through either the air or a vacuum. The radiated heat does not heat the air it moves in, but heat is generated when it strikes a receptive surface. The flow of such radiated heat is from the warmer surface to the cooler one. Fig. 17-1 gives examples of the three methods of heat transfer.

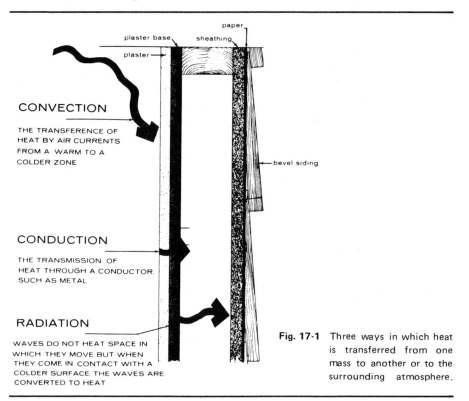

CONVECTION

THE TRANSFERENCE OF
HEAT BY AIR CURRENTS
FROM A WARM TO A
COLDER ZONE

CONDUCTION

THE TRANSMISSION OF
HEAT THROUGH A CONDUCTOR.
SUCH AS METAL

RADIATION

WAVES DO NOT HEAT SPACE IN
WHICH THEY MOVE BUT WHEN
THEY COME IN CONTACT WITH A
COLDER SURFACE THE WAVES ARE
CONVERTED TO HEAT

paper

plaster base   sheathing

plaster

bevel siding

Fig. 17-1 Three ways in which heat is transferred from one mass to another or to the surrounding atmosphere.

How Insulation Works

While all materials have some insulating qualities, there are certain ones that work better than others. Materials that have many small air spaces or pockets are good insulators since air is a poor conductor if confined to a small area that restricts movement. Such restriction eliminates the loss of heat through convection because the barrier of air cells or other insulating material effectively obstructs the flow of heat. Obviously, this means that less heat passes through the insulation, and a useful reduction in heat loss results.

Among the materials used for insulating are glass fibers, mineral fibers, organic fibers, mineral-foamed glass, foamed plastics and reflective metals. Regardless of the type used, insulation should be efficient, permanent, odor and vermin free and economical.

Insulating Terms used in describing insulation are called heat-loss coefficients. They include Btu, k, C, U and R.

Btu or British thermal unit is the amount of heat needed to raise the temperature of one pound of water by one-degree Fahrenheit.

k or thermal conductivity is the amount of heat (Btu's) that passes through one-square foot of homogenous material one-inch thick in one hour when there is an initial one-degree Fahrenheit difference between the two surfaces. The lower the k value, the higher the insulating value.

C or thermal conductance, representing the conductance or transmitting capacity of a material regardless of its thickness, shows the amount of heat in Btu's that will pass through one-square foot of material in one hour with a one-degree difference between its two surfaces. The lower the C, the higher the insulating value.

U represents the heat loss through a building section in Btu's per hour per square foot per degree-Fahrenheit difference between inside and outside temperatures. The lower the U, the higher the insulating value.

R is the measured resistance of a material to the flow of heat. All building components of a wall, roof or floor have an R value. The higher the R value, the more efficient the insulation. The R value of various wall components is listed in Fig. 17-2.

Building Material Components	R Value
1. Outside air film	0.17
2. Wood siding	0.85
3. ½" sheathing	1.32
4. Air space	1.01
5. Interior finish (½" drywall)	0.45
6. Inside air film	0.68
Total resistance (R_t)	4.48

Fig. 17-2 Table of R Values for building material components.

The U value of a construction section is the reciprocal of the R values of the various components. It is measured by the formula,

$$U = \frac{1}{R} .$$

Study of this formula clearly shows that adding insulation to a wall makes it possible to reduce the U value or heat loss. For example, the U value of a typical wood frame wall without insulation is 0.22. It is possible to reduce this to 0.10 or to 0.07—depending on which of the following insulation materials is used:

U value, without insulation	= 0.22
U value, with R 7 (2-1/4") insulation	= 0.10
U value with R 11 (3-1/2") insulation	= 0.07

In order to lower the U value, the R value (thickness) must be increased. The R value is always expressed as a numeral, and each numeral equals a resistance unit. An insulating material with an R 11 factor insulates better than one with an R 7 rating. In other words, regardless of the material or its thickness, all products with the same R factor have equal insulating value. The R value is clearly marked on insulation packages, Fig. 17-3.

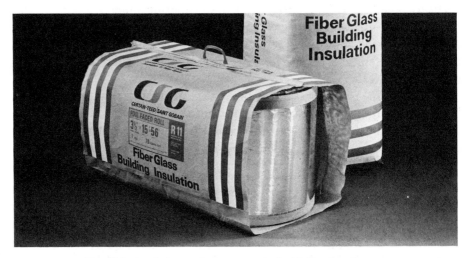

Fig. 17-3 Insulation packages are marked with R value of contents.

Cold wall effect: This phenomenon is caused by cold walls, floors or ceilings. That is, the occupants feel cold while the room temperature remains at a fairly high level. The reason for this is that the human body radiates or loses its heat to colder surfaces. To offset this, most people raise the thermostat and thus increase their fuel costs. Proper insulation would restore comfortable temperatures and eliminate this heat-loss problem.

As a reference point, comfort standards have been established for use in planning insulation. Choose insulating materials with appropriate values to achieve the desired comfort level.

Insulating Materials

Insulating materials are manufactured in a variety of forms, and generally fall into one of the following structural groups: flexible, rigid, reflective or loose fill.

Flexible insulation is made up in blankets and batts. The blankets are actually fabricated in roll form and in lengths up to several hundred feet, Fig. 17-4. The usual thickness is from one to three inches and the widths fit standard stud and joist spacing. Insulation blanket consists of fluffy mats of mineral or vegetable fibers such as rock, slag, glass wool, wood fiber and cotton. The organic fibers are treated to make them resistant to fire, vermin and decay.

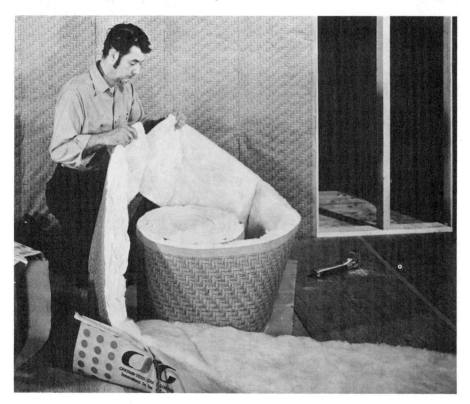

Fig. 17-4 Flexible insulation in roll form. *(Certain-teed)*

Generally, blanket insulation has paper covers on one or both sides and is provided with side tabs for fastening to studs and joists. While one covering usually serves as a vapor barrier to prevent water penetration, the other covering simply encloses the mat. Some fibrous insulations are made with only one cover. Aluminum foil, asphalt and plastic are also often used as vapor-barrier coverings.

Batt insulation is also formed of fluffy fibrous materials in thicknesses up to 6 inches or more. Widths are 15 and 23 inches and lengths are either 24 or 48 inches, Fig. 17-5. Batts are available with and without covers.

Fig. 17-5 Batt insulation without facing (covers).

Rigid insulation is manufactured from wood, cane and other fibrous vegetable materials. They are formed into large lightweight boards which combine tensile strength with thermal and acoustical insulating properties.

The material is available in a wide variety of sizes, from 12 inch squares to boards measuring 4 feet wide and 12 feet long with 1/2" to 1" thicknesses.

Where the main purpose is structural, these insulating boards are fabricated as building boards, roof decking, sheathing and wallboard. Their insulating qualities are then secondary. Such insulating sheathing is made in 1/2" and 25/32" thicknesses. The sheets are 2' × 8' for horizontal application, and either 4' × 8' or 4' × 12' for vertical use.

Other materials used for rigid insulation include polystyrene, polyurethane, cork and mineral wool. These are non-structural boards. Fig. 17-6 shows a section of tongue-and-groove rigid foam insulation.

Fig. 17-6 Closed-cell rigid foam insulation in tongue-and-groove sections. *(Amspec, Inc.)*

Reflective insulations have high reflective properties. Aluminum foil, sheet metal and reflective coatings are the most common materials used for this purpose. To function properly the insulation must have at least a 3/4" air space facing the reflective surface. In ceiling applications, the air space should be 1-1/2" or greater.

The most efficient of the reflective insulations is the accordion type which is made up of several spaced layers of foil. A reflective type in roll form is shown in Fig. 17-7. As it is applied, the material is stretched out to form the necessary air space.

Loose-fill insulation consists of bulk packed materials made of rock wool, wood fibers, shredded bark, cork, sawdust, vermiculite and other granular products.

Poured or blown into wall or ceiling spaces, it is ideally suited for use in existing structures where the installation of other forms of insulation is not practical.

Fig. 17-7   Roll of reflective accordion-type insulation.

Other Types of Insulation

Some insulations not described previously include ***corrugated paper*** in multiple layers, ***sprayed insulations*** and ***lightweight aggregates*** like vermiculite and perlite mixed with plaster.

Foamed-in-place insulation is also used extensively in existing structures. Two chemicals are mixed together to produce a urethane foam with a K factor of 0.12. Special applicators with far-reaching nozzles are used to blow it into place.

Where to Insulate

Any room that is to be cooled in the summer or heated in the winter should be properly insulated. Provide walls, floors and ceilings with a blanket of insulation. This retains heat in the winter and keeps it out during the warm weather. Place such insulation as close as possible to the rooms used as living quarters. Thus, in an unused attic, insulate the attic floor, not the roof over it. Also insulate the floors over unheated basements, crawl spaces and porches.

If basement areas are to be finished as living rooms, use thermal insulation in the walls. Also put acoustical insulation on the ceiling to reduce the passage of sound from the basement to other parts of the house. This is especially important if the basement is to be used as a playroom for children. Acoustical insulation adds to fuel savings as well. Fig. 17-8 shows the placement of insulation in various types of structures.

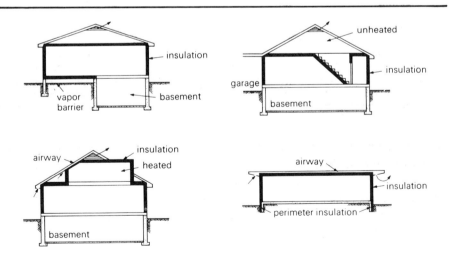

Fig. 17-8 Diagram showing areas for insulation in various structures. *(National Assoc. Home Builders)*

Thermal Insulation

Crawl spaces insulation: When homes are built without basements, many builders install a crawl space under the floor. This is better than placing the slab directly on the ground and it provides access to plumbing, heating and other services if the need arises.

The floor above the crawl space should be insulated and the area should be ventilated as well. If the crawl space is unvented, insulate the foundation walls as shown in Fig. 17-9. Over a vented crawl space insulate the floors by placing the insulation between the joists. Place it with the vapor barrier side facing toward the room. Wire mesh, laced

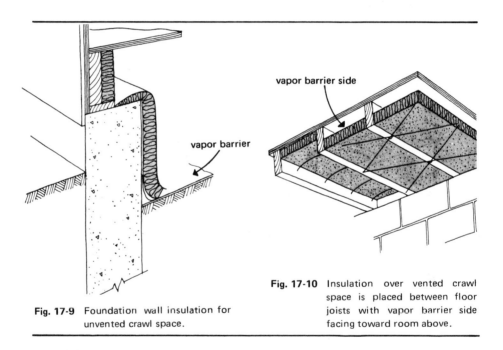

vapor barrier side

vapor barrier

Fig. 17-10 Insulation over vented crawl space is placed between floor joists with vapor barrier side facing toward room above.

Fig. 17-9 Foundation wall insulation for unvented crawl space.

wires or special bowed wires may be used, Fig. 17-10. If preferred, use polyethylene on the ground.

Use a sill sealer in basement and unvented crawl space construction to prevent air infiltration between the sill and foundation.

Cantilevered floors can be insulated as illustrated in Fig. 17-11. Cut the blanket to fit or fold it as shown.

Slab-on-ground construction may require insulation. If so, lay it only where needed, around the perimeter, since little heat is lost toward the center of the floor. Fig. 17-12 shows how rigid insulation is used between slab and foundation wall.

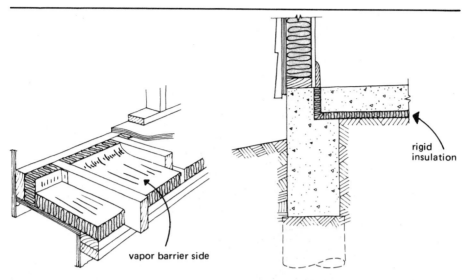

Fig. 17-11 Application of blanket insulation at floor overhang.

Fig. 17-12 Insulation around perimeter of concrete slab.

Vapor Barriers

Used with insulation to minimize the amount of vapor collecting on the cold side of the wall vapor barriers are generally part of the insulation material. They can also be a separate material however. Both batt and blanket insulation usually have one side covered with an aluminum foil barrier, but other materials may be used. Some vapor barriers consist of asphalt coated paper, plastic films or special paints. If the insulation lacks its own vapor barrier, apply sheets of polystyrene or other suitable material after the insulation is in place. The success of a vapor barrier depends on proper installation. Be sure to lap all joints and repair all tears. The only space permitted within a wall is between sheathing and insulation because that lets moisture escape to the outside. Fig. 17-13 shows how a plastic vapor barrier is applied after the insulation has been installed.

Install vapor barriers after the heating ducts, plumbing and electrical wiring are in place. Cut the material and fit it carefully around all openings.

Thermal Insulation

Fig. 17-13 Application of vapor barrier is always toward room side. *(Certain-teed)*

Installing Blanket Insulation

Install either batt or blanket insulation in open spaces in walls, floors, ceilings and roofs. Do not install either where compressive loads are present. Squeezing or compressing any flexible insulation causes it to lose its effectiveness.

Cut the batts or blankets with a handsaw or knife. Most workers prefer a knife since it is less likely to scatter particles which can be very irritating to the skin, especially fiberglass.

Make lengths slightly oversize so that a stapling flange can be formed at the top and bottom. Do this by removing some of the insulation after cutting.

Most workers cut blanket insulation on the floor. The usual procedure is to mark the required length on the floor, then roll out the material and cut on the mark. Compress the material with a straight-edge and cut with a sharp knife. Blanket insulation is being installed in Fig. 17-14.

Fig. 17-14 Method of installing blanket insulation in wall spaces.

Ceiling Insulation may be installed from below by stapling it to the ceiling joists—as in Fig. 17-15—except when unfaced insulation is used. In that case, pressure-fit the blanket between the joists. In both cases, the insulation must extend across the top plate, and it may be necessary to fill in any gap left at the inside edge of the plate.

If the specifications do not permit face stapling of the flange, staple it to the sides. Be sure to place the staples every six inches.

Insulating Masonry Walls: Masonry walls require furring strips when flexible insulation is used. The furring may be 1 × 2 or 2 × 2 stock. Spacing is either 16" or 24" OC, depending on the thickness and type of wall finish. Use either pressure-fit, batt or blanket type insulation. For the 1 × 2 furring, a 1" pressure-fit insulation without vapor barrier is generally used. Add either a separate 2 mil polyethylene or foil-backed gypsum board as the vapor barrier.

With 2 × 2 furring, use an R 7 insulation with vapor barrier, Fig. 17-16. Because of this insulation's shallow depth, the flanges must be stapled to the face of the furring.

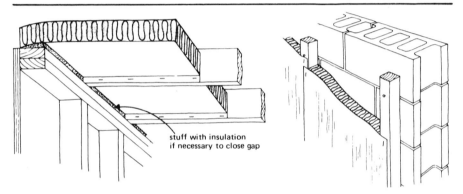

Fig. 17-15 Efficient method of applying ceiling insulation by stapling it to ceiling joists.

stuff with insulation if necessary to close gap

Fig. 17-16 Insulation fastens to furring on masonry wall.

Insulating Frame Walls: The procedure for installing batt or blanket insulation in frame walls is similar to that for ceilings. Place insulation between studs and staple the flange to the side or face. If foil-faced insulation is used, do not let the reflective surface touch the back of the drywall or paneling. In fact, inset stapling is required for this work.

Start at the top of the wall, working down toward the plate. Remove about 1-1/2 inches of insulation at the top and bottom of each piece to form a stapling flange. See Fig. 17-17 for two ways to terminate this insulation at the top and bottom plates when the inset method is used.

Use great care when installing insulation around plumbing and other outlets. If the facing is accidently torn, be sure to repair with tape. Study Fig. 17-18 to see how ducts, pipes and boxes are treated. Place the insulation so that the pipes are inward toward the room. Stuff small spaces around doors and windows with pieces of insulation. To keep

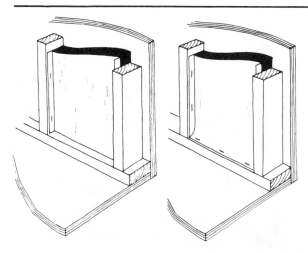

Fig. 17-17 above: Two ways of terminating ends of insulation.

Fig. 17-18 right: Proper method of insulating pipes and plumbing and heating ducts.

them effective, do not pack the pieces tightly. Finally, cover these spaces with paper or plastic vapor barrier, Fig. 17-19.

Fig. 17-19 Insulation around doors and windows must not be packed too tightly.

Thermal Insulation

Applying Fill Insulation

Granulated-type fill insulation can be poured or blown into place with special equipment. Pneumatic tools are especially useful for older houses where walls are already in place. After filling, level the surface as shown in Fig. 17-20. Note the use of a dust mask to prevent the inhalation of very fine dust particles.

Fig. 17-20   Leveling-off the surface of granulated fill insulation.

Installing Reflective Insulation

Reflective insulation is installed in the same manner as batt or blanket insulation. It is attached to either the side or the face of the framing. Be careful to maintain an air space between its reflective surface and the wall covering. Normally this is 3/4 inches for walls and at least 1-1/2 inches for floors and ceilings. Fig. 17-21 shows reflective insulation being installed in an attic.

Fig. 17-21 Application of reflective insulation in an attic.

Installing Rigid Insulation

Expanded polystyrene foam plastics have excellent thermal qualities, are waterproof and do not require the use of vapor barriers. Lightweight and easy to handle on the job, the material is available in a convenient 2 × 8 foot size and in a variety of thicknesses. It is also available in 4 × 8 foot sheets and is easily cut with a knife or saw.

Furring is not required when rigid insulation is used to insulate masonry walls, such as those in a basement. The slab may be installed with a mastic. Use mastic also to join gypsum or paneling directly to the insulation. Special mastics are made to different specifications; be sure to employ the right one for the job.

Rigid plastic foam is excellent for insulating exposed beam roof decks. Special high-density foam with high water resistance is especially suitable. An application over 2 × 6 decking is shown in Fig. 17-22.

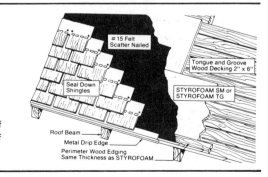

Fig. 17-22 Details for application of rigid plastic foam on roof decks.

Thermal Insulation

There is one plastic foam that can be used as a combination sheathing-insulation. A non-structural sheathing base for conventional siding, it is applied to the studs with large-headed nails. Because it is non-structural, be sure to use diagonal let-in bracing, Fig. 17-23.

Fig. 17-23 Rigid foam insulation applied to outside of diagonally-braced frame; the open-stud cavities facilitate installation of utilities. *(Amspec)*

Weather Stripping

Employ weather stripping to reduce air leakage around doors and windows. Few people realize the serious amount of heat loss that occurs because weather-stripping is either inadequate or absent. But many types are available to fit virtually all situations. Whenever possible, use adjustable weather stripping. This permits adjustments when buildings settle, doors sag or jambs warp. In addition to giving protection against drafts, weather-stripping helps control noise, dust and light. Fig. 17-24 shows several modern weather-stripping devices.

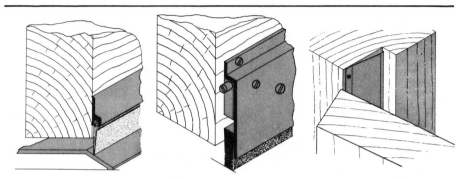

Fig. 17-24 Kinds of adjustable weather stripping: **left,** door bottom sweep; **center,** automatic door strip; **right,** spring strip. *(Pemko Mfg. Co.)*

Ventilation

Along with insulation, proper ventilation is an important factor in good construction. In summer, ventilation helps to cool a structure and in winter it helps prevent condensation. Condensation can cause considerable damage especially when concentrated in one area. Vapor barriers minimize moisture movement but some leakage often occurs through openings around pipes and other areas nevertheless. The problem can be virtually eliminated by properly ventilating the roof spaces. In the past, the cracks and openings in houses had a natural ventilating effect. However, modern techniques have resulted in tighter construction which prevents warm moist air from leaking out. Thus water vapor accumulates in the house. Left unchecked, that vapor can damage walls, ceilings, woodwork and paint.

A common problem caused by insufficient attic ventilation is the formation of ice dams at the cornice. There are several causes, but normally it happens when the attic air is warm enough so that its warm roof surface melts the snow on it. Then the water runs down to the colder surface of the cornice and freezes. The resulting ice dam causes

Thermal Insulation

water to back up and gradually work its way down into walls and ceilings. By venting the attic and keeping its temperature lower, the snow can be kept from melting. Fig. 17-25 shows how this problem has been eliminated by both increasing the attic floor insulation and installing soffit ventilation.

Louvered openings in the end walls of a gabled roof will ventilate an attic. But such installations depend on wind to circulate the air within. Adding soffit openings increases the air movement greatly, Fig. 17-26.

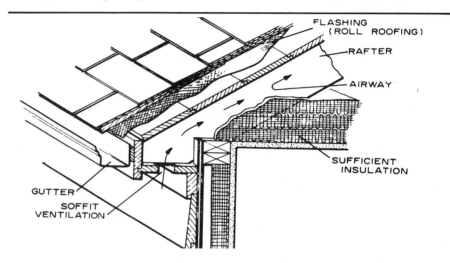

Fig. 17-25   Layout for proper ventilation and insulation of attic.

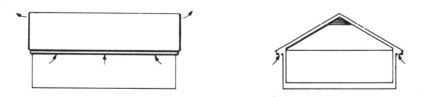

Fig. 17-26   Arrows indicate air movement through gabled roof; **right,** gable detail.

Ventilator Size depends on the roof shape and ceiling area. Normally, for a gabled roof the size of the inlet and outlet openings should be 1/900th of the ceiling area. For example, a ceiling with an area of 1800 square feet divided by 900 equals 2. Therefore, the ventilator area for inlet and outlet should be two square-feet each. Fig. 17-27 shows various inlet ventilators.

The outlet ventilators must be placed as close to the ridge as possible, with screens to keep out insects. Keep the screen size as coarse as possible. Do not use screens smaller than No. 16 since they tend to clog up with lint and dust. Precision-made wood and metal louvers are available for frame, brick veneer, and masonry constructions. Gable louvers in various pitches are made for any style of home. Manufacturers' specifications usually give the net ventilating area of each unit.

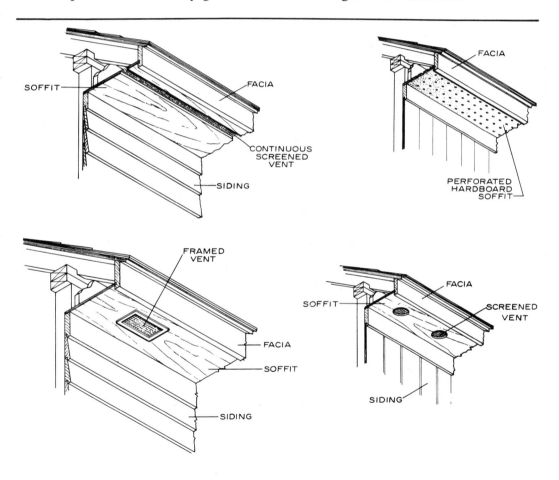

Fig. 17-27 Four types of inlet soffit ventilators.

Thermal Insulation

GLOSSARY

batt insulation: fibrous insulation material fabricated in rectangular form from 15" to 23" wide and from 24" to 48" long.

blanket insulation: a fibrous insulating material made in roll form.

British thermal unit (Btu): symbol representing the amount of heat needed to raise the temperature of 1 pound of water one-degree Fahrenheit.

C: symbol representing the conductance value of a material, showing the amount of heat (Btu's) that will flow through one square foot of material in one hour with a one-degree difference between both its surfaces.

condensation: moisture in the form of water vapor.

conduction: the transfer of heat through a solid body.

convection: the movement of heat through air.

insulation (thermal): any material used in building construction which will retard and/or prevent the flow of heat.

insulation board: a structural building board made of wood or cane fibers in various thicknesses and sizes and with insulating properties.

k: symbol representing thermal conductivity or the amount of heat (Btu's) that passes through one square foot of material one-inch thick in one hour when there is a one-degree difference Fahrenheit between both its surfaces.

R: symbol representing the measured resistance of a material to the flow of heat.

radiation: the transfer of heat by electromagnetic waves from a warmer to a cooler body.

reflective insulation: an insulating material consisting of highly reflective surfaces.

U: symbol representing the heat loss through a building section in Btu's per hour.

vapor barrier: a material used to retard the movement of water vapor into or through walls, floors, and ceilings.

vermiculite: particles of a lightweight mineral with insulating qualities, used as bulk insulation.

18

Sound Conditioning

Noise is everywhere—indoors and out. Vehicular traffic, construction tools, aircraft and trains are some of the common outdoor noisemakers. Indoors, the source of noise can be appliances, lavatories, radios, TV sets, children's play and speech (grown-ups too) and heating systems. Excessive noise is not only annoying, but it can be harmful. It causes fatigue, irritability, inefficiency and can sometimes damage the sensitive hearing nerves of the inner ear. Special construction techniques combined with sound conditioning materials can dramatically reduce irritating noise levels. Do not confuse sound conditioning with soundproofing, however. The latter is the elimination of all sound. That could be as annoying as too much sound. The aim of sound-conditioning is to reduce loud sounds to a comfortable level while eliminating or reducing unwanted noise.

Properties of Sound

Sound is a form of vibration of energy that sets certain molecules in motion and which when transmitted stimulates auditory receptors in man and animals. It travels in wave form compared somewhat to water waves. However, water waves travel in one plane while sound waves travel in all directions throughout a given medium—air, for instance. When these progressive waves or vibratory disturbances strike the ear drum they produce the sensation of hearing.

Sound waves travel at different speeds through different media. At sea level, airborne sound travels about 1,130 feet per second; in water sound travels about 4,500 feet per second; and through steel sound travels at approximately 15,000 feet per second.

Key terms used in connection with sound conditioning are defined briefly below:

Wave length: the distance a sound wave travels in one vibration.

Frequency: the number of cycles a sound wave vibrates each second.

Decibel: the unit measurement of sound; equivalent to the smallest change in sound intensity that can be detected by the human ear.

Sound Conditioning

Reverberation: a bouncing back and forth of sound waves after the sound source has stopped.

Sound reflection: the change of direction that occurs when sound waves strike an obstacle or surface so that the waves are deflected in their travel.

Sound absorption: the opposite of reflection, it is the ability of a material to absorb sound energy and convert it to heat energy within that material.

Impact sounds: vibrations transmitted by structural members such as floors and walls in a house: footsteps and door slamming are two examples.

Masking sounds: background sounds which have the effect of diminishing transmitted sounds.

Transmission loss (TL): the decrease in the sound level in decibels of airborne sound as it travels through a building construction.

Sound transmission class (STC): a number rating for evaluating the sound-insulating performance of floor and wall materials; the higher the number, the better the sound-insulating qualities.

Noise reduction coefficient (NRC): a number index of the noise-reducing efficiency of acoustical materials; it is found by averaging the sound-absorption coefficients of four common household noise frequencies which are 250, 500, 1,000 and 2,000 cycles per second respectively.

Sound Levels

Fig. 18-1 compares sound levels with common sounds. The lowest and highest numbers are extremes. Ordinary everyday sounds fall within these limits. Sounds above 120 decibels are painful to the ear and can cause damage to hearing. Jet engines are in the 140 decibel range; rocket engines may go as high as 180 decibels.

Noise Control Factors

Several things happen to sound waves heard by the persons in a room. (There is no sound in an empty room.) Some sound is reflected back into the room; some is absorbed by walls, floor, ceiling, furnishings and occupants; and some is transmitted through the construction members. The air molecules set in motion by the sound do

Noise Control Factors

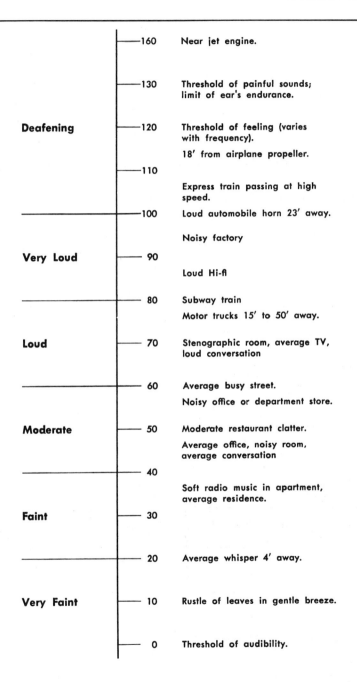

	dB	
	—160	Near jet engine.
	—130	Threshold of painful sounds; limit of ear's endurance.
Deafening	—120	Threshold of feeling (varies with frequency).
		18' from airplane propeller.
	—110	
		Express train passing at high speed.
	—100	Loud automobile horn 23' away.
		Noisy factory
Very Loud	90	
		Loud Hi-fi
	80	Subway train
		Motor trucks 15' to 50' away.
Loud	70	Stenographic room, average TV, loud conversation
	60	Average busy street.
		Noisy office or department store.
Moderate	50	Moderate restaurant clatter.
		Average office, noisy room, average conversation
	40	
		Soft radio music in apartment, average residence.
Faint	30	
	20	Average whisper 4' away.
Very Faint	10	Rustle of leaves in gentle breeze.
	0	Threshold of audibility.

Fig. 18-1 Typical sound levels compared in decibels, with common sounds. *(Acoustical Materials Assoc.)*

not actually pass through a solid wall. Instead, they cause molecules of the wall structure to vibrate. That in turn causes the air on the other side of the wall to vibrate correspondingly, thus reproducing the original sound waves but at a lower intensity. The difference in decibels on both sides of the wall or barrier is called the transmission loss (TL). If a wall has a transmission loss of 20 decibels, and the originating sound is of 60 decibels, the transmitted sound will have an intensity of 40 decibels. It means that the wall has absorbed 20 decibels of sound.

Sound-resistance ratings: To simplify the classification of building materials according to their sound resistance, the sound transmission class (STC) was established. In this system, a building element is rated according to its resistance to airborne sound. The higher the number, the better the sound barrier. Fig. 18-2 shows the degree of noise control achieved with barriers having different STC numbers. By using this system, the desired degree of acoustical privacy may easily be formulated.

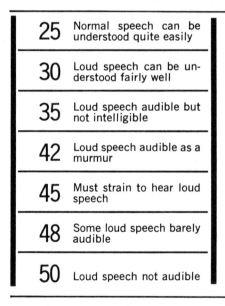

25	Normal speech can be understood quite easily
30	Loud speech can be understood fairly well
35	Loud speech audible but not intelligible
42	Loud speech audible as a murmur
45	Must strain to hear loud speech
48	Some loud speech barely audible
50	Loud speech not audible

Fig. 18-2 STC Ratings: numbers for the sound transmission classes which rate the noise-control qualities of common building materials according to their average degree of sound insulation.

Wall-Sound Transmission Ratings

The STC rating of various building materials is shown in Figs. 18-3 and 18-4. Notice that a gypsum wall nailed to studs has a fairly low or poor rating. By using thick heavy walls such as masonry, the STC number can be increased. That is not always practical, however, since

Wall-Sound Transmission Ratings

WALL DETAIL	DESCRIPTION	STC RATING
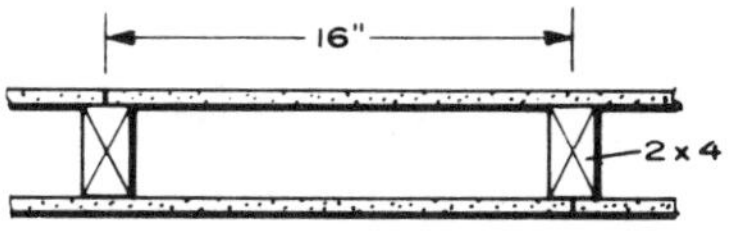 A 16" 2 x 4	½" GYPSUM WALLBOARD	32
	⅝" GYPSUM WALLBOARD	37
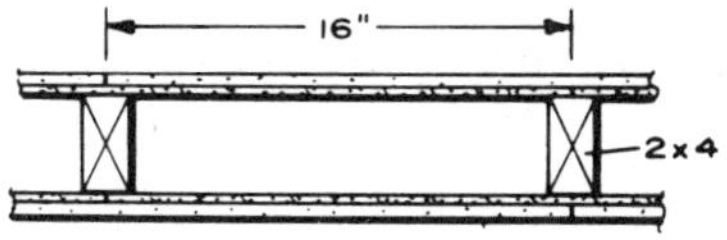 B 16" 2 x 4	⅜" GYPSUM LATH (NAILED) PLUS ½" GYPSUM PLASTER WITH WHITECOAT FINISH (EACH SIDE)	39
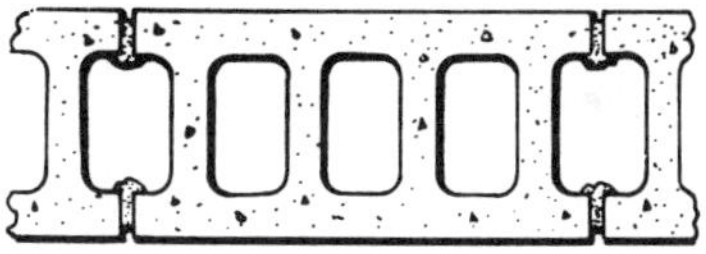 C	8" CONCRETE BLOCK	45
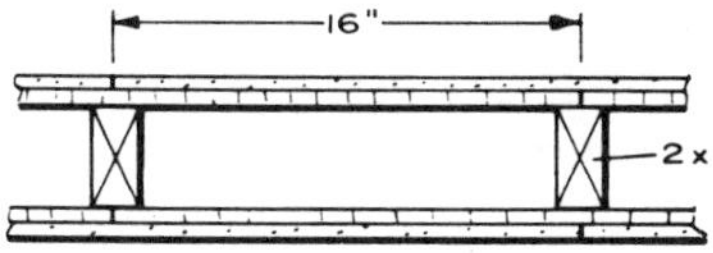 D 16" 2 x 4	½" SOUND DEADENING BOARD (NAILED) ½" GYPSUM WALLBOARD (LAMINATED) (EACH SIDE)	46
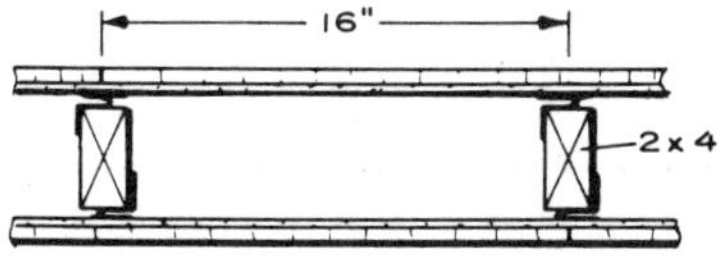 E 16" 2 x 4	RESILIENT CLIPS TO ⅜" GYPSUM BACKER BOARD ½" FIBERBOARD (LAMINATED) (EACH SIDE)	52

Fig. 18-3 STC ratings for single walls, including construction and materials data.

floor loads and cost are also factors to be considered. A more practical approach is to use materials like sound-deadening insulating board. By using the board in suitable combinations, the STC quality of a wall can be increased considerably.

A high STC rating of 52 may be obtained by fastening gypsum boards to the studs with special resilient clips and then adhesive-laminating 1/2" fiberboard to the gypsum. This method works very

WALL DETAIL	DESCRIPTION	STC RATING
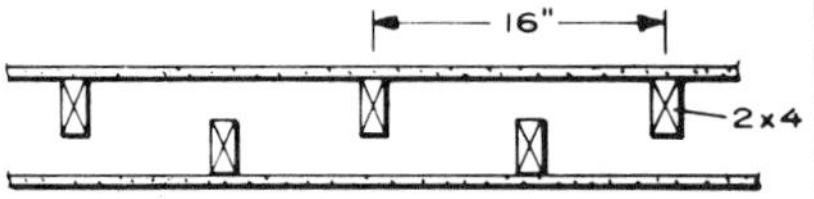 A	½" GYPSUM WALLBOARD	45
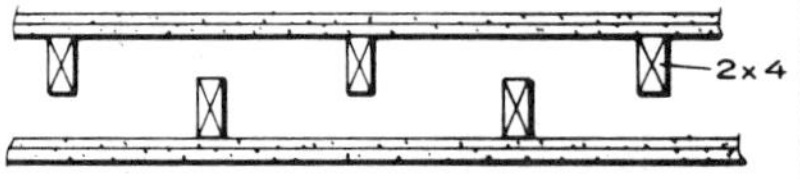 B	⅝" GYPSUM WALLBOARD (DOUBLE LAYER EACH SIDE)	45
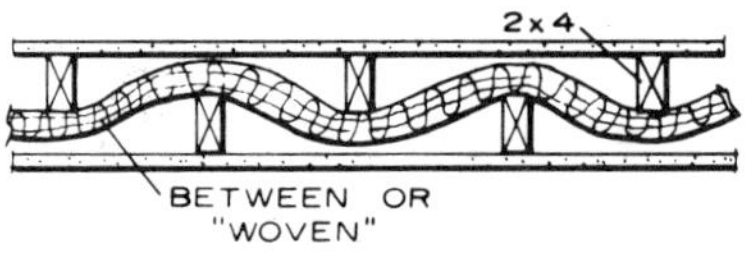 C	½" GYPSUM WALLBOARD 1½" FIBROUS INSULATION	49
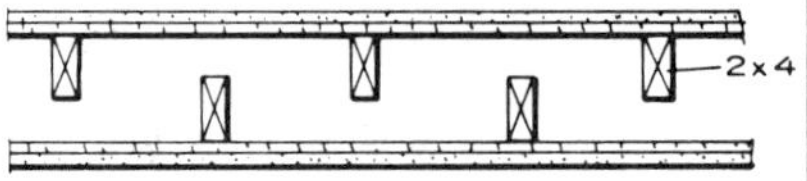 D	½" SOUND DEADENING BOARD (NAILED) ½" GYPSUM WALLBOARD (LAMINATED)	50

Fig. 18-4 STC ratings for double walls, including construction and materials data.

well because the fiberboard wall covering is isolated from the studs.

A similar system employs resilient channels to support 5/8" gypsum for an STC rating of 47. Install the steel channels at right angles to the studs. Next space the channels 24 inches apart. Then fasten them with nails or screws at each intersection. Make any needed splices by nesting over the studs—not by butting. Finally, attach the gypsum to the channels with screws.

Double walls with staggered studs on a 2 × 6 plate are sometimes used for sound deadening. Separate double walls with an air space in between are also effective. A typical double wall is spaced 2" apart with an inner facing of 1/2" sound-deadening insulation board and an outer facing of 5/8" gypsum. Such a wall has an STC rating of 50.

Floor and Ceiling Sound Transmission

In addition to airborne sounds, floors are also subject to impact sounds generated by people walking, dropping objects, moving furniture, typing and the like. Several methods of building both floors and ceilings for sound control are shown in Fig. 18-5.

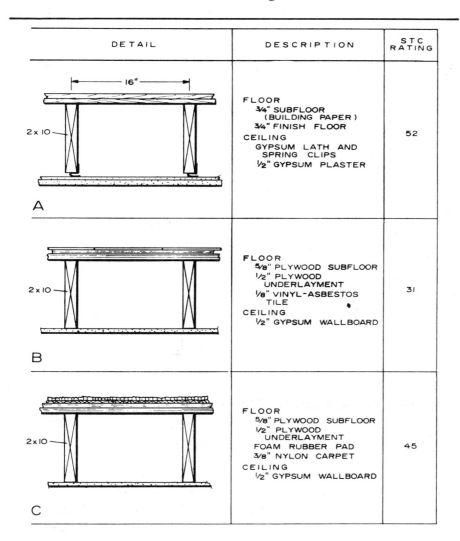

DETAIL	DESCRIPTION	STC RATING
A (2 x 10, 16")	FLOOR ¾" SUBFLOOR (BUILDING PAPER) ¾" FINISH FLOOR CEILING GYPSUM LATH AND SPRING CLIPS ½" GYPSUM PLASTER	52
B (2 x 10)	FLOOR ⅝" PLYWOOD SUBFLOOR ½" PLYWOOD UNDERLAYMENT ⅛" VINYL-ASBESTOS TILE CEILING ½" GYPSUM WALLBOARD	31
C (2 x 10)	FLOOR ⅝" PLYWOOD SUBFLOOR ½" PLYWOOD UNDERLAYMENT FOAM RUBBER PAD ⅜" NYLON CARPET CEILING ½" GYPSUM WALLBOARD	45

Fig. 18-5 STC ratings for various floors, including construction and materials data.

Sound Conditioning

The use of soft carpeting with padding is also an effective way to reduce impact sounds on floors. Much sound energy is dissipated in such fabrics so that little gets through to the room below. Sound-deadening board combined with gypsum and special clips also produces a fairly high STC rating.

Another method sometimes used to improve sound insulation is to mount the ceiling on separate ceiling joists. It works well but is costly.

Door and Window Sound Transmission

Doors and windows are poor performers acoustically but airborne sound can be damped if door and window frame areas are free of cracks and spaces. Be sure to fill up such cracks and spaces tightly with felt, rubber or metal strips.

Hollow doors transmit more sound than solid doors; nevertheless, the biggest problem lies at the door edges. Properly installed weather stripping helps considerably; and the soft-foam kind not only seals the door effectively, but also absorbs impact noise if it is slammed. A movable seal is recommended for door bottoms.

Seal window edges tightly like doors against air leaks. Window design itself also assists or retards sound transmission: fixed windows are better acoustically than operating sash since they can be sealed more tightly. Storm windows also help reduce sound transmission because the air space between the inner and outer glazing is a relatively poor conductor.

Fixture and Appliance Noises

Plumbing noises can be quite annoying. Long runs of hot water lines often produce squeaking and squealing sounds as the piping expands and contracts. Counteract this by installing S-shaped bends that allow for movement in the runs. Eliminate metal-to-metal contact of piping and supports for additional help; and minimize vibration with resilient pipe insulation. Also provide air chambers in plumbing systems to get rid of water hammer. Vibration control of household appliances also helps reduce noise levels. Accomplish that by using sound-absorbing mounts for dish washers, washing machines, clothes driers and the like.

Sound Absorption Materials

Numerous sound-absorbing materials are made for sound insulation. They are used because hard-surfaced walls and ceilings reflect most sounds striking them. However, sound-absorbing materials assimilate about 50 percent of the sound that reaches them since the sound-absorbants trap sound waves and minimize their reflection back into rooms.

Such materials have a porous structure that causes the vibrating energy or sound waves to dissipate as friction-generated heat.

Acoustical tiles commonly used for ceiling applications are made of wood fiber or similar materials. Lightweight and easily installed, their sizes range from 12" X 12" to 24" X 48" with thicknesses of 1/2" and 3/4".

Several methods have been worked out for installing acoustical tiles. You may apply them with adhesive to a smooth flat surface, staple them to furring strips, or suspend them with wires and special hangers, Fig. 18-6.

Fig. 18-6 Installation of an acoustical tile system by stapling to furring strips. *(Celotex Corp.)*

Sound Conditioning

Painting acoustical tiles does not seem to affect their efficiency ordinarily, provided the holes or pores are not clogged. Dirty tiles lose some of their efficiency because dirt tends to clog the tiny openings. Vacuum tiles occasionally to restore their sound-absorbing qualities.

GLOSSARY

acoustical tiles: soft, lightweight fibrous tiles which absorb sound waves.

acoustics: the science study of sound.

decibel: the unit of measurement of sound.

frequency: the number of sound waves per second produced by a sounding body.

noise: unwanted sound, usually at a pitch or intensity disturbing to the human ear.

sound absorption: the ability of a material, such as acoustical tiles, to absorb sound waves.

sound conditioning: the control and reduction of sound by the use of appropriate materials.

sound: a vibration of energy that sets molecules in motion which in wave form stimulates hearing.

sound transmission class (STC): a number rating that classifies the values of various materials for reducing sound transmission.

transmission loss (TL): measured in decibels, the sound insulating efficiency and non-transmitting quality of various constructions.

19

Stair Construction

Stairs consist of a series of steps which connect one level of a structure to another. When unbroken by any landing or platform, the series of steps is called a flight. When bordered by walls or framing balustrades or rails, they are called a staircase. Fig. 19-1 shows a typical main entry stairway.

Fig. 19-1 Typical residential stairs. *(Morgan Co.)*

Stair design and construction are very important to good house building. The stairs must be so built and installed that they are both safe and comfortable. Tread size, riser height, angle of the stairway and landing dimensions are important factors to be considered. Also important are headroom and step width.

Stair Construction

Steps that are too steep or too shallow are not only dangerous but also uncomfortable to climb. Fig. 19-2 illustrates the importance of properly designed stairs. Keep in mind that there is a proportional relationship between tread width and riser height. If the tread is too wide and the riser too short, the climber's body has a tendency to lean backward; and if the steps are too steep (narrow tread and high riser), his body leans forward as he climbs. The opposite tendencies prevail when the user is descending such faulty stairs.

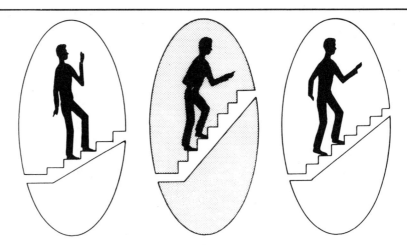

Fig. 19-2 Proper tread width and riser height are important in stair construction.

All stairs in a house must be functional but only residential stairs in living areas need be attractive. Considered service stairs, basement and attic stairs can be made of less expensive materials.

Residential stairs offer a challenge to the carpenter. Almost no other part of a structure requires the skill and patience needed to build and install a decorative and functional main stairway.

Types of Stairs

Stairs may be built straight, U-shaped, pie-shaped, or circular. Fig. 19-3 shows some common stair types. The simplest and least expensive is the straight run which leads from one floor to another without platform or landing, Fig. 19-4. Generally, such a single flight will have 13 or 14 steps.

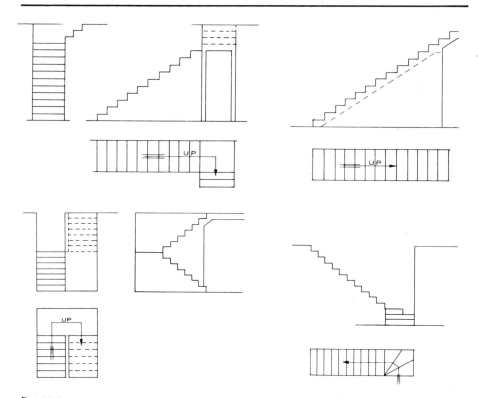

Fig. 19-3 Stair types commonly used in residential construction: **(top left)** long L; **(top right)** straight; **(bottom left)** narrow U; **(bottom right)** winder. *(Forest Prod. Lab.)*

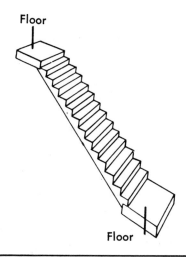

Floor

Floor

Fig. 19-4 A straight run of simple stairs.

Stair Construction

Stair Landings

A landing is the level platform either at the end of a flight of stairs or connecting one flight with another. Landings may also be used to interrupt a long flight which can be dangerous and frightening, especially to old folks and youngsters. They are also used when a stair must change direction, Fig. 19-5. The turn can either be one-quarter at 90 degrees (L-shape) or a half-turn at 180 degrees (U-shape) if the two flights are to parallel each other. When the landing is near the upper or lower level, it is called a long L; if near the center, it is a wide L. A double L has two platforms, one near the top and one near the bottom. If landings are used as a break in a long flight, make them at least 2'6" long by the width of the tread. Also be sure to keep any landing at the top of a flight with a door opening onto it a minimum of 2'6" long.

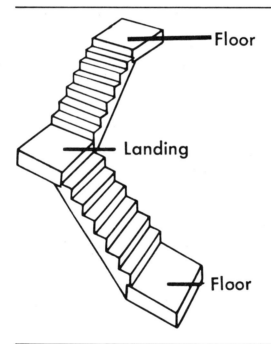

Floor

Landing

Floor

Fig. 19-5 Typical use of a landing for a turn in stairs.

When space is so limited that a landing cannot be installed, a winding or circular stair can be used instead. The steps, called winders, are pie-shaped treads whose inner width is very narrow so there is little support for the user's foot on at least part of each tread, where it fits to the inside vertical spiral. Falls are common on this type of stair. Avoid building them if possible.

Basic Parts of Stairs

Whether he designs or builds stairs, the carpenter should be familiar with the terms that define their basic parts. Fig. 19-6 illustrates them.

Stair terms: the horizontal member or part stepped on is the *tread;* the *riser* is the vertical member or front of the step; and the *nosing* is the front edge of the tread which projects beyond the riser.

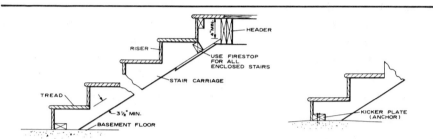

Fig. 19-6 Terms used in basic stair construction.

The *unit rise* is the distance from one step to the next; the *total rise* is the sum of all the risers, Fig. 19-7. The *unit run* is the width of a step, less the nosing projection; while the *total run* is the sum of all the unit runs.

Headroom is the vertical clearance between stair treads and the ceiling above, Fig. 19-8. The FHA minimum headroom specification is 6'8" for a main stairway, but the ideal is between 7'4" and 7'7" clearance. This allows ample room for moving large pieces of furniture.

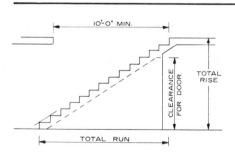

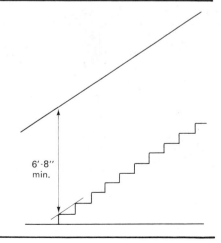

Fig. 19-7 above: Total rise and total run of stairs.

Fig. 19-8 right: Headroom is the vertical distance from tread to stairway of ceiling.

Stair Construction

Stairwell Rough Framing

Openings or wellholes for stairs are made during the floor framing since the architect's working drawings indicate the location and size of the stair wells at that stage. The openings are usually parallel to the joists. That simplifies construction, of course, but openings may also be designed perpendicular to the joists. In either case, headers and trimmers are doubled if the span is greater than 4 feet. Fig. 19-9 shows a layout with joists parallel to the opening. Fig. 19-10 shows framing detail with an opening perpendicular to the joists. The method used to secure stringers at the header also has a bearing on the wellhole opening. See Fig. 19-11.

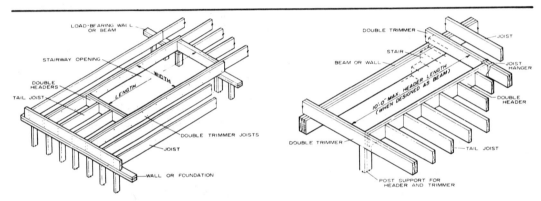

Fig. 19-9 Rough framing with joists parallel to stair opening.

Fig. 19-10 Rough framing with joists perpendicular to stair opening.

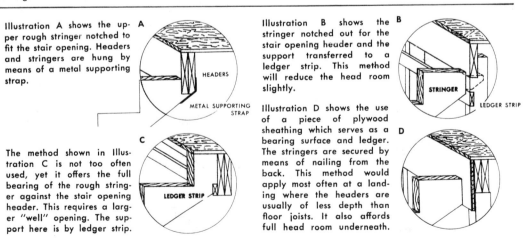

Illustration A shows the upper rough stringer notched to fit the stair opening. Headers and stringers are hung by means of a metal supporting strap.

The method shown in Illustration C is not too often used, yet it offers the full bearing of the rough stringer against the stair opening header. This requires a larger "well" opening. The support here is by ledger strip.

Illustration B shows the stringer notched out for the stair opening header and the support transferred to a ledger strip. This method will reduce the head room slightly.

Illustration D shows the use of a piece of plywood sheathing which serves as a bearing surface and ledger. The stringers are secured by means of nailing from the back. This method would apply most often at a landing where the headers are usually of less depth than floor joists. It also affords full head room underneath.

Fig. 19-11 Four methods of securing stringers to stairs opening headers. *(Morgan Co.)*

The usual rough opening for basement stairs is about 9 feet by 2 feet 8 inches. For main stairs, the RO should be at least 10 feet long by 3 feet wide. Fig. 19-12 shows how one corner of a landing is supported by a post. A load-bearing wall serves the same purpose. Fig. 19-13 details the rough framing for an interior stairway. Fig. 19-14 shows the framing for a stairway with a landing. Fig. 19-15 shows a roughed-in U-type staircase.

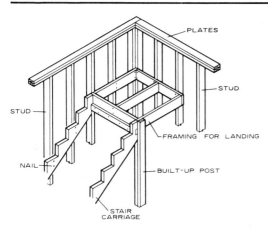

Fig. 19-12 Layout showing framing support for the corner of a stair landing.

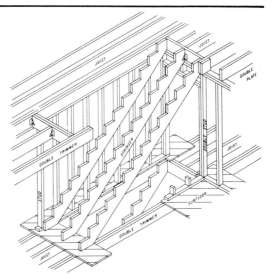

Fig. 19-13 Details of rough framing for an interior stairway.

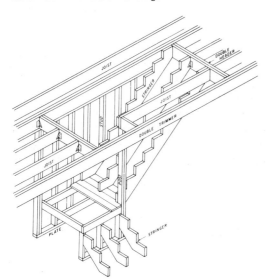

Fig. 19-14 Details of rough framing for a stairway with landing.

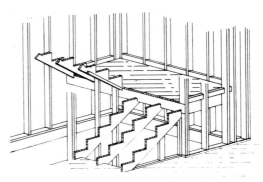

Fig. 19-15 Rough framing for U-type stairway.

Stair Construction

Work out the length of the opening after determining the rise and run of the stairway. Find the minimum length of the stairway by multiplying the number of treads in the open by the unit run. To determine the number of treads in the open, add the headroom with the upper construction thickness and divide the sum by the unit rise. This is expressed in the simple formula that follows.

To determine the length of a stairwell, proceed with the calculations represented in Steps a, b and c:

$$\text{Step a} \quad \frac{(H + UC)}{U\text{-}Ri} = O\text{-}Ri$$

$$\text{Step b} \quad O\text{-}Ri = O\text{-}Rn$$

$$\text{Step c} \quad O\text{-}Rn \times U\text{-}Rn = L$$

LEGEND

H = headroom	O-Ri = number of risers in open
UC = upper construction thickness	O-Rn = number of runs or treads in open
Ri = riser.	
U-Rn = unit run	
Rn = run	L = length of stairwell

Be sure to consider the thicknesses of all finish materials in the final calculations: for example, the risers, drywall and the like. The minimum length of the stairwell should be ten feet. If necessary, extend its length by adding one more riser to the stairs.

If ready-made stairs are to be installed, the manufacturer will furnish a framing schedule. The schedule specifies stairwell openings, support ledger sizes, and all other pertinent dimensions. Fig. 19-16 shows the rough framing for a partial wall with knee wall. A knee wall is a low wall that is about half the normal height or even less. It supports the handrail in a partially open stairway. (In a gable, the knee wall is the side wall built under the pitched roof.) The 2" × 4" cross blocking supports finish material for the stairwell's ceiling or soffit.

Staircase Width: The width of a staircase should enable two people to pass comfortably. For service stairs, the accepted minimum is 2'6" wide; for main stairs the minimum is 3'0" but 3'4" or even wider is preferred.

Give consideration to the passage of furniture when building stairs. Those with an open side are the most practical in this respect.

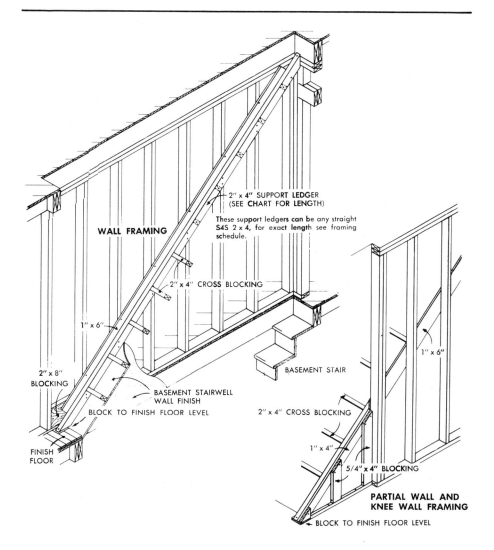

Fig. 19-16 Typical stairway framing shows cross blocking for stairwell ceiling. (*Morgan Co.*)

Remember that winders and narrow U-steps can be problems when it comes to moving large pieces of furniture from one story of a house to another.

Rail Height: The most comfortable height is 30 inches on the rake and 34 inches on landings. The hand rail should be continuous from floor to floor. Normally one hand rail is sufficient, but on wide staircases two, one either side, may be necessary.

Stair Construction

Ratio of Rise to Tread

Certain proportional rules govern the ratio between riser height and tread width and make it easy for the user to climb and descend stairs without undue strain. If that ratio is incorrect, steps are difficult to use and can be dangerous as well. It has been established that the ideal riser ranges from between 7" to 7-5/8" high. The following rules express the formula for working out correct stair dimensions:

Rule 1: The sum of two risers and one tread should be from 24" to 25".

Rule 2: The sum of one riser and one tread should be from 17" to 18".

Rule 3: The tread width multiplied by the riser height should be from 72" to 75".

Note that the ideal measurements for riser and tread, using these rules, are 7-1/2 inches for riser and 10 inches for tread. This is shown in the following equations.

Example: Assume a riser height of 7-1/2" and tread width of 10".

Rule 1: Add two riser heights and one tread width

$$
\begin{array}{l}
\text{riser } 7\text{-}1/2" \\
\text{riser } 7\text{-}1/2" \\
\underline{\text{tread } 10"} \\
\text{Total } = 25"
\end{array}
$$

Rule 2: Add one riser plus one tread

$$
\begin{array}{l}
\text{riser } 7\text{-}1/2" \\
\underline{\text{tread } 10"} \\
\text{Total } = 17\text{-}1/2"
\end{array}
$$

Rule 3: Multiply tread by riser

$$10" \times 7\text{-}1/2" = 75"$$

Likewise, a step with a 7" riser would require a tread between 10" or 11" wide, depending on which rule is used. Exception is made for basement stairs in which an 8" rise is permitted because of the normally lower height of basement ceilings.

In addition to riser-tread ratios, the slope or angle of incline of the stairway is also important. It should range from between 30 to 36 degrees above the horizontal.

Determining Riser Height and Tread Width: The number of risers in a stairway depends on the total rise. Total rise is the perpendicular distance from a finished lower floor to the top of the finished floor above. That is, it is the sum of the floor-to-ceiling height plus joist and floor thicknesses above. To find the number of risers, divide the total rise by the unit rise.

Problem: With a 7" unit rise, find the number of risers needed for a stairway with an 8'0" ceiling height and a 9" joist and floor thickness.

Solution: $\dfrac{8'9"}{7"} = \dfrac{96" + 9"}{7"} = \dfrac{105"}{7"} = 15$

Answer: 15 risers needed

If the answer includes a fraction, use the nearest whole number and discard the fraction.

Since the ideal riser is between 7" and 7-5/8", we could have divided the total rise by 7-1/2" instead of by 7". This would have given us 14 risers.

Solution: $\dfrac{8'9"}{7\text{-}1/2"} = \dfrac{96" + 9"}{7\text{-}1/2"} = \dfrac{105"}{7\text{-}1/2"} = 14$

Knowing the riser height, we can apply one of the stair rules to determine the tread width.

Problem: Using Rule #2 (the sum of one riser and one tread should equal 17" to 18"), find the tread width for a 7" riser.

Solution: (*a*) 7" + 10-1/2" = 17-1/2"
(*b*) 7" + 10" = 17"
(*c*) 7" + 11" = 18"

Answer: Any of these tread widths, 10", 10-1/2", or 11", would work well for stairs with 7" risers.

Stair Construction

The number of treads in a flight of stairs is always one less than the number of risers, Fig. 19-17. Multiplying the number of treads by the tread width gives the total run. Each tread (less the nosing) is equal to the unit run. If a stairway has 14 risers and the treads are 10-1/2" wide, the total run will be 13 × 10-1/2" or 11'4-1/2". Figure 19-18 lists tread and riser specifications.

Formula: Total Run = Unit Run × Number of Treads
Solution: 11'4-1/2" = 10-1/2" × 13

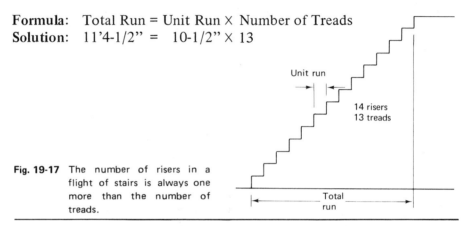

Unit run

14 risers
13 treads

Total run

Fig. 19-17 The number of risers in a flight of stairs is always one more than the number of treads.

STAIR WITH LANDING

This type is easier to climb, safer, and reduces the length of space required. The landing provides a resting point and a logical place to have a right angle turn. Landing near bottom with quarter-turn is basis of calling this type "dog-legged" or "platform" stairs.

Height Floor to Floor H	Number of Risers	Height of Risers R	Width of Tread T	RUN Number of Risers	RUN L	RUN Number of Risers	RUN L2
8'-0"	13	7-3/8 +	10"	11	8'- 4" + W	2	0'-10" + W
8'-6"	14	7-5/16"−	10"	12	9'- 2" + W	2	0'-10" + W
9'-0"	15	7-3/16"+	10"	13	10'- 0" + W	2	0'-10" + W
9'-6"	16	7-1/8"	10"	14	10'-10" + W	2	0'-10" + W

STRAIGHT STAIRS

Simplest and least costly; requires a long hallway which may sometimes be a disadvantage. May have walls on both sides (closed string) or may have open balustrade on one side (open string).

Height Floor to Floor H	Number of Risers	Height of Risers R	Width of Treads T	Total Run L	Minimum Head Rm. Y	Well Opening U
8'-0"	12	8"	9"	8'-3"	6'-6"	8'-1"
	13	7-3/8" +	9-1/2"	9'-6"	6'-6"	9'-2-1/2"
	13	7-3/8" +	10"	10'-0"	6'-6"	9'-8-1/2"
8'-6"	13	7-7/8" −	9"	9'-0"	6'-6"	8'-3"
	14	7-5/16" −	9-1/2"	10'-3-1/2"	6'-6"	9'-4"
	14	7-5/16" −	10"	10'-10"	6'-6"	9'-10"
9'-0"	14	7-11/16" +	9"	9'-9"	6'-6"	8'-5"
	15	7-3/16" +	9-1/2"	11'-1"	6'-6"	9'-6-1/2"
	15	7-3/16" +	10"	11'-8"	6'-6"	9'-11-1/2"
9'-6"	15	7-5/8" −	9"	10'-6"	6'-6"	8'-6-1/2"
	16	7-1/8"	9-1/2"	11'-10-1/2"	6'-6"	9'-7"
	16	7-1/8"	10"	12'-6"	6'-6"	10'-1"

NOTE: Dimensions shown under well opening "U" are based on 6'-6" minimum headroom. If headroom is increased well opening also increases.

Fig. 19-18 Tread and riser specifications for stairs with landings and for straight stairs.

Stringers and Carriages

Stringers and carriages are two sometimes confusing terms used in stair construction. Some authorities refer to them as separate stair members; others use them interchangeably. In this text they are used interchangeably.

A well-built stairway must have strong supporting stringers since these inclined beams hold up the risers and treads. Two main types are used in stair building; the plain stringer and the housed stringer. A *plain stringer* is of simple construction. See Fig. 19-19. Most commonly used in residential buildings, the *housed stringer* is usually made from either 1 × 10 or 1 × 12 boards. The sturdiest and most attractive wood stringer, it also gives the best finished appearance. Vertical and horizontal grooves or housings are cut into the stringer to receive the risers and treads. Then tight-fitting wedges are installed on the underside to provide a sturdy step, Fig. 19-20. Some other types of stringers used in stair construction are described below.

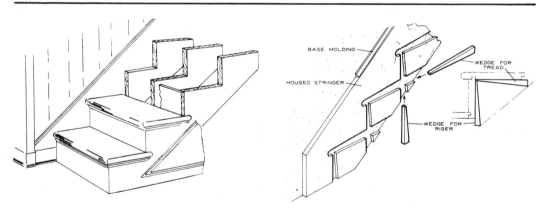

Fig. 19-19 Construction details of plain stringer.

Fig. 19-20 Housed stringer assembled with wedges.

The *cut-out carriage/open stringer* is built from 2 × 10, 2 × 12, or 2 × 14 lumber. It is used in out of the way locations where finish appearance is not of prime importance. See Fig. 19-21.

A *built-up carriage* is made from 2 × 4 or 2 × 6 lumber with triangular blocks fastened on one side to support the stair treads. For economy, a stair may have a cut-out carriage on one side and a built-up carriage on the other. In such a case, the waste or cut-off material from the cut-out carriage serves as the triangular blocking for the built-up carriage.

Fig. 19-21 Open stringers are ordinarily used for basement stairs.

The ***dadoed stringer*** is made from 2 × 6 or 2 × 8 lumber and has only treads and no risers. The treads rest on dadoes cut into the stringers. This is a simple and inexpensive type of stair, Fig. 19-22.

The ***cleated stringer*** also has only treads and no risers, but the treads are supported on cleats nailed to the inner sides of the stringers. This too is an economical stairway to build.

Cut-and-mitered stringers are made from 1 × 10's or heavier lumber and are used on open stairways where good finish appearance is desired. The treads are supported on square cuts and the risers are mitered so that no end grain lumber is exposed. A finished stairway open on only one side, may have a cut-and-mitered stringer on that side and a housed stringer on the other. See Fig. 19-23.

Construction of Stringers: Leave top and bottom ends long when cutting either carriage or stringer. Then cut them to exact size and fit only when you actually install them in the stairwell. Usually you must also shorten or drop the bottom of the carriage by an amount equal to the thickness of one tread so that all the steps have the same rise. **Note:** There is always one tread less than the number of risers in a stair because the top riser leads to a floor, not a tread.

Factory-made housed stringers complete with treads, risers and wedges come ready to install. Or to simplify your construction of job-made housed stringers, use the special ready-made router templates now available.

Stringer Layout: It is measured with a story pole. Mark off the floor to floor height or total rise with it. Be sure to add in an allowance for finish floor materials if you have not already installed them. See Fig. 19-24.

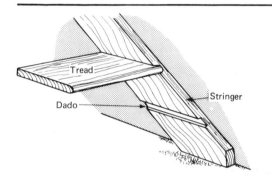

Fig. 19-22 Dadoed stringer with tread only—no risers.

Fig. 19-23 Mitered stringer on open-side of stairway; arrows indicate the mitered joints.

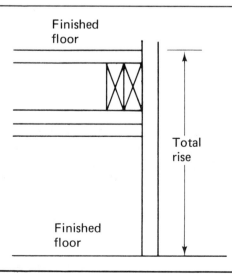

Fig. 19-24 The total rise of a stairway is the vertical distance between a finished lower floor to the top of the floor above.

Find the exact riser height with a divider. To do it, set the divider to the approximate riser height which you have already determined arithmetically. Step off the divider, adjusting as necessary, until you have marked off an equal number of divisions on the story pole. This will be the exact riser height.

To lay out the stringer, set your framing square with the riser height on the tongue and the tread width or run on the blade. Use a set of stair gauges for accuracy. Some carpenters clamp a straight edge to the square, instead of using stair gauges. See Fig. 19-25.

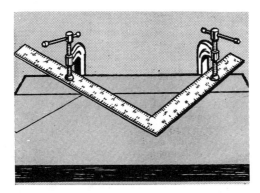

Fig. 19-25 Using the framing square to lay out stringers for a stairway.

Start at the top end of the stringer, marking the required number of treads and risers. When the layout is complete, make an adjustment at the lower end. Reduce the bottom riser in height by an amount equal to a tread thickness. You must do this because the floor line does not receive a tread. By removing that one tread thickness, you make the first step equal to all the others.

Basement Stairs

In their simplest form basement stairs may be made with two 2" × 12" carriages and 2" planks for treads. They are the easiest and the most economical to build. A better looking stairway can be made with 1-1/16" or 1-1/2" treads and risers. You can also make the risers of one inch nominal material.

A third center carriage is required if the treads are 1-1/16" thick and the steps are wider than 2'6". Three carriages are also needed when 1-1/2" thick treads make steps wider than 3 feet.

You may either fasten the carriages to the header or rest them on a ledger, Fig. 19-26. But you must fasten the bottom of the carriages to a

2 × 4 kick plate. If the stairs are installed before the basement slab is poured, use temporary blocking under the carriage.

Steel stringers are factory made for use in basement stairways. They accept wood treads and are fastened to wood carriages which are mounted to the walls. Wood treads are simply inserted into the notched stringers. Nails hold the treads in place firmly.

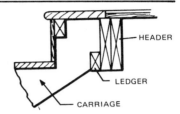

Fig. 19-26 Layout for fastening the carriage to the ledger.

Tread and Riser Details

Treads for main stairs are either 1-1/16" or 1-1/8" thick. Ordinarily a hardwood should be used; but softwoods may be used if they meet FHA standards.

Risers are usually made from 3/4" lumber of the same wood as the treads; but less expensive material may be used if they are to be covered.

If stairs are to be carpeted, plywood can be employed for the risers and treads. Because of plywood's dimensional stability, smooth surface and favorable cost, it is now widely used in stair construction.

Treads and risers are fastened several ways. Screws and glue blocks are frequently used in combination with rabbeted and grooved joints, as in Fig. 19-27. Often a decorative molding is also used to reinforce the tread nosing which the groove weakens.

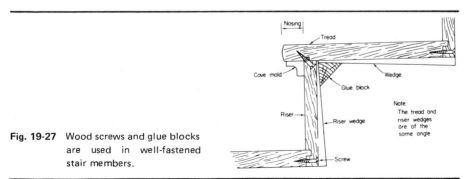

Fig. 19-27 Wood screws and glue blocks are used in well-fastened stair members.

Winder Stairs

As mentioned earlier, winder stairs are usually dangerous and should be used only when necessary. Even when the converging end is sufficiently wide, winders are difficult to tread on because the user is thrown off balance by their nonrhythmic progression. They are especially dangerous if used at the top of a stairway since a wrong step there can cause a person to fall down an entire flight.

At a distance of 18 inches from the narrow end the winder tread should be the same width as regular treads. See Fig. 19-28. You may use them in place of quarter-turn or half-turn stairs when the space is limited.

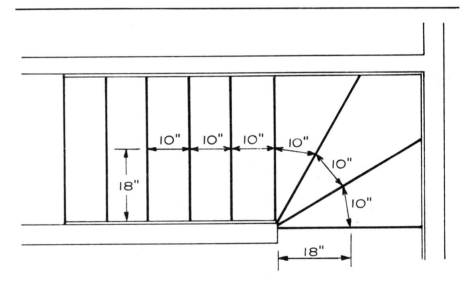

Fig. 19-28 Recommended size for winder treads.

Folding Stairs

Folding stairs are sometimes used where conventional stairs are either not practical or not wanted. They are commonly installed in seldom used attics. The stairs fold like an accordion and are held in place by counterweights or by tension springs. Various types are manufactured. Rough opening sizes, floor clearance and other specifications are furnished with the folding units.

Exterior Stairs

Exterior stair riser and tread proportions are similar to interior stairs. The standard ratios between riser and tread are generally employed, but riser height should be between 6 and 7 inches. Be sure to use only treated wood for exterior stairs.

Balusters

Balusters are the posts which support the handrail and enclose the open sections of the stairway. They are usually fastened to the treads by means of mortise and tenons. A finished return conceals the joints. The upper end fits into the underside of the handrail. The handrail is fastened to the newel post. The entire assembly is called a balustrade. See Fig. 19-29.

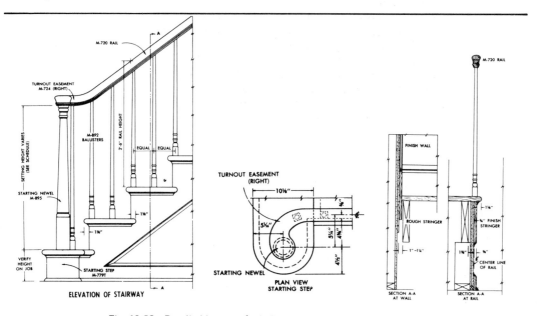

Fig. 19-29 Detailed layout of a balustrade assembly. *(Morgan Co.)*

Factory-made wood balustrades are available for speedy installation. Wrought-iron balustrade assemblies are also ready-made for use instead of wood in stair construction. Other stock parts are pre-fabricated for efficient and economical installation as well.

Stair Construction

GLOSSARY

baluster: a vertical support for a railing; the lower end may rest on a stair tread or on a bottom rail.

balustrade: a railing assembly consisting of hand rail, balusters and often a bottom rail.

carriage: the supporting member or framework of a stair; usually cut from a 2 inch plank.

flight: a series of steps between landings or floors.

headroom: the vertical clearance between the stair treads and the ceiling above.

housed stringer: a stair stringer grooved to receive the ends of treads and risers.

landing: a platform used to change the direction of stairs or to break a long run.

newel: the post to which a stair railing is fastened.

rail: a hand support parallel to the stringer and attached to the wall or to the balusters of a staircase.

rake: the incline or slope of a stairway.

riser: the vertical member between the treads of stairs.

staircase: a flight of stairs leading from one floor to another and either framed by walls or balustrades.

stairwell: the opening in which stairs are placed.

stringer (string): the sloping sides of the stair that supports treads and risers.

tread: the horizontal part of a step on which the foot is placed to ascend or descend a stairway.

unit rise: the vertical distance from one step to the next.

unit run: the width of a step less the nosing projection.

wellhole: the opening for stairs; also called stairwell.

winders: pie-shaped treads on circular or winding stairs.

20

Interior Wall Finish

This chapter covers the various methods of finishing framed interior walls. Available products for this purpose include plaster, gypsum wallboard, wood paneling, fiberboard and plywood. While plastering is a specialized trade, best done by a plasterer, the carpenter has to prepare walls for plastering so a working knowledge of plastering is essential. Fig. 20-1 shows a plasterer applying plaster to a wall.

Plaster Finishing

Plaster is made from a white mineral called gypsum. Found in the earth in rock form, it is mined just like coal, Fig. 20-2. Next, this rock is crushed, ground into a fine powder, and heated or calcined to drive off its chemically combined water. The oxidized powder is then combined with fine sand and a binder to make plaster. Mixed with fresh water, plaster then becomes an easily worked paste which when spread thinly on a flat surface dries hard and smooth. The advantages of using plaster are numerous. For instance, it is fire resistant. Fire releases the water still held in dry plaster's crystals by turning that water into steam which retards the spread of flame. Plaster also has a high resistance to sound transmission and its durable surface resists cracks and scratching. Finally, plaster can be finished with many materials.

Plaster Base: Plaster is applied in one or more coats; all but the last are called base coats. The last or top coat is the finish coat. The base coats must be applied to a surface which holds them securely. That surface is the plaster base. In the past, plaster base consisted of wood lath fastened to studding and joists so as to leave spaces between the laths. The wet plaster penetrating those spaces formed keys which held the plaster securely to the base. Such wood lath is now seldom used in modern construction, but metal and gypsum lath of various sorts are used instead.

Metal lath is made from special sheet steel which has been slit and

Interior Wall Finishing

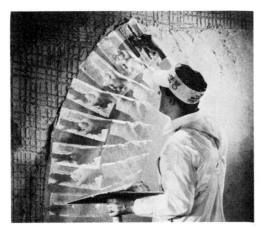

Fig. 20-1 Plaster being applied to scratch base coat. *(U.S. Gypsum)*

Fig. 20-2 Gypsum is mined underground like coal. *(U.S. Gypsum)*

expanded. In addition to providing the apertures for the necessary keying, metal lath becomes imbedded in the plaster, thus reinforcing it. Fig. 20-3 shows the types of metal lath available.

Gypsum Lath: consists of a plaster core faced with laminated paper. One paper face is porous and becomes filled with plaster crystals, resulting in a perfect bond of the two materials. Perforated lath which is similar also contains 3/4" holes spaced 4" apart. They provide mechanical keys that supplement the surface bonding, Fig. 20-4.

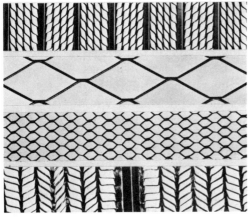

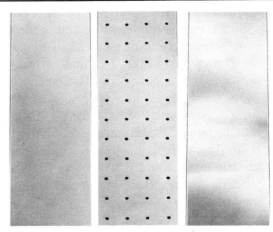

Fig. 20-3 from top to bottom: Expanded metal lath—4-mesh riblath, stucco mesh, self-furring mesh, 3/4 inch riblath.

Fig. 20-4 Types of plaster bases: one a plaster core faced with laminated paper, the other faced and perforated.

Gypsum lath is manufactured in several sizes. The standard size is 16" × 48" except on the Pacific Coast where the standard width is 16-1/5 inches. Gypsum lath is also available in 24" widths up to 12 feet long and is known as long length lath. Thicknesses are 3/8, 1/2, and 5/8 inches. For residential construction, the 3/8" and 1/2" thicknesses are used. The 3/8" lath is used for 16" stud and joist spacing; the 1/2" material for 24" stud and joist spacing.

Installing Gypsum Lath: Gypsum lath is applied horizontally with staggered joints. Fig. 20-5 shows a typical installation on a stud wall. You must not place vertical joints over jamb lines. Use either nails or staples to fasten the lath. Fastener recommendations vary among manufacturers, but 13 gauge 1-1/8" blued nails with 19/64 heads are suggested for 3/8" lath. Use 4 nails per lath in each bearing. For 1/2" lath, the nails should be 1-1/4" long, 5 per lath. Staples for the 3/8-inch lath should be 16 gauge with 7/8" divergent legs. For 1/2" lath, use 16-gauge staples with 1" divergent legs.

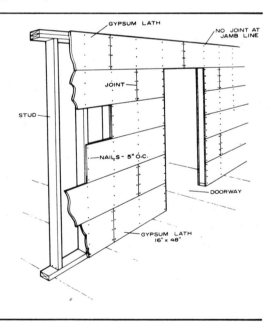

Fig. 20-5 Layout for horizontal application of gypsum lath with staggered joints.

Applying Metal Lath: Metal lath is available 27" × 96" in several mesh patterns. You fasten it to wood studs with 4d common nails bent over; and space the nails 6 inches, center to center. To fasten it to joists, use 1-1/2" 11 gauge barbed roofing nails with 7/16 heads.

Metal lath is seldom used in residential construction except in bathrooms and kitchens. Fig. 20-6 shows how to install it in a bathroom. Carefully line the area to be tiled with water resistant paper before installing the metal lath. Be sure the lath is laid horizontally with all end joints lapped; then cover it with the ceramic tile.

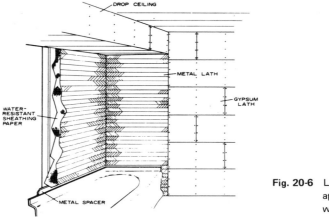

Fig. 20-6 Layout for horizontal application of metal lath with lapped end joints.

Reinforcing: The areas above doors and windows are subject to stresses which can cause cracks in plaster surfaces. To minimize the development of cracks, use reinforcing strips of metal lath above door jambs and windows. Install self-furring diamond mesh lath as shown in Fig. 20-7.

Other areas which require reinforcing are the outside corners, inside corners, and the area under flush beams, Fig. 20-8. The outside corner beads also serve as a level edge for plastering and they protect the plaster corner. Use flexible corner beads where there are irregular shapes, such as in archways.

Plaster Grounds are the wood strips placed around the perimeter of doors and windows and at the base of walls. They provide a level surface to serve as a guide for the plasterer. They also provide the necessary nailing surface for finish trim.

Grounds around doors and windows may be installed temporarily and removed after the plastering is completed. Fig. 20-9 shows how temporary grounds are installed. They are removed after the plaster has set. Permanent grounds are nailed securely, Fig. 20-10. Often the side and head jambs of doors and windows are used as grounds.

Applying Plaster is a two- or three-coat job. Two-coat work consists of one base and one finish coat. Lay it over gypsum lath and masonry like rough concrete blocks and porous brick. It is generally used in residential construction.

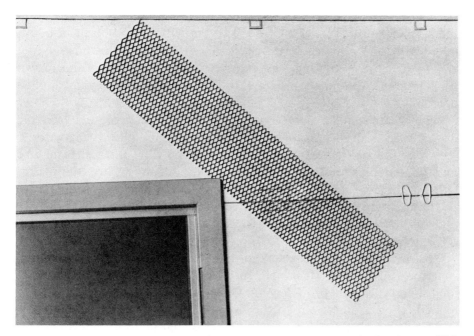

Fig. 20-7 Installation of reinforcing strip above door prevents cracks in plaster. *(U.S. Gypsum)*.

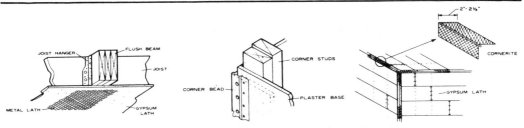

Fig. 20-8 Other areas in a room that require reinforcing before plastering.

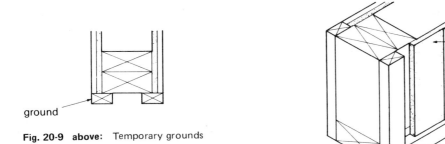

Fig. 20-9 above: Temporary grounds which are removed after plaster sets.

Fig. 20-10 right: Permanent plaster grounds which are fastened securely.

Interior Wall Finishing

The first coat consists of gypsum mixed with an aggregate of wood fibers, sand, perlite or vermiculite. They may be mixed on the job, but usually they are purchased ready-mixed. Scratch and brown is the first coat used in all plastering. In two-coat work, the scratch and brown coats are applied at the same time. In three-coat work, they are applied separately.

The scratch and brown coat fills most of the space between the plaster base and the grounds. The second coat in two-coat work is known as the finish coat and gives the wall its final surface. It may be sand floated or smooth troweled. For the sand finish, mix the plaster with special sand to give it a textured surface; for the smooth, make up the plaster without sand and trowel it to a hard smooth finish. In kitchens and bathrooms subject to moisture, a waterproof plaster such as Keene's cement is often used.

In three-coat work, the scratch coat is followed by the brown and then the finish coat. The rough surface of the scratch coat provides a good bond for the brown coat. Once it is dry, the brown coat is applied, then brought out to the surface of the grounds and trued-up. The finish coat is troweled on as in the two-coat method.

Thin Coat Plaster: Also known as veneer plaster, this method utilizes a very thin coat of veneer plaster over a special base. The coating is about 1/16 of an inch thick and may be applied in either one or two operations. Although thin, the surface is very durable. Special joint reinforcing is necessary for the base materials and special corner beads are required. Like regular plaster, the surface may be smooth or sand finished.

Gypsum Dry Wall

Dry wall construction includes paneling, plywood, fiberboard and hardboard, but generally the term refers to gypsum dry wall or wallboard. Because of ease of installation and low cost, dry wall construction is used extensively in residential construction. Other advantages are its high fire resistance and its sound insulation qualities. Made with a plain paper face, it can be painted or papered; it also comes prefinished in a variety of surface textures and colors. Numerous texturing materials are also available to apply with brush, roller, sprayer or trowel.

Gypsum wallboard, also called plasterboard, consists of a layer of gypsum plaster sandwiched between two layers of heavy paper. The face paper is folded around the long edges and the ends are cut square. Wallboard is made in four-foot widths and in lengths of 8, 10, 12, and 14 feet.

Standard thicknesses are 1/4 inch, 3/8 inch, 1/2 inch, and 5/8 inch. The 1/4" material is used for recovering old walls and ceilings. The 3/8" wallboard is generally used in two-ply construction. The 1/2" and 5/8" wallboards are used for single-ply construction for walls and ceilings. The thicker 5/8" material provides both better fire resistance and sound control.

Single-Ply Construction: In single-ply installation, the wallboard may be applied vertically or horizontally. Horizontal application is preferred since it makes for fewer joints and stronger construction. Ceilings are done first then the walls. Any end joints should be staggered.

In horizontal applications, install the top panel first, as in Fig. 20-11. This assures a good joint at the ceiling line. Any irregularities fall at the floor line and are later covered with a baseboard.

All wall frame members must be straight and true. Be sure that the studs and joists are in alignment and that the moisture content of the framing does not exceed 15 percent.

Fig. 20-11 In horizontal applications, the top panel of wallboard is installed first.

Interior Wall Finishing

Framing Defects: Any bowed stud must be straightened before you install wallboard panels. To correct bowing, make a saw cut on the hollow side of the stud. Next, drive a wedge into the kerf until the stud straightens. Then nail 1" × 4" scabs on each side, and break off the wedge flush. See Fig. 20-12.

Cutting Panels: Cut wallboard panels by scoring the face side with a sharp blade. Then snap the core by bending away from the scored side. Complete the cut by running a knife through the back of the board on the bend line, Fig. 20-13. Finally, smooth the cut edge with a rasp or coarse sandpaper.

Fig. 20-12 Straightening a bowed stud with wedge and scabs.

Fig. 20-13 Back of wallboard is slit after core has been snapped.

Fastening Methods: Wallboard panels can be fastened with nails, screws, or adhesives. Use clips and staples *only* on the first layer of two-ply construction.

Screws are superior to nails since they rarely pop and have better holding power. Make sure you drive them straight and no less than 3/8" in from the panel's edge. You will find that an electric screwdriver is essential to reduce installation time. Be sure that the tool is *properly grounded* before you use it. Special screwdrivers with a slip-clutch disengage when you drive the screw head solidly against the panel. Magnetic tips hold screws firmly to the bit. Use only approved wallboard screws; for single-ply applications, they should be 1-1/4" long. It should be noted that screws eliminate the head popping due to loose boards. They do not eliminate the "pops" caused by excessive shrinkage or warping of framing members, however. Fig. 20-14 illustrates several types of wallboard screws.

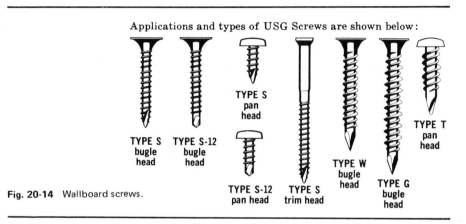

Applications and types of USG Screws are shown below:

TYPE S bugle head   **TYPE S-12 bugle head**   **TYPE S pan head**   **TYPE S-12 pan head**   **TYPE S trim head**   **TYPE W bugle head**   **TYPE G bugle head**   **TYPE T pan head**

Fig. 20-14   Wallboard screws.

Nails must be driven at least 3/8" from panel ends and edges. Start from the center of the panel and proceed outward toward the ends. Be sure to press down on panel near the nailing area. This ensures against loose boards. Use a crowned head hammer to produce a uniform dimple, as shown in Fig. 20-15.

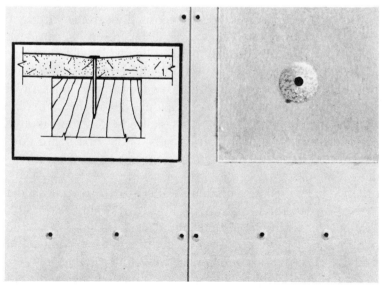

Fig. 20-15   A properly driven nail leaves a uniform dimple in the wallboard surface. *(U.S. Gypsum)*.

Interior Wall Finishing

Nail popping can be minimized by double-nailing the wallboard. Space the nails 12" on center and 2" apart, as in Fig. 20-16. Use single-nailing around the panel perimeter. Perimeter nails should be spaced 7" OC for ceilings and 8" OC for side walls. After you apply the second nails, restrike the first nails to anchor them tightly.

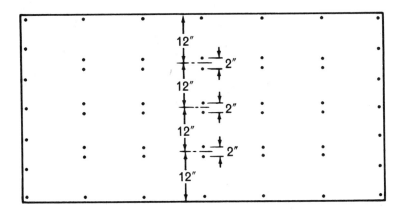

Fig. 20-16 Double-nailing pattern minimizes nail popping in wallboard.

Floating Angle Construction: You can minimize corner cracking and fastener popping by using the floating angle method. This method helps where structural stress occurs normally—at the intersection of walls and ceilings. By omitting some fasteners on the inside corners you make it possible for the wallboard to "move" under stress.

On ceilings where wallboard is applied across the joists, start the first row of nails 7" from the corner. If the wallboard goes on with the long edge parallel to the joists, nail the long edge in the regular manner.

For the side walls, drive the first nail 8" from the ceiling line; at inside corners, omit only the corner nails of the first board you apply. Nail the second board in the regular manner 8" OC. Fig. 20-17 shows you the nailing patterns.

Adhesives are used for bonding wallboard to framing or masonry. They are also used for laminating wallboard to solid backings like rigid foam, sound deadening board and other insulation. Apply the adhesive in a continuous strip. Keep the bead about 3/8" wide and center it on the framing. The adhesive will spread when the panel is pressed against it. When two panels join on the same framing member and the joint is to be taped, be sure to apply the bead in a zig-zag, Fig. 20-18. Temporary bracing ensures good contact while the adhesive sets. Use either screws or nails to provide supplemental support.

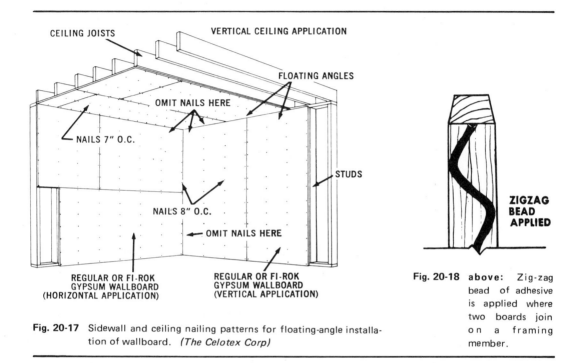

Fig. 20-17 Sidewall and ceiling nailing patterns for floating-angle installation of wallboard. *(The Celotex Corp)*

Fig. 20-18 above: Zig-zag bead of adhesive is applied where two boards join on a framing member.

Double-Ply Construction consists of two layers of gypsum panels. It offers greater strength, fire resistance, and sound insulation than single-layer installations. Walls and ceilings are less apt to crack and fastener popping is minimized.

You may fasten the panel with nails, screws, adhesive nail-on or entirely with adhesives. Since the adhesive method eliminates fasteners, it is ideally suited for use with decorator panels. Fig. 20-19 shows a two-ply installation.

The base ply of two-ply construction need not be gypsum. It may be backerboard or sound-deadening board. Double-nailing is not required for this system. Make sure that the face-ply joints offset the base joints by 10 inches or more. You may nail, screw or staple the base. If you apply the face without adhesive, use the same nailing pattern as for single-ply installations.

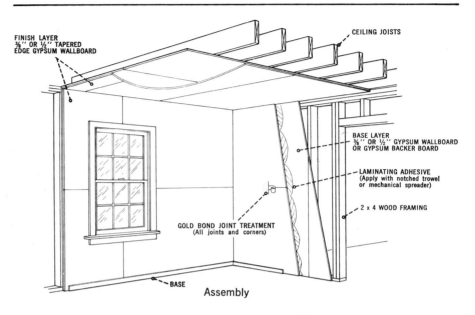

FINISH LAYER
⅜" OR ½" TAPERED
EDGE GYPSUM WALLBOARD

CEILING JOISTS

BASE LAYER
⅜" OR ½" GYPSUM WALLBOARD
OR GYPSUM BACKER BOARD

LAMINATING ADHESIVE
(Apply with notched trowel
or mechanical spreader)

2 x 4 WOOD FRAMING

GOLD BOND JOINT TREATMENT
(All joints and corners)

BASE

Assembly

Fig. 20-19 Details for two-ply application of gypsum panels.

The face-ply may be installed vertically or horizontally, except when you use decorator panels with exposed joints. Then install them vertically.

When you use adhesive for sheet lamination, spread it over the entire surface of the panel. There are occasions when spot or strip application may be desired, however. Fig. 20-20 shows how adhesive strips are applied.

Joint and Fastener Treatment: After the wallboard panels are installed, the joints and fasteners must be concealed. This is usually done with special paper tape and joint compound. Outside corners and other exposed edges require metal or other trim.

Spread joint compound in the recesses formed by the tapered edges of the joining boards, then embed the tape in the compound with a joint finishing knife. Hold the knife at a 45-degree angle to the board and apply sufficient pressure to remove excess compound. After the tape is embedded in each joint, apply a thin coat of compound over the tape, Fig. 20-21. This prevents the tape edges from curling. Otherwise, curled or wrinkled edges might later cause cracking. Fill all fastener heads with compound also. Always allow compound to dry thoroughly.

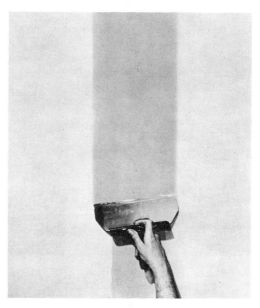

Fig. 20-20 Application of adhesive strips on face of a base panel.

Fig. 20-21 Finishing knife is held at 45-degree angle to spread compound.

Apply a second coat to the joints and fastener heads after the first is completely dry. Feather out the compound about 2" beyond the first coat. Spread the last coat after the second one is dry, again feathering the edge from 2" to 4" beyond the second coat. The last coat will be about 12" to 14" wide and should blend smoothly with the wallboard. If sanding is necessary between coats, do it very carefully with fine paper. Take care not to scratch the paper base.

Inside and Outside Corners: At inside corners, apply compound to both sides of the joint. Then fold the tape along the center crease line and press it into the corner. Otherwise follow the same steps as for flush joints. See Fig. 20-22. But, for outside corners you must nail the metal beads securely. Special tools are made to simplify installation of the beads, as shown in Fig. 20-23. Paper tape is not used since the metal flange serves the same purpose. Apply compound to outside corners just as you do for flush joints, Fig. 20-24.

After you have completed all joints and corners, allow them to dry thoroughly. Then seal or prime them with a suitable vinyl or oil base sealer/primer. That eliminates the absorption difference between wallboard and joint. You may then apply any desired decorating material.

Interior Wall Finishing

Fig. 20-22 Taping an inside corner. *(Gold Bond Building Products)*

Fig. 20-23 Crimping corner bead with a special tool. *(Gypsum Assoc.)*

Fig. 20-24 Applying compound to an outside corner. *(Gold Bond Building Products)*

Moisture-Resistant Wallboards

Bath and shower areas require the use of moisture-resistant wallboard. The installation procedure is similar to that for regular wallboard—with some exceptions. You must coat all edges of the panel with a waterproof adhesive, including holes and cutouts. Ceramic tile adhesive is suitable. Be sure to install the lower edge of the wallboard so that it is 1/4" above the tub or shower pan.

Masonry Walls

Gypsum wallboard may be applied over either insulated or uninsulated masonry walls. For *interior masonry walls* not subject to moisture, the panels may be applied directly to the masonry surface. Bond them with a suitable adhesive, such as that shown in Fig. 20-25.

For *exterior masonry walls,* the masonry must be waterproofed, and in cold climates insulation is required. You may attach the wallboard panels directly to rigid foam insulation or to wood or metal furring. You must install the wallboard 1/8" above the floor to prevent the absorption of moisture through wicking.

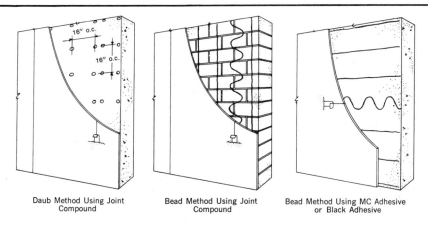

<table>
<tr><td>Daub Method Using Joint Compound</td><td>Bead Method Using Joint Compound</td><td>Bead Method Using MC Adhesive or Black Adhesive</td></tr>
</table>

Fig. 20-25 Details for adhesive application of wallboard on interior masonry walls.

Predecorated Panels

The use of predecorated gypsum panels provides an economical way to cover and decorate walls in one operation. Most predecorated panels have a fabric-backed vinyl facing laminated to a gypsum core and are available in many patterns and colors. Color-matched moldings and nails are also available. The 1/2" thickness is standard but 3/8" and 5/8" panels may be specially ordered. Remember, there may be variations in the same color if panels come from different lots. Try to use only panels with the same lot number in any one room. If this is not possible, lay out the panels from different lots so the joints fall in corners.

Various methods have been devised for joining predecorated panels. They include battens, self-edge butt joints and invisible joints. The panel manufacturers furnish detailed instructions on joint treatments.

Prefinished Paneling

Prefinished plywood paneling is widely used in residential wall finishing, Fig. 20-26. It is made in many styles, patterns, colors and textures. The special factory-applied finishes resist normal wear and require little maintenance. The standard 4' × 8' panels are 1/4" thick. However, other lengths and thicknesses are available.

Interior Wall Finishing

Fig. 20-26 Room interior finished with plywood paneling. *(U.S. Plywood)*

Panels may be fastened with nails, staples, screws or adhesives; joints may be butted, interlocked or concealed with a variety of moldings and battens. The panels can be applied directly to studs but a backer board is recommended. The backer board can be made from 5/16″ plywood sheathing or from 3/8″ gypsum board. For masonry walls, you must use horizontal furring. Place it 16″ OC, starting 1/4″ above the floor; Fig. 20-27 shows furring spacing. If necessary, use shims to ensure a flush surface; Fig. 20-28 shows various methods of attaching furring.

Keep the panels in the room to be paneled for at least forty-eight hours before installation. This allows them to reach the same degree of temperature and humidity as that in the room. Be sure that the first panel installed is plumb. Use a plumb line or spirit level as shown in Fig. 20-29. For a large expanse of paneling, leave a small space at the end for expansion. For irregular corners, scribe the panel as shown in Fig. 20-30 on page 362, at the right.

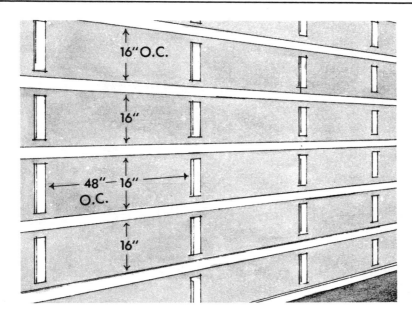

Fig. 20-27 Spacing layout for application of furring on masonry walls.

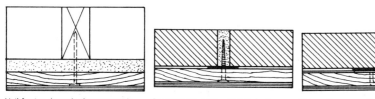

Nail furring through plaster to studs.

On masonry—screw or nail furring to shields or wood dowels inserted in wall.

On masonry—use Nail Anchors or adhesive anchors.

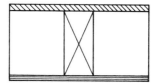

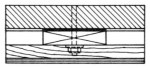

On masonry—2″ x 3″ framing wedged to ceiling and floor. (Alternate treatment—lay framing flat against wall.)

On masonry—Bolt Anchors may be used for attaching 1″ x 3″ sub-furring to wall then attach 1″ x 2″ furring to 1″ x 3″ as if to studs.

Fig. 20-28 Methods for applying furring over either plaster or masonry.

Fig. 20-29 First panel applied to room wall must be installed plumb. *(U.S. Plywood)*

Fig. 20-30 Method of scribing wall panel at an irregular joint.

If you use a hand saw, cut the panels face-up. If you use power tools such as a portable circular saw or a saber saw, you must cut the panel face-down to prevent splintering.

Fastening: Panel edges must join over a stud when nailing. Space 1-1/4" nails 6 inches apart at the edges and 12 inches apart elsewhere. Over a horizontal furring edge, place the nails 8 inches apart and intermediate nails every 16 inches. Over backer board, space 2" nails 4 inches at the edges and 6 inches elsewhere. Drive the nails with great care. Do not allow your hammer to strike the panel surface.

For adhesive application, you must cut and fit the panels very carefully. Remember, once in place they are difficult to adjust or remove. On stud or furred walls, apply adhesive with a caulking gun, Fig. 20-31. Manufacturers give detailed instructions for the correct use of their products; follow them exactly.

Fig. 20-32 shows several methods of finishing corners, ceilings and floors. Matching moldings, nails and putty are fabricated for the various panels. Some moldings are shown in Fig. 20-33.

Fig. 20-31 Panel adhesive, applied here with a caulking gun, may be used in place of nails.

Interior Wall Finishing

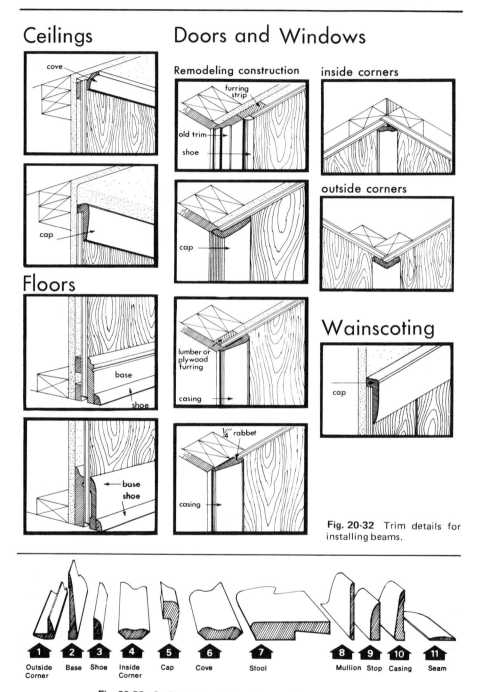

Ceilings

cove

cap

Floors

base

shoe

base
shoe

Doors and Windows

Remodeling construction

furring
strip

old trim

shoe

cap

lumber or
plywood
furring

casing

¼" rabbet

casing

inside corners

outside corners

Wainscoting

cap

Fig. 20-32 Trim details for
installing beams.

1	2	3	4	5	6	7	8	9	10	11
Outside Corner	Base	Shoe	Inside Corner	Cap	Cove	Stool	Mullion	Stop	Casing	Seam

Fig. 20-33 Prefinished moldings for use with wood paneling.

Hardboard Panels

Decorative hardboard panels are made to resemble wood, marble, stone and many other materials. They are printed, embossed and laminated so skillfully that it is often difficult to distinguish them from the "real" product. Fig. 20-34 is an example of hardboard made to resemble eighteenth century hand-carved wood blocks.

Hardboard panels are applied in much the same manner as wood panels. Most wood grains have random grooves which simplifies installation. V-joints are butted; square-edge panels may be concealed with moldings or battens. If it is necessary to nail into a panel surface, use color-matched nails. Adhesive applications are similar to those for wood.

Some hardboard panels have tongue-and-groove edges. Others are slotted to engage metal strip fasteners. In both cases, face nailing is entirely eliminated. Use special nails to fasten the metal strips to furring.

Install tongue-and-groove panels over solid backing with adhesives. Over open studding or furring, use special clips. They are not visible when properly installed.

If you use hardboard below the site grade, humidity may be a problem. Certain precautions are necessary. Waterproof all walls and insulate the outside walls as well. In the summertime, the use of dehumidifying equipment by the homeowner is recommended.

Solid-Wood Paneling

Solid-wood paneling is a distinctive wall covering which may be installed horizontally, vertically or diagonally; Fig. 20-35 shows an effective use of patterned panels on wall and cabinets. Panels are made with smooth or rough surfaces and with various edge treatments. Boards with surface defects such as wormy chestnut, knotty pine and pecky cypress are very popular today. Likewise, the rough-sawn boards once used principally for exterior walls are now used for interior wall coverings as well.

Use only well-seasoned panels. For horizontal application, nail the boards directly to studs. For vertical installations, use furring strips as with prefinished paneling. The panels must also be left unwrapped for a few days until their moisture content is in equilibrium with that of the surroundings.

Interior Wall Finishing

Fig. 20-34 Hardboard panels made to resemble carved wood blocks. *(Masonite Corp.)*

Fig. 20-35 Solid wood paneling applied to walls and cabinets *(Western Wood Products Assoc.)*

Tongue-and-groove patterns under 6" wide are blind-nailed to eliminate the need for countersinking and give a smooth blemish-free surface. This is especially important when natural-finish wood panels are applied. If face nailing is used, countersink the nail heads and fill them with wood putty. Figure 20-36 shows how tongue-and-groove boards are installed on horizontal furring.

If paneling is to be installed on an outside wall, you will have to lay down a vapor barrier.

Herringbone Paneling: Tongue-and-groove panels can be installed diagonally to form a herringbone pattern. Because the joint is exposed, you must do the work with great care. Cut the boards at 45-degree angles and install them with tongue up. Fasten the boards by blind-nailing into the ends as shown in Fig. 20-37.

Fig. 20-36 Applying redwood lumber strips to horizontal furring.

Fig. 20-37 Blind-nailing ends of diagonal panels. *(Western Wood Prod. Assoc.)*

Other Wall Covers

Walls can be covered with *artificial stone, bricks,* and *brick panels* which are exact reproductions of the real products. Because of their light weight they can be used without concern for floor loads. Each such brick weighs only 4 ounces and is 1/4" thick. These bricks can be applied to any flat interior surface with a special adhesive mortar. It is troweled onto the surface to be covered and the brick is simply pressed into place.

Another realistic brick is available in 12" X 28-3/4" panels. They have interlocking ends and are installed with nails on any flat surface. These lightweight panels are made with crushed stone imbedded in a fiberglass backing. Do the grouting with a caulking gun.

Plastic laminates are also used for wall finishing. Very durable, they resist scratching and denting far better than wood or hardboard. For wall installations, you must use backed laminates.

Cedar shingles add a decorative rustic accent when used indoors. They are practical in bathrooms since they are unaffected by moisture. Install them as you would for external coverings, except that waterproofing building paper is not necessary. A backer board of 3/8" or 1/2" plywood serves as a nailing base. Use two nails in each shingle. See Fig. 20-38.

Fig. 20-38 Applying red cedar shingles to an interior wall.

Interior Wall Finishing

Ceiling Tiles

In addition to covering exposed ceiling joists and beams, ceiling tiles serve other purposes. They help solve noise problems, can retard the spread of fire, and provide an effective way to decorate old and new room ceilings.

Many sizes, styles and finishes are made; some acoustical, others strictly decorative. For residential construction, the 12" × 12" square tile with tongue-and-groove is very popular. Others available include the 12" × 24", 24" × 24", and 24" × 48" sizes. Thicknesses range from 1/2" to 3/4".

All tiles deaden sound to some degree. To be considered acoustical, however, a tile must absorb at least 50 percent of the sound striking it. When you choose any ceiling tile, be sure to consider its primary function.

Most tiles are made to fit together with a V-groove seam, Fig. 20-39. However, they are also made with a square edge to eliminate the "block" appearance.

The tiles' surfaces are also important. For instance, tiles to be used in a kitchen should be chosen for washability and grease-resistance. Such tiles are usually plastic-coated today. See Fig. 20-40.

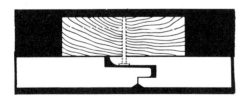

Fig. 20-39 Cross-section view of ceiling tile with V-groove seam.

Fig. 20-40 **right:** Plastic-coated tiles applied to a kitchen ceiling. *(The Celotex Corp.)*

Many methods have been worked out for installing tiles. They can be applied directly to a flat ceiling or to furring, or they may be suspended. They may be held with cement or staples, or they may be dropped into place in an open grid system.

Tiles may be cemented directly to a flat level ceiling. If the ceiling is in poor condition however, it is best to apply the tiles to furring strips which are nailed to the ceiling joists. That provides a solid base for stapling the tiles.

Regardless of the application method, lay out the ceiling tiles so that the border tiles on opposite sides of the room are of equal width. This improves the appearance of the ceiling. To determine the width of the border tiles, measure one of the short walls of the room. If this measurement is not an exact number of feet and you are installing 12" square tiles, add 12" to the inches over the exact footage. Then divide that total number of inches by 2 to obtain the width of the border tile needed for the long wall.

Problem: Find the border tile width for the longer wall, where the shorter wall measures 10'4" and you are installing 12" × 12" tiles.

Solution: For a 10'4" wall there are 4 extra inches.

$$12" + 4" = 16"$$

$$\frac{16"}{2} = 8"$$

Answer: Border tile width for the longer wall should be 8 inches.

Use the same procedure to figure the size of the border tiles for the shorter walls. *Note:* If you are installing 24" × 24" tiles, you add 24" to any extra inches of the wall measurement and divide the sum by 2 to get the size of border tile.

Installing Furring Strips: If furring strips are to be used, you must place the first two strips accurately in order to properly align the border tiles. Place the first furring strip against the wall at right angles to the ceiling joists. Fasten it with two 8d nails at each joist location. Then place the second furring strip parallel to the first strip and space it so that its center is the width of a border tile away from the wall. Install succeeding furring strips by spacing them 12" OC (for 12-inch square tile), until the entire ceiling is covered. *Note*: For 24" square tiles, the furring strips would be 24" OC, etc.

After all furring strips have been installed, check them for levelness, using a carpenter's level. If necessary, install shims, as shown in Fig.

20-41. Because of the unevenness of joists, some shimming is usually necessary.

If pipes, wires or other obstructions are located below the ceiling, it will be necessary to box them in. If they project less than 1-1/2" below the ceiling, a double row of furring may be used to eliminate projections altogether. Apply the first course of furring directly to the center of each joist. Then apply the second course perpendicular to the first, spaced at correct intervals for the ceiling tiles.

Before tiles are installed, snap two chalk lines as reference lines to align the first row of ceiling tiles. Since most walls and ceilings are not perfectly level and square, these chalk lines are necessary for proper installation of all square or rectangular tiles. Make sure that your chalk lines are at right angles to each other; one line parallel to the long wall, the other parallel to the adjacent shorter wall. Space the chalk lines away from the walls at a distance equal to the width of the border tile plus 1/2" (to allow for flange).

Fig. 20-41 Wedges are inserted to correct high spots between furring and joists.

Installing Tiles: You must cut border tiles individually. Measure carefully as explained previously in this chapter, then remove the tongue edge for the first row of tiles. This leaves the wide flanges for stapling. Cut the tiles either with a coping saw or sharp knife. When you cut tiles with a knife, use a straight edge as a guide and be sure to keep your fingers well away from the knife blade. Always cut tiles from the face side—except when you use a saber saw. With a saber saw, cut the tiles from the back side.

Most likely you will have to cut the first corner tile on two edges, as in Fig. 20-42. Use 9/16" staples to attach each tile, Fig. 20-43. Use three staples on each tile if you are installing 12" × 12" tiles, and more staples if the tiles are larger. Install the tiles snugly so there is no gap between them.

Most ceiling tiles are fragile. Use care in handling them and be sure not to mar their edges with the staple gun.

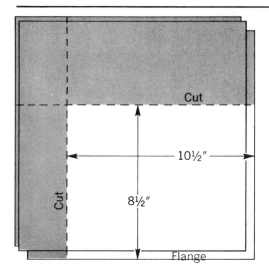

Fig. 20-42 above: Tongue edge is removed if border tiles must be cut.

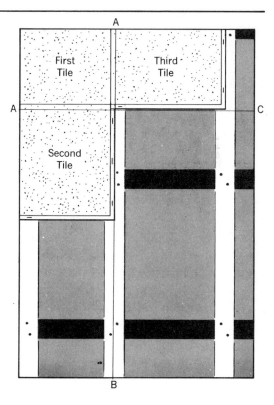

Fig. 20-43 right: Stapling pattern for installation of 12" ceiling tiles.

Interior Wall Finishing

When using elongated tiles, as in Fig. 20-44, be sure to stagger the joints. Drive in a staple at each furring strip in this type installation.

Cementing Tiles: You may use adhesive mastic to install ceiling tiles to any fairly flat and reasonably sound ceiling. Place daubs of mastic on the back of the tile at its four corners and center, Fig. 20-45. Then position the tile on the ceiling and press it firmly in place. Place two staples in each exposed flange of the tile to hold it in place while the cement sets.

Fig. 20-44 Fastening ceiling plank tiles so joints are staggered.

Fig. 20-45 Daubing cement on back of ceiling tile at corners and center.

Metal Furring Channels: A new development in ceiling tile installation replaces the wood furring strips with a self-leveling metal channel. These metal channels eliminate warping that can sometimes occur with wood furring. Another advantage you have with metal channels is ease and rapidity of installation. You can also use these channels on drywall or exposed joists. There are three basic steps: nail molding to all four walls, 2" below the level of the existing ceiling; then install metal furring channels perpendicular to the direction of the joists and nail them directly to the joists; finally, install the ceiling tiles, using metal crosstees for additional support, Figs. 20-46 through 20-48.

Suspended Tile Ceilings: Suspended ceilings may be used to lower a high ceiling or to conceal overhead piping, wiring, and heating ducts. One great advantage of suspended ceilings is that the area directly above the tiles is readily accessible, for repair or maintenance work. The tiles merely rest on a metal grid and may be raised to gain access to the area above.

Fig. 20-46 Furring channels are installed perpendicular to the ceiling joists.

Fig. 20-47 Crosstee is slid along the channel so that it engages edge of tile.

Fig. 20-48 Self-leveling metal channel system which can be used with either rectangular or square ceiling tiles.

Using rectangular tiles, follow these basic steps to install a suspended ceiling. See Figs. 20-49 and 20-50: first nail metal molding on all walls at the desired ceiling height. (This molding will support the tiles around the perimeter of the room.) Then attach hanger wires to the joists at 4-foot intervals and fasten the main runners to these hanger wires. Then snap crosstees into place between the main runners. Finally lift up the ceiling panels and drop them into place in the grid.

Fig. 20-49 Hanger wires are attached to the ceiling joists at 4-foot intervals for a suspended ceiling system.

Fig. 20-50 Ceiling panels are dropped into place in the grid formed by the main runners and crosstees. (*Armstrong Cork Co.*)

Recessed lighting units can be simply installed by eliminating several ceiling tiles and substituting translucent plastic panels in the grid.

GLOSSARY

crosstee: a crosspiece used between the main runners of a suspended ceiling system.

drywall: an interior wall covering of gypsum board or plywood; generally refers to gypsum board.

floating angle: a system used in gypsum wallboard installation to prevent cracks at wall and ceiling joints.

gypsum: a pure white mineral used for making plaster.

hanger wire: steel wire used to hold the framework of a suspended ceiling.

lath: a plaster base made of wood, metal, gypsum or insulating board.

plaster: a combination of lime, sand and water used for wall and ceiling finishing.

plasterboard: a rigid board made of plaster sandwiched between paper coverings.

plaster grounds: wood guides placed around doors, windows and wall base.

runners: longitudinal supports for suspended ceiling tiles.

scratch coat: the first coat of plaster applied to a wall; its surface is scratched to provide a grip for the following coat.

wallboard: a wall covering made of either gypsum, wood fibers or other materials.

21

Interior Doors & Windows and Room Trim

The two basic types of interior doors are flush and panel. For both the standard thickness is 1-3/8'' but novelty doors like the cafe door shown in Fig. 21-1 are 1-1/8'' thick.

Doors should only be ordered for delivery when the carpenter is likely to be ready to install them. The doors must be handled and stored carefully until actual installation. Common-sense precautions will ensure that they remain in tip-top condition until you are ready to hang and finish them.

It is important that you store doors flat on a level surface to prevent warping. You should store them in a clean, dry room; never in moist or freshly plastered areas. Wear gloves when you do handle the doors to keep dirty fingerprints off them; and always try to carry them from one location to another. If it is absolutely necessary to drag very heavy doors, do so only after you place a scuff strip at the bottom end for protection.

Door Frames

Interior door frames and trim are usually installed after the finish floor is in place. If the framing is done first, be sure to allow for the finish floor thickness. Regardless of which is done first, the carpenter should always work very carefully since any tools he might drop onto a finished floor could easily damage it. Likewise, door frame edges can be easily dented by a careless carpenter who is installing a floor.

The door frame which forms the finish opening of the whole doorway is sometimes referred to as the door jamb. But, the frame actually consists of a head jamb and two vertical side jambs. Its parts are shown in Fig. 21-2. They form the support for the door, door stops

Interior Doors & Windows and Room Trim

Fig. 21-1 Cafe doors installed between kitchen and dining room.

and casing. Interior jambs are made of 3/4" lumber and in widths to accommodate different wall thicknesses. They are also constructed as two- and three-piece units, as shown in Fig. 21-3. They can then be adjusted to fit various wall thicknesses. Jambs for plaster walls are 5-3/8" wide. For 2 × 4 walls with 1/2" drywall finish, the jamb is 4-1/4" wide.

Door jambs are manufactured with the side jambs dadoed at the top to receive the head jamb. Edges are usually given a 1/16" bevel so that the casing forms a good tight joint at the edge. Kerf cuts are often made to lessen the possibility of warping. Not all jambs are kerfed, however.

Door frames are usually shipped knocked down (KD) and are assembled on the job. If it is necessary to cut the jambs of KD units to fit the rough opening, cut them before assembly. Door frames are also available pre-assembled.

Frame openings are usually made 3/16" wider than the door width

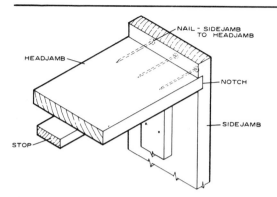

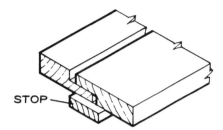

Fig. 21-2 Parts of framing for an interior door.

Fig. 21-3 Details of three-piece interior jamb.

and from 1/2" to 1" longer than the door height. The extra clearance at the floor line is needed for carpeting. See Fig. 21-4.

Connect the side jambs to the head, using three 7d coated nails per side. Place the assembled frame in the opening either by nailing temporarily or by using wedges between the studs and jamb. Make sure that the header is level. If it is not, measure the amount to be cut off and carefully remove the frame from the opening. Trim the bottom of the too-high piece, then replace the whole frame in the opening. Next insert wedges at top and bottom and add temporary braces to the frame. See Fig. 21-5. Use a straight edge to check the side jambs to learn where to add additional wedges. Place them at the hinge and latch levels. Then fasten the jamb to the rough framing with two 8d finish nails. Be sure to drive the nails into the jamb through the wedges and on into the studs. Hammer in the nails where the door stop molding will later cover them over.

You can cut the door stop moldings and install them temporarily at this time. Miter them at the top if the stop has a molded edge. If the stops are square-edged install them with butt joints.

Door and Window Trims

The trim used around the tops and sides of door and window frames is called casing. Made in various shapes and sizes, casings are hollow-backed so they will lie flat. The casing covers the joint between

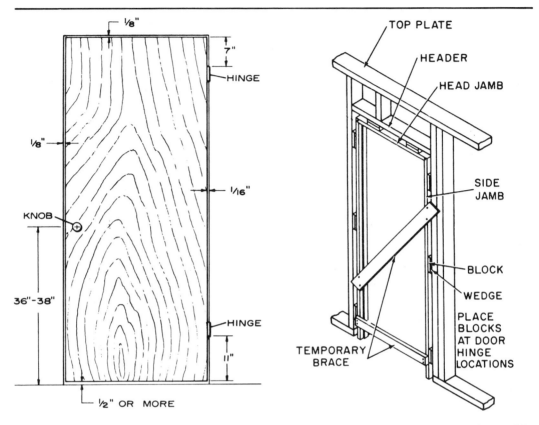

Fig. 21-4 Typical door clearances. *(Forest Products Lab).*

Fig. 21-5 Door frame installation details, here wall is not covered.

the whole door jamb and the rough framing and adds rigidity to the frame. Usually it is installed before the doors and windows are hung, but it may be installed afterwards. The casing is set back 1/8" to 1/4" from the edge of the jamb. If you use shaped casing, you must miter the top joints, as in Fig. 21-6. Before installing any casing, be sure to ease the corners of the door jamb with sandpaper so there won't be any sharp edges.

To simplify trim installation, draw a light pencil line on the edge of the jamb so it corresponds to the edge of the casing. Cut the bottoms of the side casings square. Then take one piece and hold it in place on the jamb and mark it for the miter. Cut the 45-degree miter. Then nail it temporarily in place, using the pencil line on the jamb as a guide.

Miter one end of the head casing, then hold it against the casing just installed. Mark the opposite end, again miter and install. Hold the second side casing in place, mark and cut. Your pieces should fit perfectly. If not, make the necessary adjustments and nail permanently in place. Use 4d finish nails at the jamb edge and 8d nails along the outer edge. Space the nails 16" OC (on center).

Many builders use pre-cut door casings. They are factory mitered and assure perfect joints. See Fig. 21-7.

Fig. 21-6 Mitered casing being installed.

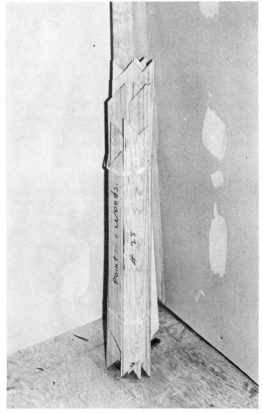

Fig. 21-7 What is wrong with the way this bundle of pre-cut casings is stored?

Pre-hung Doors

Many manufacturers make pre-hung doors, some with factory-applied finishes. They can ordinarily be installed complete in less than an hour. Fig. 21-8 shows the jamb detail of one type with a vinyl surface. Its split jamb allows adjustment for walls from 4-1/4" to

5-3/8" thick. The first installation steps are shown in Fig. 21-9. Doors are made 1-3/8" and 1-3/4" thick in both the hollow and solid core styles.

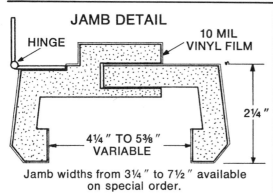

JAMB DETAIL
HINGE

10 MIL VINYL FILM

2¼"

4¼" TO 5⅜" VARIABLE

Jamb widths from 3¼" to 7½" available on special order.

Fig. 21-8 Detail of pre-hung door jamb.

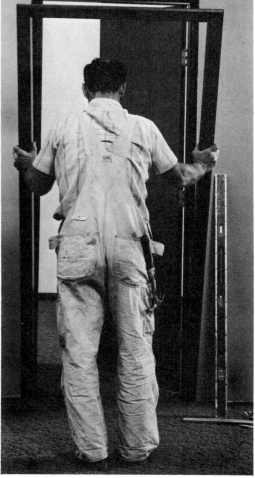

Fig. 21-9 right: Jamb/casing is fitted to frame.

Interior Door Types

Among the many interior doors used in residential construction are swinging, sliding, folding and by-passing types. They are also made in various materials and sizes.

Flush Doors

Flush doors may be made with either a solid or hollow core and with wood, plastic or hardboard faces.

Hollow core doors are widely used in residential construction because of their low cost and superior performance. Light in weight, they are strong and durable nevertheless. They weigh about one-third as much as solid doors and are warp resistant. One called the ladder-core is made up with an interior grid of struts. These 1/2" wood struts are set at a slight angle 6" apart and support the face panels. Paper is also used to form grids. Several core designs are shown in Fig. 21-10. All hollow doors have lock blocks which are at least 4" wide and 20" long. They are located midpoint on each side of the door.

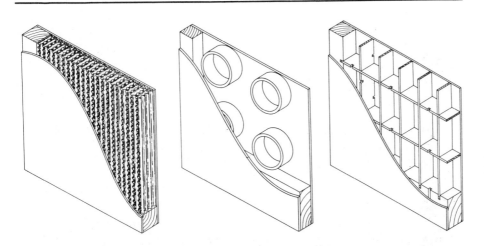

Fig. 21-10 Typical hollow door cores: **left,** honeycomb; **center,** tube; **right,** lattice.

Hollow core doors to be used for exteriors must be made with waterproof adhesives. The glue used for interior doors must be at least water resistant to prevent damage from dampness.

Solid core doors are usually made with narrow wood blocks of random length. Other core materials are particleboard, mineral board, and other compositions. Both hollow and solid core flush doors are usually covered with two or more cross-banded wood veneers. The outer veneer always runs vertically.

Interior Doors & Windows and Room Trim

Panel Doors

Panel doors consist of stiles and rails which form sections or divisions called panels that are raised above or lowered below the door's frame. Usually only decorative, the panels are often made of wood or hardboard. If some of a door's panels are made of glass and admit light and/or air, it is referred to as a sash door. When all the door's panels are glass, it is called a casement or French door, Fig. 21-11.

Panel doors are made in many styles and sizes for both interior and exterior use. The principal parts of a panel door are shown in Fig. 21-12. The cross-section through a panel door is shown in Fig. 21-13.

Fig. 21-11. French doors with frame made of ponderosa pine.

Standard Door Sizes

Interior passage doors are generally 1-3/8" thick. Widths range from 1'6" up to 3'0". Heights of interior and exterior doors are generally 6'8" but 7'0" doors are also available. FHA minimum standards require 2'0" widths for bathrooms and 2'6" widths for bedrooms.

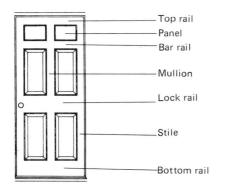

Fig. 21-12 Principal parts of an interior door.

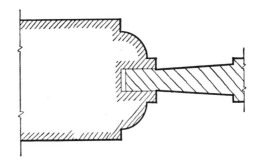

Fig. 21-13 Cross-section through a panel door.

Installing Doors

Exterior doors are quite heavy and usually require three hinges, while lighter interior doors can be hung with two. Your first step in hanging a door is to fit it correctly. Trim the door to fit the opening with the following allowances: 1/16" on the hinge side, 1/8" at the top and lock side and from 1/2" to 1" at the bottom. Install doorstops temporarily at this time.

You must make a three-degree bevel on the lock side to permit the door to swing without striking the jamb, Fig. 21-14.

Do the trimming with either a hand or power plane. Be sure to hold the door firmly while trimming. Special door jacks are made for this purpose. Or you can easily make simple wood holders on the job. These consist of a horizontal cross member nailed to a vertical board which contains a long V notch. One end of the door is placed in the notch. The other end rests on the floor. Some carpenters like to use two such holders, one at each end.

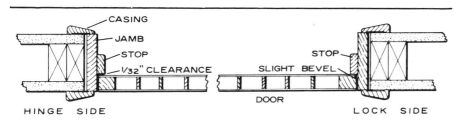

Fig. 21-14 Clearance specifications for hanging an interior door. *(Forest Products Lab.)*

After you fit the door, you can round the edges slightly. However some carpenters prefer to leave this rounding until after the hinge grooves have been mortised—especially if the gain or notching is laid out and cut by hand.

Hinges: Loose-pin butt hinges are used on residential doors. For 1-3/4" doors 4" × 4" hinges are recommended. For 1-3/8" doors, use 3-1/2" × 3-1/2" butts. Unless the architect's plans call for specific hinge location, place them 7" from the top edge of the door and 11" from the bottom.

Butts may be mortised by hand. They can also be cut with a router using a hinge template. The use of router and template cut down installation time considerably; and the templates can be adjusted to fit most door sizes. If you use a template, follow the manufacturer's instructions.

When you cut the gain by hand, follow this procedure: Separate the two halves of the hinge by removing the loose pin. Using the half which fastens to the door itself mark its outline on the door's stile with a 3/16" to 1/4" setback, as in Fig. 21-15. Next trace the outline of the hinge with a sharp knife or scriber. Or you may use a butt gauge instead. This tool is designed to locate a hinge outline accurately. After you cut in the gain, check the fit of the hinge. It should countersink so as to rest flat and be flush with the back edge of the door. Install the screws carefully.

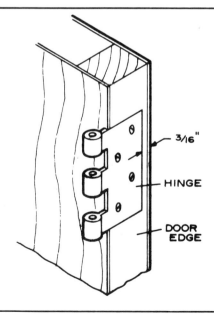

Fig. 21-15 Details of setback for door hinge.

Place the door in the opening and block the bottom to obtain the proper top clearance. Mark the hinge locations on the jamb, then mortise for the second half or frame-mounted side of the hinge.

If you install the hinges correctly, the door should fit with proper clearances all around. Some adjustment can be made by using shims or by deepening the gains, of course.

Lock Installation: Some door locks must be chosen according to the hand of the door. The hand of the door is always determined by standing outside and facing the door. If the hinge is at the right and the door swings inward, it is a right-handed door (RH). If the hinge is on the left and the door swings inward, it is a left-handed door (LH).

If the hinge is at the right and the door swings outward, it is a right-handed reverse (RHR); if the hinge is at the left and the door swings outward, it is a left-handed-reverse (LHR). See Fig. 21-16.

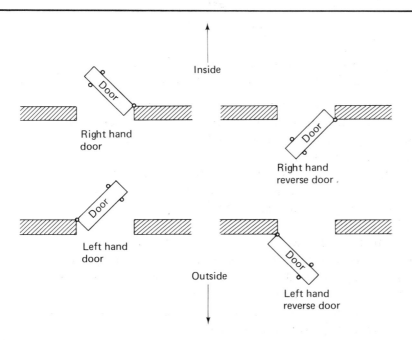

Fig. 21-16 Determining "hand" and swing of doors before installation of locks. (Observer on outside.)

Cylinder locks are generally used in residential construction. They are easy to install and ordinarily provide ample security. Fig. 21-17 shows a typical cylinder lock. To install this type of lock you have to bore two holes: one in the face of the door, the other in the edge.

Special boring jigs and faceplate markers are useful for installing

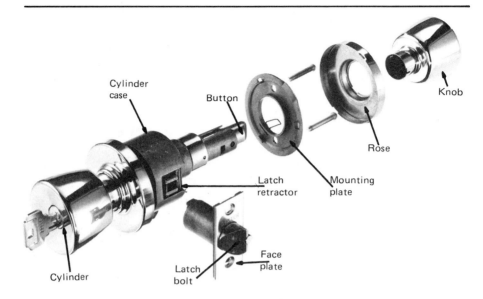

Fig. 21-17 Parts of a typical cylinder lock. *(Dexter Industries Inc.)*

locks, Fig. 21-18. They locate and guide the drill bit accurately, eliminating the chance of error. Most jigs are made for use with power tools; however, they can be used with a hand brace. The face plate markers are used by hand.

Lock sets are furnished with complete instructions and templates. But if you use a boring jig, you will not need the templates. Fig. 21-19 shows a latch bolt hole being bored.

Mark a line on the lock edge of the door, 36" above the floor. Fold the template and place it at the edge of the door. Align it with the mark on the door, then mark the centers for the holes and carefully drill them. Bore into the door for the large cross hole until the point of the bit breaks through, then withdraw the bit and complete the hole from the other side. This eliminates splintering of the wood. Finally bore the edge hole through to the cross hole.

Insert the latch unit, mark the outline, then mortise. A faceplate mortise marker may be used for this purpose. After you finish the mortising, attach the latch with screws. The lock case can now be installed. Then locate the height of the strike plate on the door jamb using the lock latch as a guide. Next mortise the jamb and attach the strike with screws.

Doorstops may now be installed permanently. With the door closed, place the lock-side stop tightly against the door and nail with 1-1/2" finishing nails. Use two nails every 16 inches. See Fig. 21-20.

Install the hinge stop next. Place it about 1/32" away from the door so it won't rub against the door. Lastly, install the top or head stop.

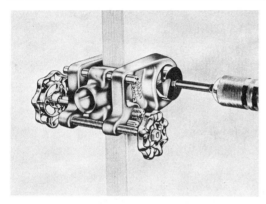

Fig. 21-18 A boring jig cuts down on installation time and assures accuracy in drilling. *(Dexter Industries, Inc.)*

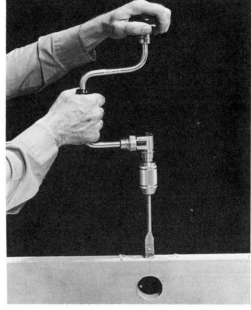

Fig. 21-19 Boring the latch bolt hole in the edge of a door.

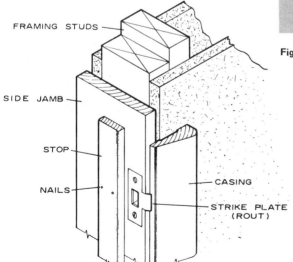

FRAMING STUDS

SIDE JAMB

STOP

NAILS

CASING

STRIKE PLATE (ROUT)

Fig. 21-20 Details for installation of a door-stop.

Interior Doors & Windows and Room Trim

Pocket Doors

Sliding pocket doors are used where swing space is a problem since they slide into partition openings. Made in various sizes, pocket doors come complete with door frame, stops, track, guides and hardware. They are made as single- and two-door units. The carpenter installs their framework during rough framing.

Most pocket door units are factory assembled with split jambs. Both door and RO sizes are shown in Fig. 21-21. Adjustable hangers permit the doors to be raised so cutting at the floor line is not necessary.

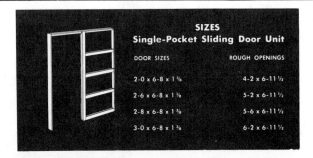

SIZES
Single-Pocket Sliding Door Unit

DOOR SIZES	ROUGH OPENINGS
2-0 x 6-8 x 1 ⅜	4-2 x 6-11 ½
2-6 x 6-8 x 1 ⅜	5-2 x 6-11 ½
2-8 x 6-8 x 1 ⅜	5-6 x 6-11 ½
3-0 x 6-8 x 1 ⅜	6-2 x 6-11 ½

SIZES
Twin-Pocket Sliding Door Unit

DOOR SIZES	FRAME SIZES	ROUGH OPENINGS
2-0 x 6-8 x 1 ⅜	4-0 x 6-8	8-2 ½ x 6-11 ½
2-6 x 6-8 x 1 ⅜	5-0 x 6-8	10-2 ½ x 6-11 ½
2-8 x 6-8 x 1 ⅜	5-4 x 6-8	10-10 ½ x 6-11 ½

Fig. 21-21 Door and opening sizes for single- and twin-pocket door units.

Folding Doors

Folding or bi-fold doors hang on overhead tracks and slide on rollers or glides. They operate in pairs and are generally used for closets, wardrobes, and room dividers. The overhead track permits them to swing flat against the door frame.

Folding doors are installed on jambs with hinges, usually three per side. The weight of the doors is supported by hinges and pivots. Fig. 21-22 illustrates the hardware mounting used in a typical unit. Self-lubricating bushings allow the doors to operate smoothly. The doors are made in a number of sizes and styles in metal, plastic or wood.

Accordion folding doors are made of flexible material, usually vinyl, and close into a series of accordion-like pleats. The folds are maintained by full slats of steel, fiberboard or thin wood, and are suspended from an overhead track. These doors may be used singly or in pairs and come in widths of 32", 36", 48" and up. They are relatively inexpensive, take up very little floor space, and are excellent where room space is at a minimum.

Fig. 21-22 Installation of pivot hardware for folding door.

Sliding Doors

Sliding doors are suspended on overhead tracks and operate on rollers. The overhead tracks may be surface mounted or recessed. Floor guides keep the doors in alignment. Some manufacturers furnish frame units consisting of side jambs and header assembly with the track already installed. Standard doors are 3/4" and 1-3/8" thick but those 1-3/4" thick can be special ordered. The advantage of sliding doors is that they eliminate door swings; a disadvantage is that they do not give full access to the opening.

Interior Doors & Windows and Room Trim

Finish Trim

Various trim members are used in finishing the interior of a house. Applied when all other phases of construction are completed, trim is used to cover seams and joints around door and window openings as well as at floors, ceilings and corners. It is also used in other areas for decorative purposes.

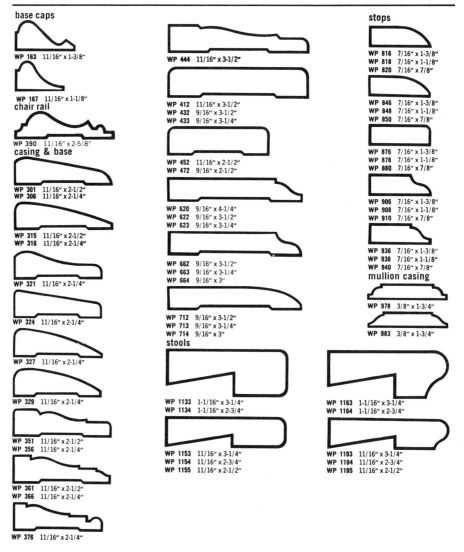

Fig. 21-23 Typical molding shapes. *(Western Wood Moulding & Millwork Producers)*

Moldings used for trim can range from the simple to the very ornate. They are produced prefinished or unfinished and in wood, plaster, composition, metal, and plastics. Some prefinished moldings are stained or painted, others are clad in vinyl. Some of the more common shapes are illustrated in Fig. 21-23.

At one time moldings were made by hand. Now they are manufactured with high speed equipment in modern factories. A cutterhead used to make fingercuts for wood moldings is shown in Fig. 21-24.

Fig. 21-24 Cutterhead used for making finger cuts in wood molding.

Interior Window Trim

Interior window trim includes stool, casings, mullions, aprons and stops, Fig. 21-25. The window casing in a room should match that used for the doors. The stool and apron are not used in one method of trimming a window. Instead, casing alone is used for all trim, as in Fig. 21-26. The trim is generally installed in this order: stool, side casing, mullion (for multiple windows), head casing, apron and stops.

The bottom of the stool is rabbeted and rests on the sill. Viewed from the top, the front edge of the stool extends to the sash. The ends are notched so that they bear against the wall.

Hold the stool centered and level against the sill, then carefully mark the window opening (side jambs). Extend these lines with a

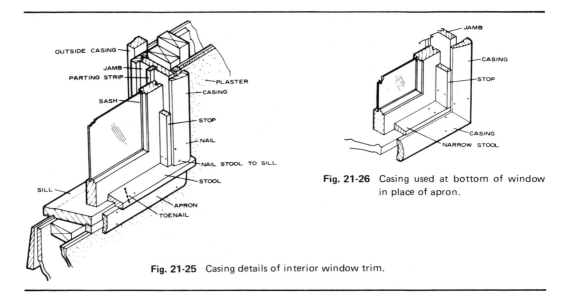

Fig. 21-26 Casing used at bottom of window in place of apron.

Fig. 21-25 Casing details of interior window trim.

square. Also mark the depth of the notch; it will equal the measurement from the sash to the face of the wall less 1/16". The 1/16" setback allows a clearance between the window sash and the front edge of the stool.

Use a backsaw or other fine toothed saw to notch out the corners. Make sure that the piece fits snugly between the jambs. Cut the length of the stool so that the ends extend 3/4" beyond the side casings. The end of the stool is usually coped to conform to its shaped edge. Fasten the stool to the sill with 8d finishing nails. If it is required, as in masonry walls, bed the sill in caulking. Check to make sure the stool is level; then toe-nail the ends into the rough frame. Place the nails so their heads will be concealed by the side casing. To avoid splitting the stool at the ends, predrill the nail holes.

Install the side casing next. Cut the bottom edges square and miter the tops. The side casing is usually set even with the corner but it can be set back like the door casings. Check the miters carefully then nail. If mullions are used, fit and install them next.

Measure the distance between the side casings at the outside edges, then cut the apron to this size. On quality work the ends of the apron are coped to match its profile.

Add the stops last. To prevent binding, leave a slight space between the stop and the sash. The front edge of the stop should lap over the casing. Complete all nailing, then set the nail heads.

Baseboard Moldings

Molding for the base of a wall is one of the last items of trim installed in a room or hallway. Base molding is applied around the perimeter of the room at the intersection between walls and floor. A shoe molding is often used at the bottom of the base molding. It is flexible and conforms to any unevenness of the floor. If it is used, fit it but do not fasten it permanently until all surface finishes are applied or resilient tile is installed. It is not used with carpeting.

Baseboard molding is mitered on the outside corners and coped at all the inside corners. Cut the first piece of baseboard to fit from corner to corner or from corner to door casing. Be sure to cut the pieces so they fit snugly. If you must piece any sections, cut a scarf joint (diagonal cut); and be sure to locate it in front of a stud.

The second piece of baseboard at an inside corner is coped because that makes a better joint than a miter. To make the cope, saw a 45-degree miter at the end of the second piece. This should be an inside miter. The remaining profile is recut with a coping saw. A slight undercut assures a tight joint. Fig. 21-27 shows baseboard molding details.

Use nails long enough to penetrate into the wall frame. Place two nails at every stud location; and angle the lower one into the sole plate. Sink all nail heads. If a base shoe is used, you must miter it where it meets the door casing.

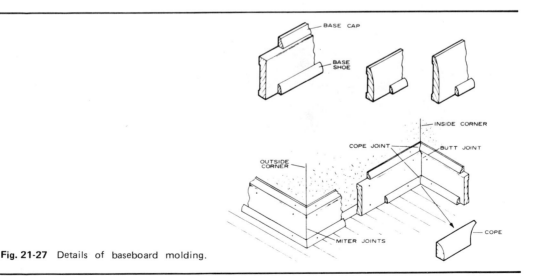

Fig. 21-27 Details of baseboard molding.

Ceiling Moldings

Ceiling molding is generally used when walls are paneled. It gives the room a finished appearance and covers irregularities between wall and ceiling. As in baseboard applications, ceiling moldings are coped and mitered. When you work with crown moldings, you must cut them in the same position as they are to appear on the wall. Some carpenters nail a temporary stop in the miter box to hold the molding at the proper angle.

Cut the molding at a 45-degree angle and then cope it. The profile serves as a guideline for the coping saw. The wedge cut is also cut at a 45-degree angle, Fig. 21-28. After you install all moldings, sink all nail heads and fill.

Fig. 21-28 Crown molding being coped with saw held at a 45-degree angle.

Installing Beams

Beams fabricated of plastic or wood are in wide use today as a decorative trim on walls and ceilings. The plastic types are usually molded foam. They are lightweight and have a realistic appearance.

Box type beams are generally used in remodeling work. Solid beams are commonly used in new construction. Fig. 21-29 shows a hollow beam used on a wall. Fig. 21-30 shows solid beams built on the job. Hand-hewn beams are easily shaped with an axe, Fig. 21-31. Be sure to keep your hands and feet well away from the axe when making these. The top edges should be relieved or beveled slightly. This makes it easier to refinish the beams when necessary.

Fasten the beams to the ceiling joists with nails or lag screws. If lag screws are used, set the heads below the surface. For a realistic effect you can use dummy pegs on the beams. Cut these pegs from 1" dowels and fasten with finishing nails driven through the center of the peg. Details of various installations are shown in Fig. 21-32.

Fig. 21-29 Hollow box beam applied to an interior wall.

Fig. 21-30 Hand-hewn beams with fake pegs.

Fig. 21-31 Shaping a solid beam with an axe.

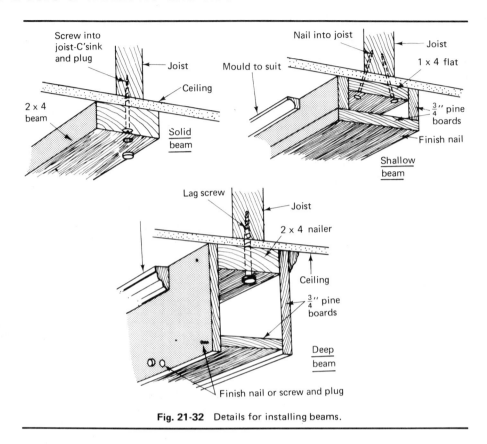

Fig. 21-32 Details for installing beams.

GLOSSARY

baseboard: flat trim nailed to the bottom of a wall just above the floor line.

butt gauge: a tool for marking the outline of hinges.

caulking: a flexible material used for making watertight joints.

coping: a method of cutting joints in molding.

flush door: in construction, a door with a smooth level surface.

gain: the mortising or notching of a cavity in a piece of wood so it can accept a hinge, hardware or another piece of wood.

hollow core door: a door made with a flush surface and with an interior grid which supports the faces.

kerfing: the cutting of grooves to permit bending; also, the longitudinal cuts made in millwork to prevent warping.

latch: a device for holding doors shut.

mortise: a hole, groove or slot cut into wood to receive a tenon or another piece.

panel: in house construction, a thin material such as wood or plywood framed with stiles and rails as in a door; also, a wall or ceiling covering of wood, plywood or other materials.

rail: the horizontal crosspiece members of a door or window; also, the upper (handrail) and lower sloping members of a balustrade.

scarf joint: a diagonal lap joint made by splicing two pieces of stock.

shim: a thin, often tapered strip of wood, metal or other material; commonly used for setting door and window frames.

solid core door: a door with a solid interior; usually made with wood blocks bonded together with end joints staggered.

stile: the vertical member of a door or window.

22

Finish Floor Coverings

Finish flooring is the final covering placed on a floor construction. Many materials are used, including asphalt, wood, linoleum, vinyl, asbestos, rubber, ceramics, fabrics, stone and cork—and many others. See Fig. 22-1.

Wood floors are made in strip, plank and block form. Other materials are made up in sheets or tiles. Recent developments in product and installation techniques have greatly increased the variety and choice of flooring. In the past, only asphalt tiles could be used in below-grade construction. Now many products, including wood, may be used on below-grade slabs.

The choice of a particular flooring depends on several factors. Appearance, comfort and cost are important. Function must also be considered. A wood floor is not practical in a bathroom, nor is a brick floor a wise choice for a bedroom. Hard-wearing floors should be used in areas of heavy traffic. Regardless of the type of flooring selected, you must install it properly and carefully so that it will be long lasting and attractive.

Install the finish floor only after walls and ceilings are completed. Do it, however, before you install door jambs, casings and floor trim.

Wood Floors

Wood is an excellent material for floors. Strong, durable and decorative, it will give years of trouble-free service if installed properly. The most common hardwoods used for floors are oak, maple, and birch. Some harder softwoods like Douglas fir, western larch and southern yellow pine are also used but they are less resistant to wear and abrasion. Vertical grained woods are the best for flooring since they shrink and swell less than flat-grained woods.

The wood flooring used for residential construction is available in strip, plank or block form. The most common thickness is 25/32-inches. For heavy-duty use, the 31/32, 41/32, and 53/32-inch

Fig. 22-1   Low maintenance vinyl floor does not require waxing.   *(Armstrong Cork Co.)*

thicknesses are recommended. For light duty on sound underfloors, 3/8-inch and 1/2-inch stock is quite satisfactory.

Strip flooring is made of narrow strips of hardwood or softwood in various thicknesses. For residential floors, 2-1/4" widths are generally used. It is usually side and end matched or tongue-and-groove and has a hollow or grooved back. Look at it in cross section and you will see that the face width of T&G flooring is slightly greater than its bottom width. This assures a very tight joint between the strips. If you plan to fasten it with mastic instead of nails, make sure that the backs are flat.

Strip flooring is also made with square edges. This type must be face nailed. Often used for budgetary reasons, it is cheaper than T&G stock and easier to install, Fig. 22-2.

Finish Floor Coverings

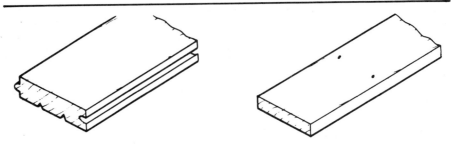

Fig. 22-2 Strip flooring: **left:** tongue and groove; **right:** square edge.

Plank flooring is similar to strip flooring. It resembles the floors of early America and is very popular today, especially in colonial-style homes. The planks are available in 3-1/2" to 8" random widths, with and without a bevel at the top edge. The plugs that conceal the installation screw heads add to the antique look since these plugs resemble the pegs used by the colonists to fasten the flooring to the subfloor. See Fig. 22-3.

Install plank flooring in a random pattern, end to end, and stagger the joints. Be sure to leave a 1/32" space between planks to allow for dimensional changes as seasonal weather changes occur.

Block flooring: There are several types to choose from. One consists of a unit block made of short lengths of strip flooring joined at the edges to form square blocks. Another is the laminated or plywood block, made by gluing several layers of wood to a desired thickness. Laminated blocks may have either tongue-and-groove edges or grooves all around. The grooved blocks are assembled with special metal splines. Both types of block flooring are installed with a mastic over wood or on concrete slab.

Slat blocks, a third type, are made up of thin slats of hardwood, generally assembled into basic squares. The slats measure 3/4" to 1-1/4" wide and 4" to 7" long and can be laid in many interesting patterns. Figure 22-4 lists oak flooring details.

Do not store finish flooring in the building unless the building is thoroughly dry. But the flooring must be on hand in the building at least three days before you install it. That gives the wood a chance to acquire the same on-site moisture content. Temperature is important too. Store and install wood flooring only in room temperatures ranging from 50° to 70° Fahrenheit.

You must have a subflooring in good condition in order to lay

Fig. 22-3 Plank flooring with pegs. *(E.L. Bruce & Co., Inc.)*

Finish Floor Coverings

Plain-Sawed	Quarter-Sawed
Clear	Clear
Select	Select
No. 1 Common	
No. 2 Common	
1¼ -Foot Shorts	

OAK FLOORING SIZES

	Thickness	Width
	Tongued and Grooved	
STRIP	Most popular: 25/32" Standard thicknesses: ½", ⅜" and 25/32"	Most popular: 2¼" Standard widths: 1½", 2", 2¼" and 3¼"
PLANK	25/32"	3" to 8"
UNIT BLOCK	½", 25/32" and 33/32"	Individual pieces are 1½", 2" and 2¼"
	Dimensions of unit block are in multiples of the widths of component pieces. Example: squares from pieces 2¼" wide are 6¾" x 6¾" or 9" x 9" or 11¼" x 11¼". Rectangles from pieces 2¼" wide are 6¾" x 12½".	
	Square Edged	
STRIP	5/16"	1⅓", 1½" and 2"

Fig. 22-4 Oak flooring: grades, sizes, and styles.

finish flooring well. If you use boards, select a softwood species of 1" stock 4" to 6" wide. Lay them diagonally with 1/4" space between the boards. Face nail with two 7d threaded or 8d common nails at every bearing.

If you use plywood subflooring, lay it with the outer grain plies at right angles to the joists, as shown in Fig. 22-5. Use 3d nails, 6" apart along every joist. Instead of nailing, you may glue the subfloor to the joists.

Fig. 22-5 Installation of tongue-and-groove underlayment plywood flooring. *(American Plywood Assoc.)*

Installing Strip Flooring

Strip flooring should be laid to run continuously from room to room. Plan it so that the strips run lengthwise in halls or other narrow spaces. It should also extend from one room to the next without a change of direction. You must be sure to place the starter strip very accurately. If you do not set it exactly parallel to the long wall it adjoins, it will throw off alignment of the flooring in the other rooms.

Scrape and sweep the subfloor clean. Be sure to remove all traces of plaster and hammer down any nails that may be protruding from the floor. Then cover the floor with a layer of waterproof building paper, with 4" lapped seams, Fig. 22-6. Felt is necessary on board subfloors but is not required on plywood subfloors. Locate the joists on the felt with chalklines. This simplifies nailing.

Nailing: Lay flooring at right angles to the floor joists with the starter strip parallel to the side wall. Place it so the edge groove faces the wall. But leave at least 1/2" expansion space between the wall and the edge of the strip, Fig. 22-7. The base molding covers this later.

Face nail the edge of the strip nearest the wall, locating these nails so they will be hidden when you apply the shoe molding. Blind-nail into the tongues at an angle of 45 to 55 degrees, Fig. 22-8. Refer to Fig. 22-9 for proper nail size and type. Special nailing devices are sometimes used to install flooring. Exercise great care when you are nailing to prevent damage to the strip edges. Drive the nails to a point where the nail heads are still slightly above the wood surface. Then complete the

Finish Floor Coverings

Fig. 22-6 Laying building paper over subfloor.

Fig. 22-7 Starter strip is placed 1/2" from the wall. *(National Oak Flooring Mfg. Assoc.)*

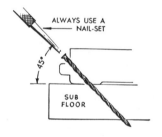

Fig. 22-8 Proper nailing angle for strip flooring.

ALWAYS USE A NAIL-SET

45°

SUB FLOOR

NAIL SCHEDULE

Flooring Size	Nail Size	Spacing
25/32" x 3¼" 25/32" x 2¼" 25/32" x 1½"	7d or 8d cut steel flooring or screw-type nails	10" to 12" on center
½" x 2½" ½" x 2" ½" x 1½"	5d or 6d cut steel flooring or screw-type nails	8" to 10" on center
⅜" x 2" ⅜" x 1½"	3d or 4d cut steel flooring or screw-type nails	6" to 8" on center
Machine driven barbed fasteners, of the size recommended by the manufacturers, are acceptable. Space as shown above.		

Fig. 22-9 **Right,** Specifications for nail sizes and OC spacing for strip floors.

1½" NO. 1 SPIRAL FLOOR SCREW-NAIL

4d CUT STEEL FLOOR NAIL

3d FINISHING NAIL

2¼" NO. 5 SPIRAL FLOOR SCREW-NAIL

6d CUT STEEL FLOOR NAIL

7d CUT STEEL FLOOR NAIL

8d CUT STEEL FLOOR NAIL

nailing with a nail set. Toenail the first nail in each strip toward the preceeding strip. Toed nails (slanted vertically and horizontally) ensure tight joints, Fig. 22-10. Strip flooring must be nailed with tight joints.

When tapping the end or edge of a strip, be sure to use a scrap piece of flooring as shown in Fig. 22-11. This prevents marring of the strip. If splitting is a problem, especially near the strip ends, you may have to predrill pilot holes. Use a drill bit the same size as the nail shank.

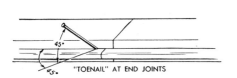

Fig. 22-10 To ensure tight joints, the first nail in each strip is toenailed toward the preceding piece.

Fig. 22-11 Using scrap to tap flooring strips into place and to tighten joints before nailing.

You must lay the secondary strips so the end joints are staggered at least 8" apart. When you cut strips for an end wall, use the leftover piece to start the next course at the opposite wall. Use very short leftovers in closets.

As you work on the floor, lay out the strips a few rows ahead of the nailing. This allows you to work faster and also gives you a chance to arrange the color blending and grain pattern of the wood flooring, Fig. 22-12.

The end joints of strip flooring are pulled tight with the aid of an "L" shaped tool to complete the work. Fig. 22-13.

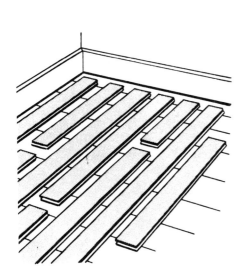

Fig. 22-12 Lay out strips before nailing in order to arrange color blending and grain patterns.

Fig. 22-13 Pulling flooring joints tight before nailing.

Installing Wood Floors Over Concrete

When you lay wood floors over concrete, you must take certain precautions. If the slab is on the ground, install a suitable vapor barrier to resist ground moisture and vapor. If a barrier was used under the slab when it was poured, another is not required. Likewise, slabs with an air space between them and the ground do not require a vapor barrier; however, air circulation should be provided.

There are several ways of waterproofing a slab before you install wood flooring. For instance, you may coat the entire slab first with an asphaltic mastic, then put a layer of polyethylene film over that, and then apply another coat of mastic. Next, you embed overlapped screeds (also called sleepers) in the top layer of mastic and finish up with still another coat of mastic. The screeds serve as your nailing base for the flooring.

A simpler method is shown in Fig. 22-14. Nail treated 1 × 4 wood strips to the slab 16" OC after you apply a waterproof coating. Next

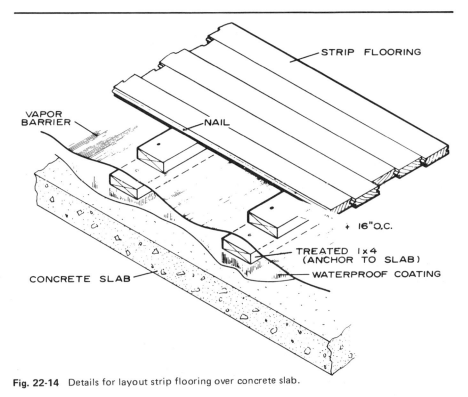

Fig. 22-14 Details for layout strip flooring over concrete slab.

place a polyethylene film over the 1 X 4's, then nail another set of 1 X 4's to the first series. These second strips are not treated; use 1-1/2" nails staggered 12" OC.

If strip flooring is to be installed, nail it directly to the wood strips. If other types of flooring are to be installed, you must nail a plywood base to the strips first. Figs. 22-15 through 22-19 show the steps in strip flooring installation.

Fig. 22-15 Application of flooring. Slab is swept clean and primed; when primer is dry, chalk lines are snapped 16" apart, then covered with mastic in 2" width strips. (Adhesive can be asphalt mastic or any suitable adhesive designed for bonding wood to concrete. If heating elements are in the slab, the adhesive must be a heat-resistant type.) (*National Oak Flooring Mfg. Assoc.*)

Fig. 22-16 Installation of bottom sleepers: Bottom sleepers should be made of 1 X 2 lumber treated with wood preservative; imbed this first course of strips in adhesive and secure to slab with 1-1/2" concrete nails, approximately 24 inches apart. (The 1 X 2's should be random lengths, laid end to end with slight space between ends—not butted together.)

Fig. 22-17 Application of polyethylene film: After all bottom sleepers have been installed, .004 polyethylene film is laid over the first course of strips; the polyethylene sheets are joined by lapping their edges over the sleepers.

Fig. 22-18 Laying the top-nailing sleepers: The second course of 1 x 2's do not have to be preservative treated and should be nailed with 4d nails, 16 to 24 inches apart; nails should go through top sleepers and polyethylene into bottom sleeper.

Finish Floor Coverings

Fig. 22-19 Installation of strip flooring: Install strip oak flooring at right angles to sleepers by blind-nailing to each sleeper; drive nails at angle of approximately 50 degrees, using either threaded or screw-type nails, cut nails or barbed fasteners. No two adjoining floor strips should break joints in same sleeper space; each strip must bear on at least one sleeper, and a minimum 1/2 inch clearance must be left between flooring and wall to allow for expansion.

Installing Block Floors

Unit and laminated block flooring are installed in a similar manner. Most can be nailed or set in mastic; however, some are made for mastic application only.

Generally block flooring requires an expansion clearance at the walls. But some laminated oak blocks do not expand noticeably and normally do not require any expansion allowance. Manufacturers literature will specify. When expansion strips are necessary, cork is ordinarily used at the walls. See Fig. 22-20.

When you install blocks on wood subfloors, nail them at the tongue edges. But the base must be prepared: first give it a thin coat of special mastic; then follow with a layer of 30-lb. asphalt-saturated felt; and finally apply another thin coat of mastic over the felt. You then lay the blocks in the top coat. For installations on concrete, either on or below grade, you must use a damp-proofing membrane in addition to the felt.

Self-Adhesive Blocks

Factory-applied adhesive strips are a fairly new development for block flooring application. Two pressure-sensitive adhesive strips are fixed to the backs of each block. You remove the protective covering on the strips just before you are ready to install the blocks. Then you position the blocks carefully and press them into place.

You can install these blocks over wood or concrete, as well as over old vinyl and asphalt tile floors. Make sure that all the floor surfaces are free of dust, wax, and moisture. If concrete floors are very dusty, you may have to seal them with a good grade of latex or acrylic primer.

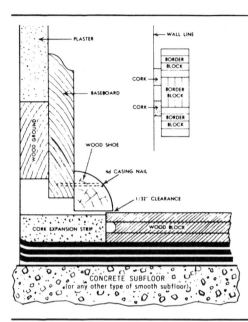

Fig. 22-20 Details for expansion allowance when installing block flooring. *(E.L. Bruce & Co., Inc.)*

Resilient Tiles

Resilient flooring comes in sheet form and in squares or tiles. Tiles vary in size but the most common for residential use are either 12" × 12" or 9" × 9" square. In new work the tiles are installed on one of the underlayment products described in detail below. If the tiles are installed over old finished wood floors as in remodeling, an asphalt-saturated felt underlayment must be used first to prevent the old flooring joints from showing through the tiles.

The range of resilient flooring materials is almost unlimited. Some require special underlayments, adhesives and methods of application. Always follow the manufacturer's recommendations in that regard. The technique for laying the tiles is fairly standard, however.

Underlayment for Resilient Flooring

Resilient flooring must not be laid directly on plank subfloors. To do so would invite trouble since any irregularities in the subfloor—knots, cracks, and other defects—would soon telegraph or show through to the surface. A suitable underlayment of plywood, hardboard, or particleboard is necessary to provide a smooth even surface for the flooring materials. It should be applied before you lay the finish floor.

Plywood underlayment is ideal. It is dimensionally stable, strong and warp resistant and can be used for tile, carpeting, linoleum and other nonstructural flooring. Not all plywood can be used. Use only the grade that is so stamped, certifying that the plywood is an underlayment grade. Figure 22-21 shows underlayment grade marks.

Fig. 22-21 Examples of underlayment grade marks.

Plywood underlayment is available from 3/8" to 1-1/8" thick and in 4'0" widths and in lengths of 8, 10 and 12 feet. It is made with exterior and interior glue lines. The grade chosen depends on its ultimate use. If floors are to be subject to unusual moisture conditions, you must use panels made with waterproof glue. When thin resilient flooring is to be laid, the underlayment panel must be fully sanded. Install the panels with 1/32" space to allow for expansion. Be sure to countersink all nails 1/16" before you lay the finish floor. For staples a 1/32" countersink suffices.

Hardboard underlayment is used like plywood—to provide a smooth surface for resilient coverings. You can use it either over interior wood floors or subfloors. It is made in two standard sizes (3' X 4' and 4' X 4') with a nominal 1/4" thickness. Unwrap the panels and stand them around the room for at least twenty-four hours before you use them. If the weather is very humid, hold up installation until conditions are normal.

Check the subfloor carefully. Nail any loose boards and plane high spots. Start at a corner and fasten each panel before going on to the next. Leave a 1/8" space between underlayment edges and walls, Fig. 22-22. Otherwise follow the same procedure as with plywood underlayment including the 1/32" space between panels.

You can fasten hardboard underlayment with 1-1/4" ring-grooved underlayment nails or 4d cement-coated sinker nails. If you use staples, make sure they are the 7/8" long divergent type. Start the fastening at the center of the panel and work out toward the edges. This eliminates the possibility of buckling. Space nails 4" OC throughout. Some underlayment panels have the nailing guide printed on their surfaces. A nailing machine will facilitate the installation additionally. See Fig. 22-23. Space staples 6" OC throughout the body of the panel and 3" OC at the edges.

Fig. 22-22 Starting underlayment installation at corner of room.

Fig. 22-23 Using a nailing machine for installation of underlayment.

Particleboard underlayment is composed of wood particles which have been compressed and bonded under high pressure. The resulting boards are exceptionally flat, free of knots and voids. They are made in 4' × 8' sheets and from 3/8" to 3/4" thick. Particleboard is intended for interior use only and is fastened either with ring-grooved or coated box nails or 16 gauge staples. The fasteners must be long enough to enter the subfloors to a depth of from 1" to 1-1/2". Space the nails 10" throughout the field and 6" at the edges. Space staples 6" throughout and 3" at edges. *All fasteners* should be placed at a distance of 1/2" to 3/4" from the edges.

Tile layout: Start by finding the center of each of the two end walls. Drive a nail at each of these points, then carefully strike a chalk line, Fig. 22-24. Be sure the line is stretched tight; then locate its center and strike another chalk line at right angles to it. Use a square to locate the right angle at this point, Fig. 22-25. The floor will now be divided into quarters as in Fig. 22-26.

To locate the true centerlines, place a row of tiles along the perpendicular chalk lines as shown in Fig. 22-27. Measure the distance between the last full tile and the side wall. If that space is less than 2" or more than 8", move the centerline 4-1/2" closer to the wall,

Finish Floor Coverings

Fig. 22-24 Striking a chalk line at center of floor.

Fig. 22-25 Using a square to draw a line at right angles to the first.

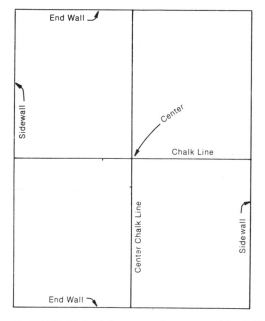

Fig. 22-26 Method for quartering room with chalk lines.

Fig. 22-27 Method for locating true centerlines.

assuming that the tiles are 9" × 9" square. Be sure you keep the new centerline parallel to the original chalk line. The new centerline will give wider and more attractive border tiles on opposite sides of the room.

Repeat this procedure for the end walls. The point where the two new chalk lines meet is the actual starting point for laying the tiles. See Fig. 22-28. Wipe away the original chalk lines to prevent mistakes.

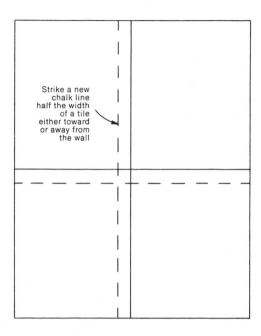

Strike a new chalk line half the width of a tile either toward or away from the wall

Fig. 22-28 Layout showing new chalk lines; the original lines to be wiped out.

Spreading adhesive: Be sure the floor is free of all foreign matter. Sweep it several times; if a vacuum is available use it to remove all traces of dust and dirt. Spread whichever adhesive is recommended by the tile manufacturer on one segment of the floor. Work up to the chalk lines but do not cross them. Spread the adhesive carefully and avoid bare spots or lumps. Some adhesives are made to be brushed on while others must be trowelled on. Follow the manufacturer's instructions precisely.

Allow the adhesive to set before laying the tiles. The time of set depends on both the material used and the room temperature. Adhesive is set when it is tacky to the touch but not wet, Fig. 22-29.

Installing Tiles: Start placing tiles at the center point. Lay the first tile carefully since all other tiles align with it. Butt each tile tightly against the adjoining tiles. Lower, *do not slide*, the tiles into place. Sliding may cause adhesive to ooze up over the surface of tiles. Press each tile down firmly as it is installed. Some tiles may also require rolling.

Fitting and cutting: Border tiles are installed after the main area has been completed. To cut and fit tiles at the walls, place a loose tile directly over the last full tile. On top of this, place another tile and slide it until it touches the wall. Use the edge of the uppermost tile as a guide

Finish Floor Coverings

Fig. 22-29 Testing that adhesive is set—when tacky to the touch but not wet.

and mark the tile below it. See Fig. 22-30. Cut the tile along this mark. It should fit the space exactly.

To cut around pipes and other irregularities, make a pattern of paper. Check and if the fit is O.K., trace the pattern onto the tile and cut. Some tiles may be cut with shears or a sharp knife at room temperature. Others may have to be heated before cutting. If you use a torch or heat gun to soften the tile, be sure to follow safety rules. Keep flame away from walls or other flammable objects. Do not leave a heat gun or torch unattended.

Wall base may be either the set-on or the butt-cove type. The set-on base is the easiest to use. It is placed against the wall and on top of the tiles. Little cutting and fitting is required. Butt-cove wall base is installed flush with the tiles. Fitting must be perfect as all joints are exposed.

Fig. 22-30 Fitting border tiles after main area of floor is laid.

Intensive research in the industry has produced more new and better materials for flooring than ever before. Sheet flooring is an example. Tough, easy to clean and highly recommended for high traffic areas such as kitchens, it is very popular—especially because it is seamless. But the installation of sheet flooring is not discussed here since it is generally done by flooring specialists rather than by carpenters.

GLOSSARY

asphalt: a by-product of petroleum; used for waterproofing roofs, floors and walls.

block flooring: flooring in square shapes, usually 9" X 9" and made with glued or laminated blocks; edges are tongue-and-groove or grooved all around.

border tiles: tiles used at the edges of a room; usually they must be cut to fit.

end matched: boards with tongue-and-groove joints at the ends as well as the sides.

expansion strip: in flooring, a cork or other resilient strip placed at the edge of flooring to permit expansion.

mastic: a material used as a cement or as a waterproof coating.

plank flooring: wood flooring in from 3-1/2" to 8" random widths; edges may or may not be bevelled.

plug: a piece used to fill a hole.

pressure-sensitive: an adhesive backing which sticks upon contact when slight pressure is applied.

resilient flooring: any vinyl- or asphalt-base floor covering with a certain amount of resistance to denting or deformation; usually installed over plywood subflooring and underlayment.

sleeper: wood strip laid on or in concrete to support a subfloor or finish flooring.

strip flooring: flooring of narrow strips usually 2-1/4" wide and tongue-and-grooved.

sub-floor: boards or plywood laid on joists to serve as a base for finish flooring.

vinyl: a tough, durable plastic.

23

Chimneys & Fireplaces

Chimneys create the draft necessary for combustion and vent the gases and smoke produced by heating units and fireplaces. To operate efficiently, they must be designed and built properly. A poorly made chimney can cause discomfort and be a fire hazard as well.

Ordinarily other trades build chimneys and fireplaces. But you, as a knowledgeable carpenter, should master the fundamentals of their techniques.

A masonry chimney is perhaps the heaviest part of a residential structure and must be built on a solid foundation. It must not depend on or rest on other parts of the structure for support. Its footings must be below the frost line and should support the chimney load amply and without settlement. For one-story houses the footings should be 8" thick; for two-story houses, 12" thick. Footings should extend at least 6" beyond the chimney on all sides.

Chimney size depends on the number, size and arrangement of flues; the number of flues depends on the number of heating units and fireplaces. Flues are the tubes or ducts that serve each heating unit that uses air and discharges smoke and gases.

The height of a chimney should be sufficient to prevent downdrafts caused by surrounding obstructions, Fig. 23-1. Consult local building

Fig. 23-1 Chimneys must be free of obstructions to function properly.

Chimneys and Fireplaces

codes of course,—but remember that a chimney should be at least 2'0"
above any ridge or wall within 10'0" of it, Fig. 23-2. Furthermore, flat
roofs require a 3'0" minimum chimney (Fig. 23-3) and the space
between framing and chimney must be filled with a noncombustible
insulating material. This serves as a fire stop. Fig. 23-4 shows a detail of
a chimney serving a heating unit and a fireplace.

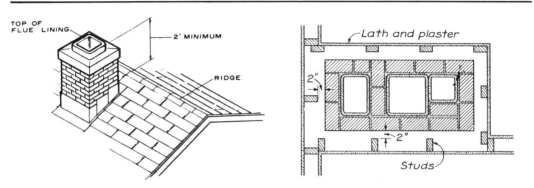

Fig. 23-2 Chimney height must be a minimum of 2 feet above the ridge.

Fig. 23-3 Proper clearance between chimney and framing members.

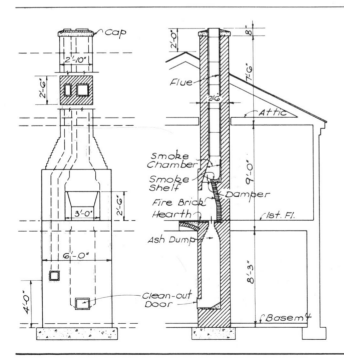

Fig. 23-4 Details for construction of a residential chimney. *(U.S. Dept. of Agriculture)*

Flues

Each fireplace and furnace should have its own flue for proper operating efficiency. The flue area is governed by both the fireplace opening and the chimney height. Generally, the area of the flue should be 1/10th the area of the fireplace opening when the chimney is 15'0" high. Lower chimneys require larger flues. Refer to the table in Fig. 23-5 for proper flue sizes. For example, a fireplace opening measuring 30" × 30" has an area of 900 square inches. One-tenth of 900 equals 90. Reference to the table indicates that an 8" × 12" or 96 square-inch flue is recommended. This is the standard size nearest to the figure derived from the fireplace opening area. It is suitable because it is slightly larger than actually needed. Flue size may be more but must *never* be less than the calculated size.

Although some chimneys are made without flue linings, most codes require that they be used as a safety measure. The linings prevent the escape of smoke and hot gases through cracks in the masonry.

| Finished Opening | | | | | Dimensions Rough Masonry | | | | | | |
| A | B | C | D | E | F | G | H | I | K | | |
Width	Height	Depth	Back	Throat	Width	Depth	Smoke Shelf Height	Smoke Chamber	Vertical Back	Inside of Brick Flue	Standard Flue Lining
24	28	16	16	9	30	19	32	11	14	8½x 8½	8½x 8½
26	28	16	18	9	32	19	32	11	14	8½x 8½	8½x 8½
28	28	16	20	9	34	19	32	11	14	8 x12	8½x13
30	30	16	22	9	36	19	34	11	15	8 x12	8½x13
32	30	16	24	9	38	19	34	11	15	8 x12	8½x13
34	30	16	26	9	40	19	34	11	15	12 x12	8½x13
36	31	18	27	9	42	21	36	11	16	12 x12	13 x13
38	31	18	29	9	44	21	36	11	16	12 x12	13 x13
40	31	18	31	9	46	21	36	11	16	12 x12	13 x13
42	31	18	33	9	48	21	36	11	16	12 x12	13 x13
44	32	18	35	9	50	21	37	11	17	12 x12	13 x13
46	32	18	37	9	52	21	37	11	17	12 x16	13 x13
48	32	20	38	9	54	23	37	15½	17	12 x16	13 x18
50	34	20	40	9	56	23	39	15½	18	12 x16	13 x18
52	34	20	42	9	58	23	39	15½	18	12 x16	13 x18
54	34	20	44	9	60	23	39	15½	18	16 x16	13 x18
56	36	20	46	9	62	23	41	15½	19	16 x16	18 x18
58	36	22	47	9	64	25	41	15½	19	16 x16	18 x18
60	36	22	49	9	66	25	41	15½	19	16 x16	18 x18

Fig. 23-5 Schedule of flue sizes which are determined by dimensions of fireplace openings.

Chimneys and Fireplaces

Fire-clay linings are available in round, square or rectangular shapes. They range in size from 8" × 8" to 20" × 20" and their wall thickness increases correspondingly. The minimum recommended for a flue is a 5/8" thickness. They are usually made in 2-foot sections, and round linings are made from 8" to 24" in diameter.

If a chimney contains more than one flue, the flues must be separated from each other. This is done by setting brick spacers at least 4" thick, between the flue liners. They are called withes.

Flue liners ought to be installed in a straight line, but if it is necessary, they can be offset. The slope should not exceed 45 degrees; and an angle of 30 degrees or less is preferable. See Fig. 23-6. Lining can be cut with a chisel, but some masons prefer to use a portable saw fitted with a masonry blade. Be sure to use safety goggles while you are doing such work.

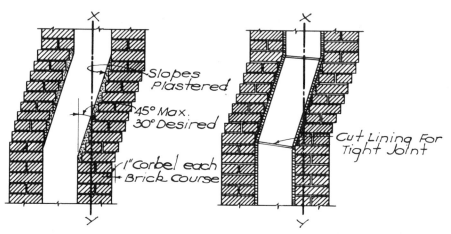

Fig. 23-6 Chimney offset: center of upper flue, line *XY*, must not fall past center of lower flue wall; chimney on left has no flue liner—slope is plastered.

Install the liners slightly ahead of the brick work. It gives you the opportunity to make a good mortar joint between sections. Strike the inside of the joint smooth, then lay the brick work in cement mortar.

Chimneys are often enlarged in girth just before they emerge from the roof. Although it is usually done for the sake of appearance, it also strengthens them. The practice is called **corbeling** and is done by offsetting the bricks so each course projects outward slightly more than the one below it. See Fig. 23-7. The offsets should not exceed 1" per course.

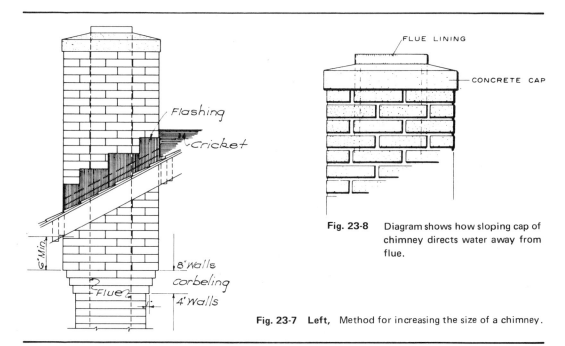

Fig. 23-8   Diagram shows how sloping cap of chimney directs water away from flue.

Fig. 23-7   **Left,**   Method for increasing the size of a chimney.

The flue lining must extend at least 4" above the top course of bricks. It is then capped with two inches of mortar, sloping away from the top, as in Fig. 23-8. This allows water to drain away from the flue. It also provides an upward draft when air currents strike it.

Brick and masonry chimneys must be protected against water entry and damage. If water penetrates mortar it causes deterioration and crumbling, and eventually both bricks and mortar fall. Water that penetrates the joints at the rooftop can also gain entry to interior walls and ceilings and cause extensive damage. Capping, flashing and counterflashing all furnish protection against water leakage in and around chimneys.

Capping is the placing of a waterproof top over a brick or masonry chimney. Ordinarily it is a concrete cap, but sometimes a metal cap, usually made of lead, is used. Unlike other metals, lead is not affected by the gases discharged through the flue.

Flashing is the installing of metal sheets at the joints of chimneys and roofs. The metal (tin, lead or copper) is nailed under the shingles at the joints and extends up the side of the chimney walls. See Chapter 11 on roof framing for details.

Counterflashing is the installing between the bricks of the chimney wall of metal sheets which extend downward over the flashing, Fig. 23-9. Together the two make a snug waterproof joint.

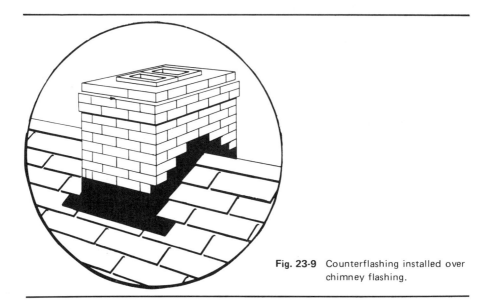

Fig. 23-9 Counterflashing installed over chimney flashing.

Fireplaces

At one time, fireplaces were essential in a house because they were the only source of heating. Today the fireplace is an optional item in house construction since more satisfactory and efficient central heating systems have been devised.

Masonry fireplaces are not efficient sources of heat because most of the heat produced is lost up the chimney. The heating efficiency of prefabricated metal fireplaces is considerably greater. Nevertheless, whatever its heating value, many people want a fireplace for aesthetic reasons. Besides being decorative, the fireplace often functions as a focal point of interest in a household.

The main parts of a masonry fireplace and typical construction details are shown in the cutaway view of a fireplace, Fig. 23-10.

Hearth: The fireplace hearth is the lower part or base of the firebox that extends outward into the room. It should be made of fire brick, stone or other fireproof materials. The hearth may be made flush with the finish floor or it may be raised on a platform. It should rest on a reinforced concrete slab at least 3-1/2" thick.

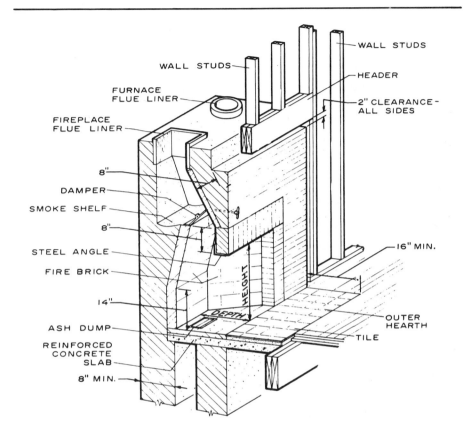

Fig. 23-10 Details for construction of typical masonry fireplace.

An ash dump in the floor of the hearth lets ashes drop into an ash pit in the basement—or at the bottom of an outside wall if there is no basement. See Fig. 23-11. The ash pit should be of tight masonry and have a snug fitting cleanout door.

Be sure that the floor framing around a hearth is strong and substantial. If the headers are more than 4'0" long, they should be doubled. See Fig. 23-12. If more than four tail beams are supported, use metal joist hangers.

Walls: The back and side walls of fireplaces must be at least 8" thick if they are of solid masonry or reinforced concrete. Stone construction should be a minimum of 12" thick. Line the back and sides with fire brick at least 2" thick.

The rear and side walls of the firebox must slant in order to reflect

Chimneys and Fireplaces

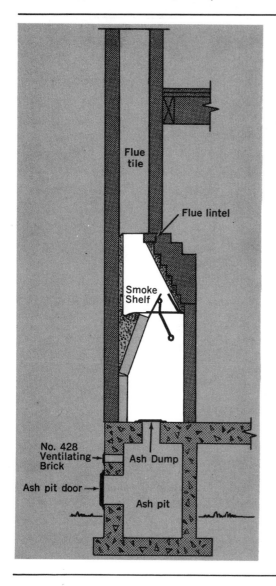

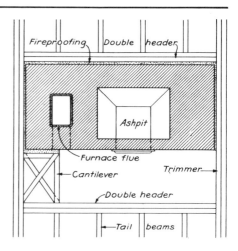

Fig. 23-12 Framing details around hearth.

Fig. 23-11 **Left,** Details of ash dump and pit.

heat back into the room. The width should always be greater than the height in a correctly proportioned firebox.

Lintel: The lintel supports the masonry above the fireplace opening. It may be made of heavy steel bars but usually 3-1/2" × 3-1/2" × 1/4" angle iron is used. Heavier lintels are required for openings more than 4'0" wide.

Throat: The throat is the area directly above the firebox. It forms the passageway from the firebox to the smoke chamber. It must be carefully designed and built, if the fireplace is to function properly. The throat has tapered sides which start 6" to 8" above the lintel.

The area of the throat should be no less than that of the flue. Also, its length should be equal to the fireplace opening, Fig. 23-13.

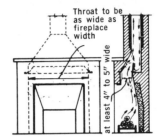

Throat to be as wide as fireplace width

at least 4" to 5" wide

Fig. 23-13 Throat above firebox should be same width as fireplace opening and at least 4 to 5 inches deep.

Smoke shelf: The purpose of the smoke shelf is to prevent down drafts from entering the firebox. It is the horizontal surface directly behind the throat, and the *smoke chamber* is the area above the smoke shelf. The smoke chamber's upper opening leads into the flue so its surface must be smooth in order for smoke and gases to pass upward freely.

Damper: The damper consists of a metal framework with a hinged lid. It is adjustable and is used to regulate the draft by varying the size of the throat. Most dampers are made with flanges which set into the masonry.

Some dampers have flanges designed to support the masonry above the opening, thus eliminating the need for a lintel. Figure 23-14 illustrates two types of damper controls. The rotary unit extends through the fireplace wall; the other unit is operated with a poker.

Multi-Opening Fireplaces

If a fireplace has more than one face or opening, the flue sizes given for the single face fireplace do not apply since cross drafts and other problems occur. So stronger drafts and larger flues must be provided. Tables listing maximum opening heights for various hearth and flue sizes are available from fireplace manufacturers.

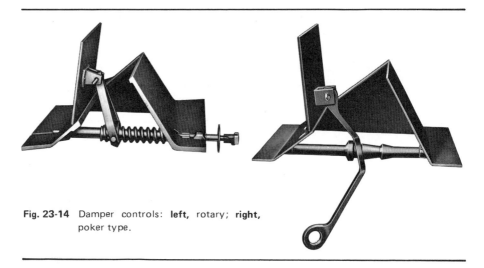

Fig. 23-14 Damper controls: **left,** rotary; **right,** poker type.

Prefabricated Fireplaces

Many builders use prefabricated fireplaces. They can be installed in a relatively short time without extra masonry support, or footings and foundation. By utilizing special angles and other components, several fireplaces may even be stacked one above the other. See Fig. 23-15.

Packaged units are available complete with triple-wall chimney and roof projections, Fig. 23-16. Local building codes should be consulted concerning the installation requirements. Fig. 23-17 shows the framing of a prefabricated unit.

Triple-walled chimney components permit cool outside air to circulate down and around the firebox. This improves the draft which controls the burning process. The advantage of the system is that a minimum amount of room air is used for combustion since air is mainly brought in from the outside.

Circulator Fireplaces

Circulator fireplaces are factory-made units which are designed to send warm air convection currents into a room. In the process cool air enters intakes near the floor, is heated, and reenters the living area through outlet vents. See. Fig. 23-18. Specially designed blades in the

Fig. 23-15 Multi-floor stacking installation.

Fig. 23-16 A modern prefabricated fireplace complete from hearth to chimney.

Fig. 23-17 Left, Framing of prefabricated fireplace; note flue from basement heating unit. *(The Majestic Co.)*

double-walled construction directs the flow of air over the hottest parts of the unit for maximum efficiency Small fans may be used to convert the system to forced-air operation.

The integral parts of a circulator include firebox, damper, smoke dome and smoke shelf. The use of circulator fireplaces eliminates the calculations and labor involved in installing conventional fireplaces. Of heavy steel, the units are designed to be concealed behind the usual brick or other masonry constructions.

The installation of a typical circulator involves the pouring of a footing and foundation. Fire brick is laid out on the hearth area. The circulator is then set in position over the fire brick. Angle seals at the sides provide a starting place for the finishing masonry work.

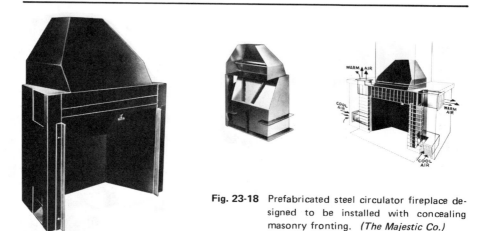

Fig. 23-18 Prefabricated steel circulator fireplace designed to be installed with concealing masonry fronting. *(The Majestic Co.)*

Fiberglass wool is used to cover all outer surfaces of the unit. Masonry can then be applied in the conventional manner. It must not come closer than 1/2" to 3/4" to the circulator. The manufacturers' instructions must be followed carefully. See Fig. 23-19 for an installation without masonry and intake outlet grilles in the side walls.

Free-Standing Fireplaces

Free standing wall-hung fireplaces are made in many shapes and sizes. Some burn wood, others are gas fired or electric powered. Fig. 23-20 illustrates a wall-hung electric unit.

Fig. 23-19 Prefabricated circulator fireplace without masonry; warm air enters from upper duct.

Fig. 23-20 A wall-hung fireplace powered by electricity. *(The Majestic Co.)*

When a free-standing fireplace is installed on a combustible floor, a protective covering is required. Slate, stone, ceramic tiles and asbestos are among the materials used to make such a bed or base.

Mantels

A mantel is the decorative facing around a fireplace. It may consist of a single shelf or it can be very ornate and include lintel or pilaster and facings. Wood is widely used for this construction, but other materials may be used. If flammable materials are used for the mantel, they must not be closer than 3-1/2" to the fireplace opening. The mantel shown in the photo is factory-made of ponderosa pine. It is cut to final size on the job by the carpenter; Fig. 23-21 shows construction details of the unit.

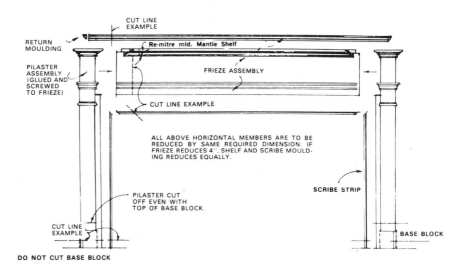

Fig. 23-21   Construction details of a factory-made mantel.

GLOSSARY

capping: the top of a chimney or other structure.

circulator: a prefabricated fireplace designed to circulate heated air into a room.

corbelling: a stepped projection from a vertical surface; in chimneys, an offset in the masonry.

counterflashing: the chimney flashing used at the roof line to cover shingle flashing.

damper: a device used to control the draft in a chimney.

down draft: the flow of air down a chimney.

firebox: the chamber in a fireplace where fuel burns.

flashing: any material used, ordinarily metal, to prevent water seeping or leaking into a building.

flue: a passageway in a chimney to carry off smoke, gases and fumes.

flue liner: a lining usually of fire clay used to cover the interior of a chimney.

hearth: the inner and outer floor of a fireplace.

lintel: a horizontal structural support over openings such as doors, windows, and fireplaces.

mantel: the shelf above a fireplace, including the decorative facing around it.

non-combustible material: one that will not support combustion—will not burn.

smoke chamber: the area directly above the smoke shelf in a fireplace.

smoke shelf: the horizontal surface behind the throat of a fireplace; it eliminates down drafts.

throat: the area in a fireplace above the firebox; it forms the passageway from the firebox to the smoke chamber.

withe (wythe): the partition between flues in a chimney.

24

Cabinets & Woodwork

The modern carpenter, while not a cabinet-maker, is often called upon to make or install kitchen cabinets, counter tops, shelves, closets and similar units. So he must have a good working knowledge of the various materials and procedures used in cabinetry, as well as the basic skills to do the necessary woodwork.

Although most kitchen cabinets today are custom-built, as in Fig. 24-1, some are built on the job by the carpenter, either in part or wholly. The built-in unit shown in Fig. 24-2 was constructed almost entirely on the job. Whether building a unit or installing a factory-made one, you as the carpenter must work with care. Make all cuts accurately and all joints true.

Plywood is widely used in cabinetry and appears in many of the illustrations in this chapter. In most cases however, what you are learning also applies to solid lumber.

Fig. 24-1 Installation of custom-built kitchen cabinets. *(U.S. Plywood)*

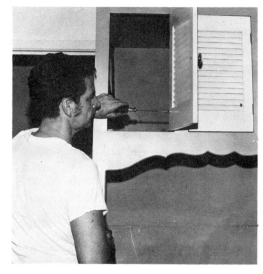

Fig. 24-2 Built-in unit constructed by the carpenter on the job.

Joints

A number of special joints are used in carpentry and cabinet work. Some of the most common are shown in Fig. 24-3.

HALF LAP

CROSS LAP

END LAP

MIDDLE LAP

TONGUE & GROOVE

BUTT

RABBET

DADO

DADO & RABBET

DADO TONGUE AND RABBET

MITRE

THRU MORTISE TENON

STUB MORTISE TENON

BLIND MORTISE TENON

OPEN MORTISE TENON

LAP DOVETAIL

DOVETAIL DADO

THRU SINGLE DOVETAIL

THRU MULTIPLE DOVETAIL

STOPPED LAP DOVETAIL

LAP DOVETAIL OR HALF BLIND DOVETAIL

BLIND MITRE OR SECRET DOVETAIL

Fig. 24-3 Details of various wood joints.

Cabinets & Woodwork

The *butt joint* is the simplest to make; it is also the weakest. For thin stock, add reinforcing cleats. To further strengthen the joint, glue it in addition to the regular nailing or screwing. If you also want to make a joint without fasteners showing, use blind dowels, as in Fig. 24-4.

Rabbet and *dado joints* are similar, except that the rabbet occurs at the edge of a piece of wood, the dado along the surface. Both are used in the construction of case work and drawers. If the edge grain is objectionable, combine a miter with the rabbet, Fig. 24-5. The dado is generally used for shelves and drawers. Both can be cut by hand, but are much easier to do with power tools. Either fit the router with a gauge to cut a dado, or use the table saw.

In order to reduce the weight and cost of cabinets, you can make the framework of 1" stock with butt joints and then cover it with thin plywood.

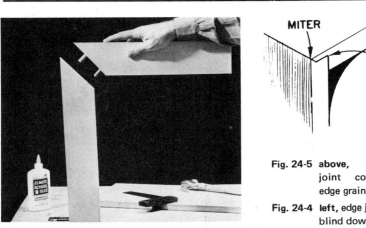

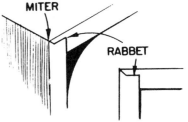

Fig. 24-5 above, Mitered rabbet joint covers unattractive edge grain.

Fig. 24-4 left, edge joint made with blind dowels

The *stub tenon* is another joint widely used in cabinetry. It is used to join stiles and rails, is fairly strong and not difficult to make. The tenon fits into the groove or stub mortise in the frame. A stub mortise is a cut which extends only partly through the wood—deep enough to receive the tenon securely.

Dovetail joints are generally used in construction of quality drawers. They are perhaps the most difficult to make by hand. A router used with a special jig and template greatly simplifies the making of this fine joint. See Fig. 24-6. The finished joint is shown in Fig. 24-7.

Fig. 24-6 Using a dovetail template to make drawer joints.

Fig. 24-7 Completed dovetail joint in finished drawer.

Miter joints can be assembled in several ways for frame facing and other things. A special tool is designed to set corrugated fasteners accurately and safely. Such fasteners can also be used to reinforce miter joints on case work.

Screw or nail assembly: Cabinets can be assembled with either. But you must always predrill a pilot hole for screws. If you must place nails close to an edge, predrill for them also. Countersink all heads and fill with a suitable putty. Generally, you can space the nails about 6" apart. When you work with thin wood, closer spacing may be needed to prevent buckling.

Countersink screwheads as shown in Fig. 24-8.

Glued joints should be used where additional strength is needed. Choose the proper glue for the job you are doing. If the piece will be subjected to dampness, use a moisture-resistant glue. Edge-grain wood absorbs glue quickly, so you must give the edge a thin prime coat. Allow it to set for a few minutes, then apply glue to both surfaces and join.

Base and Wall Cabinets

Base and wall cabinets if made on the job are usually cut and assembled before installing. This procedure is recommended to assure accuracy. Building the cabinets directly on walls and floors can cause problems because corners can be off-square and walls be out of plumb.

Cabinets & Woodwork

A simple base cabinet is shown in Fig. 24-9. The sides and back are made of plywood; the facing and web frame of solid lumber. Factory-made cabinets are usually made this way. The four compartment unit in Fig. 24-10 has a plywood top, but particleboard may also be used—especially if the surface is to be laminated. Figure 24-11 has construction details for a typical kitchen remodeling. Complete

Fig. 24-8 Drilling countersink holes for screw heads. *(Black & Decker)*

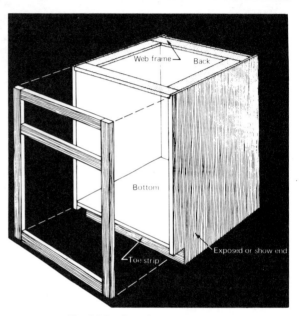

Fig. 24-9 A typical base cabinet.

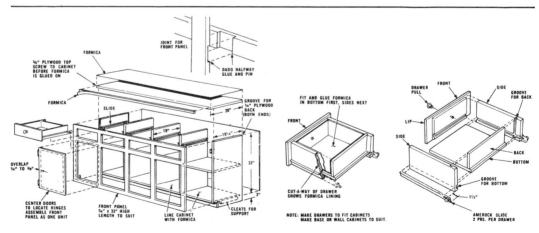

Fig. 24-10 Detailed layout for typical base cabinets.

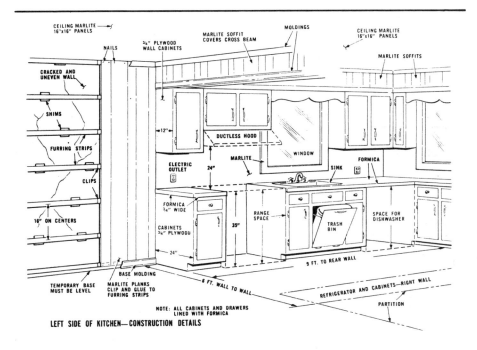

LEFT SIDE OF KITCHEN—CONSTRUCTION DETAILS

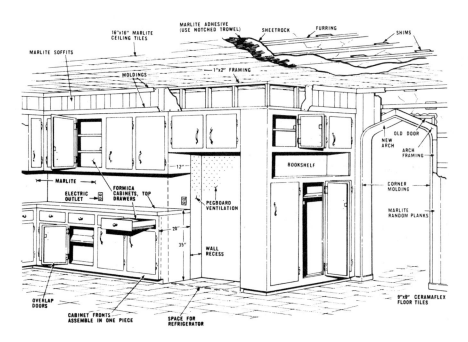

Fig. 24-11 Construction details for remodelling a kitchen.

Cabinets & Woodwork

construction details of two-compartment base and wall cabinets are illustrated in Fig. 24-12.

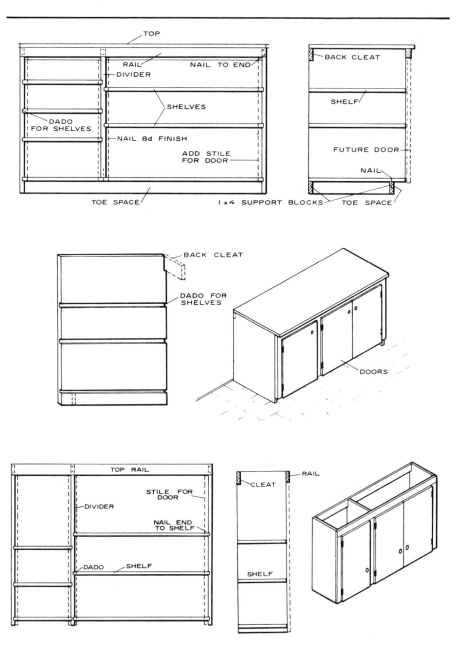

Fig. 24-12 Details for constructing a base cabinet.

If built-in appliances are to be included in the cabinets, make proper allowances so that the cabinets will fit; Fig. 24-13 illustrates the proportions of standard kitchen cabinets.

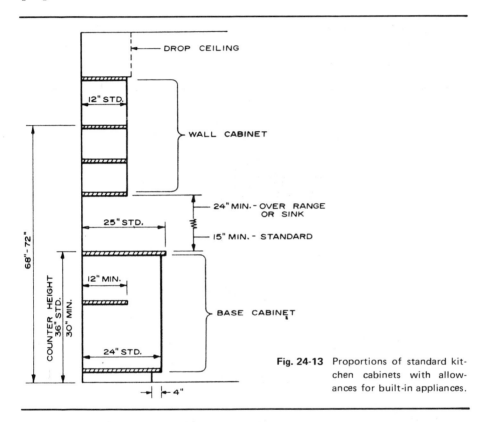

Fig. 24-13 Proportions of standard kitchen cabinets with allowances for built-in appliances.

Assemble cabinet members with clamps whenever possible since that guarantees a good tight joint.

There are several ways to apply cabinet backs. The usual method is to rabbet them either flush or set-in. Large units are usually set-in to allow for walls that may not be perfectly flat. The lip that remains after the back is installed may be trimmed as necessary to get a good fit between the wall and the cabinet. If the back will not be seen, use 1" headed nails, otherwise use 1" brads.

Drawers

Drawers are of two basic types—flush or overlap. The *flush drawer* requires greater accuracy in construction and fitting since any discrepancies between frame and drawer edges are clearly visible. In *overlap-*

ping drawers the space between the drawer sides and frame is concealed. The bottom overlap is sometimes beveled to allow for a finger grip, thus eliminating the drawer pull.

A simple drawer is shown in Fig. 24-14. The joints may vary in different drawers, but basically most drawers are made this way. Drawers have five parts—front, two sides, rear and bottom. Construction details for front and side joints are shown in Fig. 24-15.

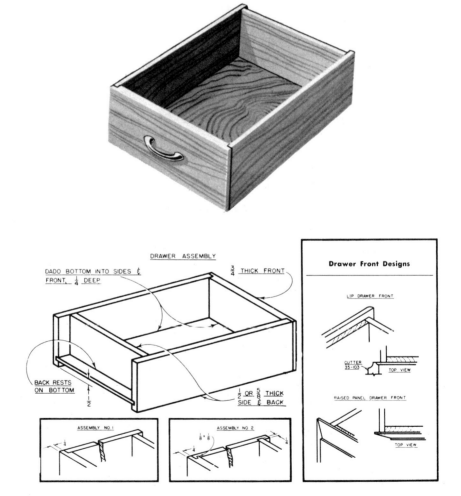

Fig. 24-14 A typical drawer with assembly details below.

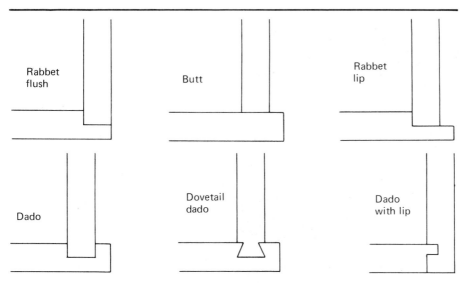

Fig. 24-15 Details for front and side in drawers.

Drawer Fronts are usually made with 3/4" solid lumber, however lumber core or plywood may also be used. Prefabricated drawer fronts are available in many styles and can be added to a basic drawer. Some are covered with plastic laminates, Fig. 24-16.

Drawer Sides are normally made of 1/2" thick materials. Small drawers may be made with 3/8" stock; and some with side guides may require 3/4" thick material. Solid wood or plywood are both used in construction. The sides of quality drawers are usually made of hardwood. Grooves are cut near the lower edge to receive the drawer bottom.

Drawer Backs are usually made with 1/2" thick materials. They may be grooved like the sides or simply rest on the bottom panel; Fig. 24-17 shows several kinds of corner joints that may be used.

Drawer Bottoms are usually made of 1/4" material—either plywood or hardboard. In very simple constructions, the bottom may be nailed to the sides and back. Then a strip of wood is glued to the front to reinforce the bottom panel.

Drawer Guides are runners or tracks on which drawers slide in and out without jamming. The three basic types are the side, corner and center guides, shown in Fig. 24-18.

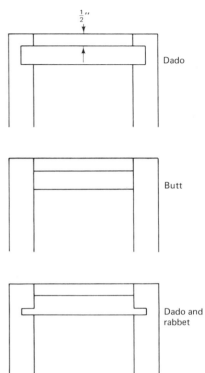

Fig. 24-17 Joints for drawer backs.

Fig. 24-16 **left:** Kitchen cabinet doors and drawers covered with plastic laminate. *(Formica Corp.)*

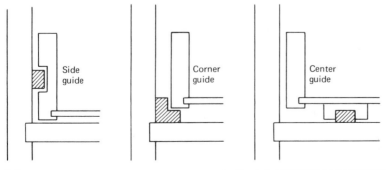

Fig. 24-18 **left:** types of drawer guides.

In the *side guide* type, grooves are cut in both drawer side panels slightly above center. Hardwood strips are then fastened to both sides of the cabinet interior so the drawer rides on them.

In another version of this style, the grooves are cut in the cabinet sides, and the hardwood strips are added to the drawer's side panels.

Both versions work equally well, but the grooved drawer type is more often used. It is easier to cut the grooves or plows in the smaller drawer section than in the larger cabinet sides.

The *corner guide* has an L-shaped strip which supports the bottom and sides of the drawer. A kicker is usually put at the top underside of the cabinet frame to keep the drawer from tipping downward when open.

The *center guide* is a grooved strip of wood attached and centered to the drawer bottom. It is made so it rides on a runner attached to the framing. On very wide drawers, two guides may be used instead of a single center one.

Commercial Slides and Guides are made of metal and plastics, others all of plastic. These factory-made units operate at the sides or bottom of the drawers. Several types are shown in Fig. 24-19. Most of them have automatic stops; some include a self-closing feature.

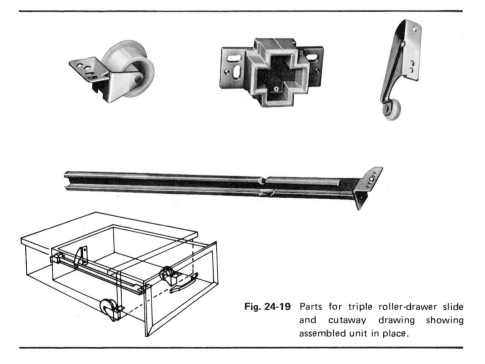

Fig. 24-19 Parts for triple roller-drawer slide and cutaway drawing showing assembled unit in place.

Cabinets & Woodwork

Cabinet Doors

Cabinet doors are of two general types—sliding or hinged. The hinged swinging door is widely used for kitchen and bathroom cabinets. Sliding doors are useful when swinging doors are not practical because of space limitations or other reasons.

Hinged doors may be purchased ready-made or they can be constructed by the carpenter. Fancy raised-panel doors are not practical to make on the job. Solid doors are the easiest to construct on the job. They are generally made of lumber-core plywood and may be flush, lipped, or have a full overlap, Fig. 24-20.

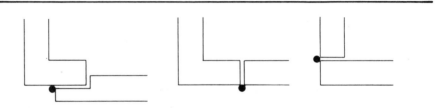

Fig. 24-20 Hinged-door types: **left**, overlapping (lipped); **center**, flush; **right**, full overlap.

Lipped doors are easier to fit than flush doors because they cover part of the face frame of the cabinet and no cracks show.

Flush doors must be fitted into the framework of a cabinet so that all surfaces are level. The carpenter must work very carefully to get an accurate fit so that no cracks or spaces show around the doors. Only the pin loops on semiconcealed hinges are visible. Except for the pivot, a concealed pin hinge is not visible from the face. See Fig. 24-21. Fully concealed hinges are also made, but they are not generally used on kitchen cabinets.

The *full overlap door* is used mostly in contemporary furniture. The door does not fit into a frame but fits flush with the front edges of the case. Or it can be made to cover 2, 3 or 4 front edges of the case. With this type of door, pin or pivot hinges are ordinarily used. When the door is closed only the pin or pivot shows from the front.

Sliding Doors may be made of various materials. Wood, plastic, glass and hardboard are most commonly used for them. They slide in grooves cut in the cabinet frame or in tracks mounted on the edge of the frame. The tracks can be factory-made or you can make them with square strips and quarter-round molding. The square strips are used between

Cabinet Accessories and Hardware

Fig. 24-21 The concealed pin hinge is used when the appearance of a hinge is objectionable; large doors require at least three hinges. (*American Plywood Assoc.*)

the doors, and the quarter-round molding is placed at the front and rear of the doors, parallel to the narrow strip, to form a groove or track. The upper grooves for the sliding doors are deeper than the lower ones: the lower are usually 3/16" deep, the upper 3/8" deep. The doors are installed by pushing them upwards so the top edges enter into the upper grooves first; then the doors are lowered so that the bottom edges drop into the lower tracks.

Shelves

Cabinet shelves may be hung with ***cleats*** or ***shelf supports*** or they may be set into ***dadoes***. The dadoed shelf is the strongest, but the shelf is fixed and cannot be moved. When you cut dadoes with a router, you may have to make several passes to obtain the proper width. In Fig. 24-22 the width of the dado cut is being checked with a piece of scrap. The use of shelf supports makes the shelf adjustable in height. Peg the supports into blind holes drilled 3/8" deep into the cabinet sides at desired intervals.

Cabinet Accessories and Hardware

Often the back shelf space in corner cabinets is wasted because it is difficult to reach. That space can be utilized by using revolving doors with shelves attached to their backs. Prefabricated units are available

Fig. 24-22 Checking the size of dado cut for shelf.

for both upper and lower cabinets. Manufacturers instruction sheets are fully detailed. Follow them carefully for installation. Various catches, pulls, and other cabinet hardware are shown in Fig. 24-23.

Laminated Tops

Kitchen and bathroom sinks and counter tops are usually surfaced with plastic laminates. Durable and heat resistant, these materials are not affected by most household chemicals.

Laminates are made from layers of resin-impregnated paper bonded under very high pressure and heat into rigid sheets. Standard sizes are .031 and .062 thick (1/32" and 1/16"); 24, 30, 36, 48, and 60" wide; and 72, 84, 96, 120 and 144" long. Laminates come in a wide range of colors and patterns, including simulated woodgrains.

For cabinet work the laminates should be bonded to a core material, such as particleboard or plywood. Particleboard is preferred for counter top cores because of its smooth surface. The heavy grain patterns of plywood underlayment may telegraph or show through the surface.

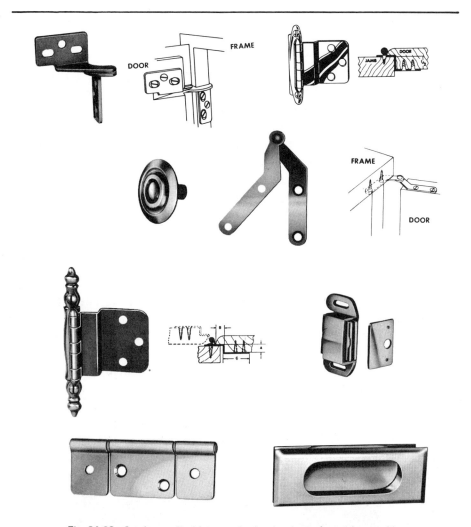

Fig. 24-23 Catches, pulls, hinges and other hardware for cabinet making.

In addition to the general purpose grades described above, other laminates are also made for backing and for cabinet interiors. Backing material is nondecorative and serves only as a moisture barrier to minimize warping. The cabinet liner grade is designed for cabinet interiors. It is similar to backing laminate but looks better. Both types are less expensive than general grade laminates.

Cutting Laminates: You can cut with either hand or power saws or with tin shears. The tin shears can only be used to cut the .031 or

Cabinets & Woodwork

1/32" laminate; they are not practical for thicker material. Regardless of the type of saw you use, the saw must have fine teeth and the laminate must be supported along the cutting line. Pieces must be cut from 1/8" to 1/4" oversize. Then trim after you glue the sheet to the base material. When you use a portable circular saw or saber saw, be sure to place the sheet face down. But when you cut with a router, handsaw or table saw, you must keep laminate sheet face up. If you work with a handsaw, be sure to hold it at a very low angle.

Applying adhesives to laminates: Before you apply adhesives, make sure that the surface to be covered is clean and free of bumps. If necessary, sand the surface smooth. Fill all voids or holes.

Before you apply any adhesive, take note of all instructions and cautions on the container. If the contact cement is flammable, extinguish all pilot lights, cigarettes and other flames. Do not operate electrical switches while any traces of vapor remain in the room. Open all windows while you work with the adhesive and avoid breathing its vapors. If nonflammable cement is used, be careful. The vapors are usually toxic. It also must only be used with adequate ventilation.

Apply the cement by brush, roller or notched applicator to the back of the laminate and to the face of the piece being covered. Allow ample time for the glue to dry; bare wood usually requires two applications of glue. Be sure the first coat is dry before applying the second; and if possible, do the gluing outdoors, Fig. 24-24.

Fig. 24-24 Applying contact cement to cabinet case before positioning laminate sheet.

Final Pressure Technique

If the edges are to be laminated, glue and finish them first. The reason is fully explained in the section on ***Edge Banding*** on page 452.

When you bring the two cemented surfaces into contact, they will bond immediately and cannot be repositioned. Therefore the laminate must be positioned accurately before the glued surfaces touch. So the following procedure is recommended: Apply a coat of cement to both surfaces (cabinet top and back of laminate) and allow to air dry about 45 minutes. Next, cut several sheets of kraft (wrapping) paper and place them on the cabinet's surface so that they overlap slightly. Then position the laminate on top of that paper and in exact position over the cemented cabinet top. The paper will not stick to the dried cement. Pull one sheet out slowly and gently press on the laminate. Now remove the rest of the paper carefully. As the paper is removed, the two cemented surfaces make contact and adhere to each other in the proper position.

Final Pressure Technique

When you have removed the paper completely, apply heavy pressure to the surface. Use a small roller or a smooth block of wood, tapping it with a hammer as shown in Fig. 24-25.

Fig. 24-25 With sheet in place, use a roller or hammer and block to apply heavy pressure.

Cabinets & Woodwork

Edge Banding: For best appearance, you should make the exposed edges of sink and counter tops at least 1-1/2" thick. Usually this requires that you build up the edge of the counter top with strips of wood or particleboard, also called flakeboard. The build-up strips should be 3/4" thick and from 1-1/2" to 2" wide. Be sure the edges are aligned perfectly. Apply these strips with nails and glue; then sand smooth.

Next, cut a strip of laminate 1/4" wider and longer than the edge to be covered, Fig. 24-26. Apply glue to both the wood build-up strips and to the back of the laminate strip exactly as before. Allow both to dry and then apply the laminate and roll or hammer the edge with firm pressure.

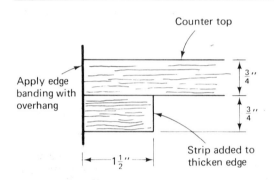

Fig. 24-26 Details for edge banding of exposed edges of sinks and counter tops.

Edge Trimming: Trim the excess so the laminate is flush with the top and bottom of the wood surfaces. Use a router (Fig. 24-27) or use a block plane and finish with a file. The special trimmer shown can be adjusted to make straight and bevel cuts.

After the edge band has been trimmed, apply the laminate sheet to the top as described on page 451. Trim off any excess laminate with a 22-1/2 degree bevel-trimming bit.

A standard counter top is shown in Fig. 24-28. The laminated back splash is finished as a separate unit. Together with the cove molding, it is added to the trimmed top. Fig. 24-29 illustrates a fully-formed top. This type is factory-made since special equipment is required to heat and form the laminate. The carpenter usually has to make the sink cutouts in these tops, with either a saber saw or router. The router method is much faster. Use a hardboard template and clamp it to the

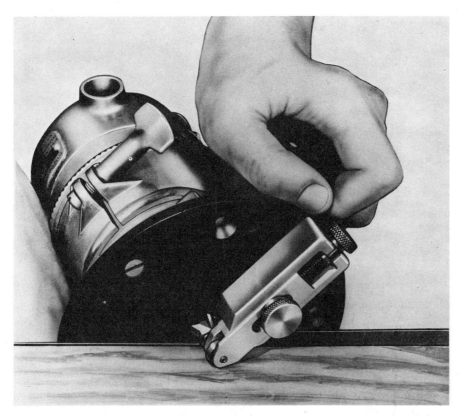

Fig. 24-27 Trimming off excess laminate at edge with router fitted with a special cutter.

Fig. 24-28 Standard counter top; a type that can be built by the carpenter on the job site.

Fig. 24-29 Factory-built fully formed counter top ready for installation.

counter top in exact position. For this, fit the router with a combination panel bit and template guide. Do the cutting as shown in Fig. 24-30.

Cabinets & Woodwork

Fig. 24-30 Using a router to make a sink cutout. *(Black & Decker)*

Cabinet Installations

Install all cabinets carefully according to the specifications on plans. Wherever possible, fasten the cabinets to wall studs, Fig. 24-31. Do all fastening with screws or bolts. But when you mount cabinets to hollow masonry walls, you should use toggle bolts or "Molly" fasteners. Make holes with either a star drill or masonry drill bit. Be sure to protect your eyes when you make the holes. If cabinet fasteners must go into gypsum walls, make sure that the wall will support the load. For solid masonry walls, you may use either expansion shields or lead plugs. Special anchor bolts held with mastic may also be used. Press the bracket into mastic so that it squeezes through the holes for a firm bond.

Fig. 24-31 Wherever possible, wall cabinets are fastened to studs. *(American Plywood Assoc.)*

If walls and floor are not plumb and true, shim as illustrated in Fig. 24-32. After the cabinets have been shimmed, they are joined as necessary with flat head wood screws, Fig. 24-33. Countersink for screw heads. Secure any ceiling-hung cabinets with wood screws driven into the ceiling joists, Fig. 24-34.

Install blind corner-base and wall cabinets with filler strips, Fig. 24-35. If necessary, such cabinets may be pulled away from the wall to about 3″ distance for additional length. Set and fill all exposed nail heads.

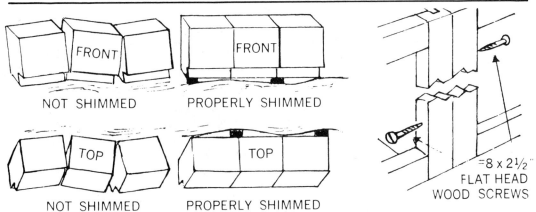

Fig. 24-32 Cabinets must be shimmed to highest point to ensure proper alignment.

Fig. 24-33 Front frames must be flush at front surface and bottom edge and fastened with screws.

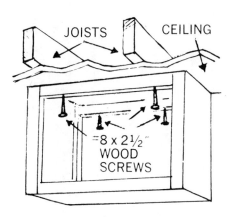

Fig. 24-34 Ceiling-hung peninsular cabinets are fastened to ceiling joists.

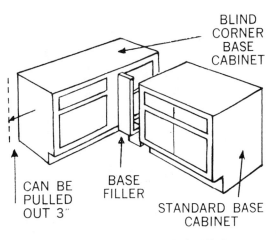

Fig. 24-35 Cabinet details for blind corner.

Cabinets & Woodwork

Cabinet doors can be decorated by applying decorative molding as illustrated in Fig. 24-36. Be sure to miter all straight lengths of molding. To prevent chipped edges, use a fine toothed saw when you cut the miters. Corners are available precut; and pieces can also be glued to save time. Apply a bead of glue along the center of each piece. Do not use too much glue. A fast drying glue is recommended.

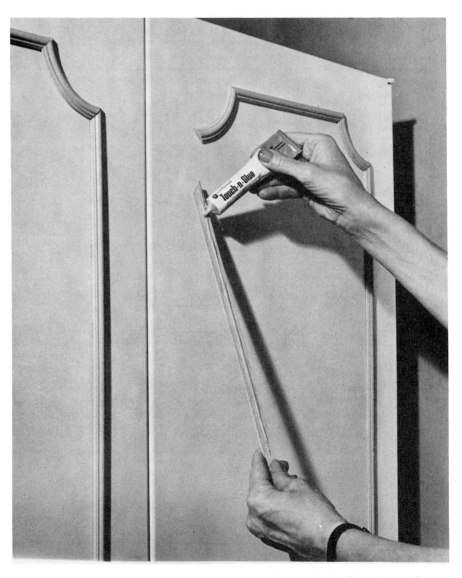

Fig. 24-36 Application of fast drying glue to a molding strip. *(U.S. Plywood)*

GLOSSARY

blind dowel: concealed dowel usually used to strengthen butt joints.

butt: to bring things together squarely so they touch or abut, but do *not overlap.*

butt hinge: square or rectangular hinge usually with a loose pin; generally used for hanging doors.

cleat: a strip of wood used for reinforcement or to support some objects such as shelves or stair treads.

corrugated fastener: also called wavey nails; it is a corrugated metal strip with one edge sharpened; used to join two pieces of wood.

counterbore: a depression or cavity made in the surface of wood or metal, usually to receive the head of bolts and washers; the sides of the depression are straight and perpendicular to the surface.

countersink: a depression similar to a counterbore, but with the sides of the depression tapered—usually to match the shape of a screw head; also the sinking or driving of nail heads below the surface.

dado: a rectangular groove cut across the grain of a board.

edge band: a narrow strip of wood or other material used to conceal an exposed edge as of plywood, particleboard, and the like.

groove: a narrow channel cut on the surface of a board along its length.

guide: in cabinetry, a member used in cabinet construction to keep drawers in alignment.

joint: the place or part where two even surfaces or members meet; they can be either movable or immovable.

kicker: in cabinetry, a low flat member used above a drawer on the underside of the cabinet to keep the drawer from tipping downward when extended.

miter: the cutting of the ends of two pieces of wood or the like at any angle other than 90 degrees.

plow (plough): a lengthwise groove cut in a piece of wood.

rabbet joint: a joint formed by cutting a groove along the edge or end of a board.

25

Plank & Beam Construction

Plank–and–beam construction is a simplified method of framing floors and roofs. Also called post-and-beam framing, it has been used commercially for many years, and is still widely used in residential construction, Fig. 25-1.

Plank-and-beam framing offers many advantages when compared to conventional framing. This system uses fewer and larger pieces than conventional framing and the pieces are spaced farther apart—all making for faster construction at a lower cost. Fig. 25-2 dramatically illustrates the difference between the two systems. Compare especially the differing quantities of materials used in the two; note also the absence of headers over doors and windows in plank-and-beam work. Cross-bridging of joists is also eliminated and fewer nails are required. All this adds up to substantial savings in materials and labor.

Fig. 25-1 Plank-and-beam construction. *(U.S. Plywood)*

Plank-and-Beam Construction

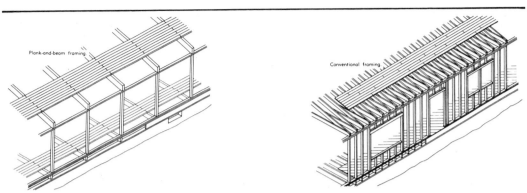

Fig. 25-2 Layout of plank-and-beam framing compared to that of conventional framing. *(National Forest Products Assoc.)*

In plank-and-beam framing 2" thick (nominal) lumber is generally used for subfloors and roofs. They are supported on beams spaced up to 8 feet apart. Posts or piers support the beam ends.

The plank-and-beam system allows the architect to design unusual structures. Exposed beams and ceiling planks can be used effectively for dramatic interiors. By increasing the size of roof beams, extra wide overhangs are possible. The wide spacing between posts permits the large expanses of glass which is preferred in contemporary styling.

Another advantage of the system is illustrated in Fig. 25-3. The

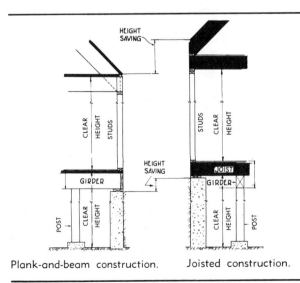

Plank-and-beam construction.   Joisted construction.

Fig. 25-3 Plank-and-beam construction cuts down overall height of building.

ceiling height is measured to the underside of the plank and not to the bottom of the joist. The result is less overall height in both interior and exterior wall.

Functional Limitations

Although plank-and-beam framing does have some limitations, they can be largely overcome with proper planning and design.

For instance, the plank floors are not strong enough for heavy concentrated loads. So the weight of refrigerators, bath-tubs and partitions must be carried by some additional framing.

Insulation provided by the 2" planks is usually adequate in moderate climates but not in colder climates. Consequently additional insulation is then needed with plank-and-beam framing. The amount depends on prevailing weather conditions. This insulation may be applied above or below the planks: if below, its appearance is a consideration; if above, the rigid type must be installed. Then a vapor barrier also has to be placed between planks and insulation.

Because of the limited amount of concealed spaces in ceilings and walls, the placement of electrical wiring may also be a problem. Spaced beams with adequate blocking is one solution; another is to rout a groove along the top edge of solid beams so they can hold the wiring.

Post Specifications

In plank-and-beam framing, roof and floor loads are distributed to the beams by the planks; the beams in turn transmit their load to the posts. The exterior walls are non-loadbearing and only serve to brace the frame.

The foundation for the posts may be a continuous wall or it may consist of individual piers resting on adequate footings, Fig. 25-4. Ordinarily the posts are spaced a maximum of 8 feet apart, however, spacing varies according to the loads involved. The posts should be strong enough to carry the load and large enough to fully support the ends of the beams. Generally, posts should be made from 4 × 4's (nominal size). When two beams join over a post, blocks should be added to increase the bearing surface, as shown in Fig. 25-5. The posts may be solid or built-up. Several pieces of 2-inch lumber are spiked

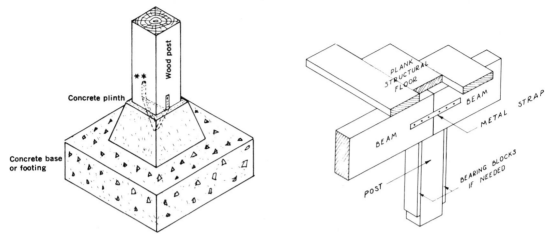

Fig. 25-4 Pier construction detail for post support.

Fig. 25-5 Construction of bearing beam over basement post.

together for built-up parts. Allowable loads for posts can be found in "Wood Structural Design Data" published by the National Lumber Manufacturers Association.

Beam Specifications

Beam sizes vary according to the span, spacing, loads and other construction factors. These factors are all determined by the architect to serve the structure he designs. The tables in the Appendices list some beam sizes for simple spans.

Beams may be solid, glued, or built-up lumber. Fig. 25-6 illustrates several methods of finishing exposed beams. Framing anchors fasten the

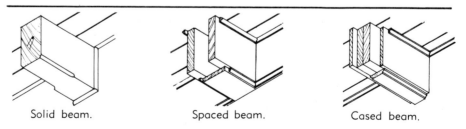

Solid beam.   Spaced beam.   Cased beam.

Fig. 25-6 Standard methods of finishing exposed beams.

beams to the posts; two types are shown in Fig. 25-7. Angle clips are also used for fastening.

Beams may be set on sills as in conventional platform framing, Fig. 25-8. If a low house silhouette is desired, the beams may be set in notches provided in the foundation wall. See Fig. 25-9. The result is a lower floor level.

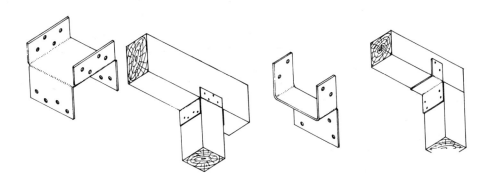

Fig. 25-7 Details of two post-and-beam anchors and typical usage.

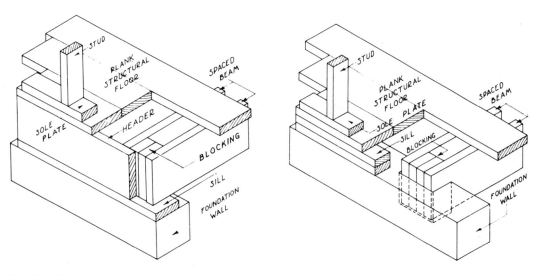

Fig. 25-8 Beam set on sill.   **Fig. 25-9** Beam set in notches in foundation wall.

Plank-and-Beam Construction

Wall Specifications

The framing walls are non-loadbearing in plank-and-beam construction. But they do give lateral support to the posts. They also serve as weather barriers and help prevent racking. The framing may be installed vertically. or horizontally, depending on the type of exterior siding to be applied later.

Partition walls are usually non-loadbearing. If they are meant to be loadbearing, they must be placed over beams proportionally enlarged to carry the load. No additional framing is needed for non-loadbearing partitions at right angles to a plank floor. If the partitions run parallel to the planks, however, extra support is required. Fig. 25-10 shows two methods of support. If the partition has openings, the partition is rested on a sole plate with a small beam placed on the underside of the planks. Otherwise, the small beam, made of two 2 × 4's on edge, can be placed directly under the studs.

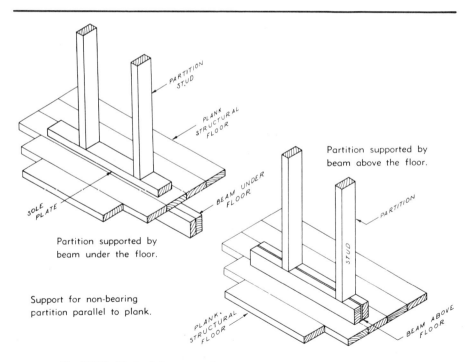

Fig. 25-10 Methods for supporting non-bearing partitions parallel to planks.

Plank Specifications

Planks for plank-and-beam flooring and roofing are usually 2 × 6's or 2 × 8's. Heavier stock up to 4 × 6 may be used if the span size warrants it. The planks serve as the subfloor and may have tongue-and-groove or spline grooves. If the planks are end-matched, the joints need not occur over a beam.

Planks usually serve as the finished ceiling for the room below. It is therefore important that their appearance be considered when they are selected for use.

For greater strength and stiffness, planks should be made continuous over more than one span. A plank which is continuous over two spans is about two and a half times as stiff as one over a single span. See Fig. 25-11. Planks that are 2" thick should be fastened to supports with two 16d nails; but thicker stock should be face-nailed, toenailed and edge-nailed.

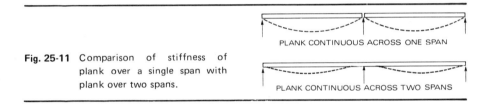

Fig. 25-11 Comparison of stiffness of plank over a single span with plank over two spans.

PLANK CONTINUOUS ACROSS ONE SPAN

PLANK CONTINUOUS ACROSS TWO SPANS

Roof Beam Specifications

In plank-and-beam construction, the roof beams may be either transverse or longitudinal. Transverse beams are placed at right angles or perpendicular to the length of the building. They extend from the ridge to the sidewalls. This means that, except for the posts or partitions supporting the ridge beam, the interior is unobstructed.

Longitudinal beams run parallel to the ridge and are usually heavier than transverse beams because of the greater spans involved. See Fig. 25-12.

Several approved methods of supporting roof beams are shown in Fig. 25-13. Metal connectors provide ample holding power and eliminate toe-nailing. For the greatest strength, lag screws or bolts should be used whenever possible.

Plank-and-Beam Construction

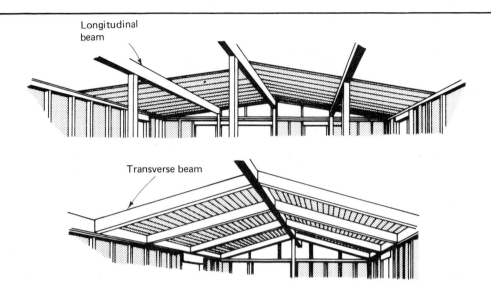

Fig. 25-12 Plank-and-beam construction: **top,** longitudinal beams; **bottom,** transverse beams.

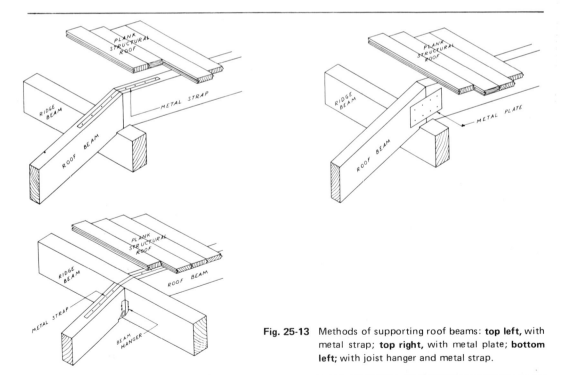

Fig. 25-13 Methods of supporting roof beams: **top left,** with metal strap; **top right,** with metal plate; **bottom left;** with joist hanger and metal strap.

GLOSSARY

lateral support: sidewise support for any structure; particularly true of non-loadbearing walls which give some to the posts in plank-and-beam frames.

longitudinal beam: a structural member running lengthwise or parallel to the ridge of a building.

plank-and-beam construction (post-and-beam construction): a system of construction in which fewer but larger framing members are used than in conventional framing.

racking: the forcing out of shape of a structure due to high pressure or impacts, such as those from gale force winds.

transverse beam: a structural member running at right angles to the ridge.

26

Prefabrication

In the building trades, prefabrication refers to the construction of various components, sections, or complete homes in modern well-equipped factories. These ready-made units are then transported to the building site. See Fig. 26-1. Doors, windows, stairs, and cabinets are typical examples of the prefabricated components now much used in home construction.

The advantages of prefabrication over on-site building are many. The use of specialized tools and machinery under factory-controlled conditions increases production at lower cost since bad weather has no effect on the progress of the work, as it often does on-site. Volume purchasing that encourages efficient materials stockpiling is another important advantage. Also, the finished parts can be better protected until their installation is scheduled.

Prefabricated roof trusses are an excellent example of manufactured components that are widely used in commercial and residential construction. In Fig. 26-2 a load is ready for delivery to a job site.

Self-supporting floor trusses that permit wide unsupported spans are also prefabricated. The large openings between such truss members permit the concealment of plumbing and electrical lines as well as heating and cooling ducts, Fig. 26-3.

Fig. 26-1 Sections of prefabricated house being assembled. *(American Plywood Assoc.)*

Fig. 26-2 Prefabricated trusses ready to be trucked to site. *(Automated Building Components)*

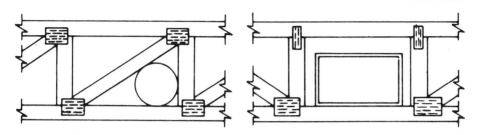

Fig. 26-3 Layouts of typical flat-floor truss sections: **left,** with off-center panel duct; right, with open-center panel duct.

Trusses can also be prefabricated to include wall sections, as in Fig. 26-4. In fact, the configuration of trusses is almost unlimited; Fig. 26-5 illustrates some of the many types made. They are designed by structural engineers who specialize in this field.

Fig. 26-4 Prefabricated roof and wall trusses in combination are being used in this commercial structure. *(Automated Building Components)*

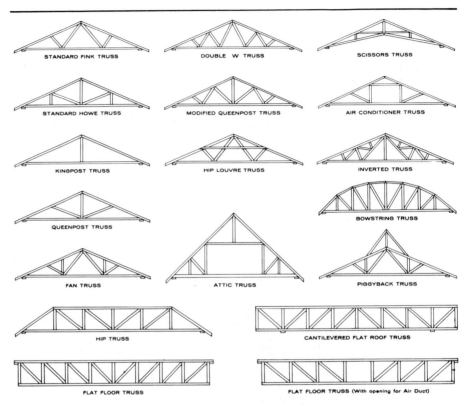

Fig. 26-5 Some of the many types of trusses manufactured for on-site installation. *(Automated Building Components)*

"Prefab" trusses are often designed and built to cover very wide spans. These trusses are assembled from accurately cut members which are first clamped in fixed positions on jigs, Fig. 26-6. They are next fastened with connectors that have great holding power. Gang nails are then firmly pressed into the lumber on hydraulic presses, Fig. 26-7. These specially designed points penetrate without splitting or cutting the wood fibers.

Types of Prefabrication

Precut components for on-site assembly are made at the factory, then shipped and put together on the job site. The members are cut to size and identified with markings, as in Fig. 26-8. This is the simplest

Fig. 26-6 Truss section assembly: clamps are used to hold parts securely while the "prefab" construction workers apply the fasteners. *(Automated Building Components)*

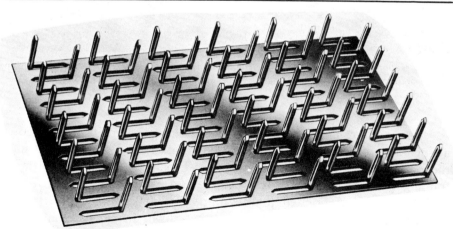

Fig. 26-7 Gang nail used to assemble prefabricated trusses. *(Automated Building Components)*

Prefabrication

type of house prefabrication. But it ensures accuracy since all the pieces are cut on jigs and fixtures. Carpenters erect these structures in the usual manner but with a minimum of cutting and fitting.

Prefabricated Shells are factory-built structures that consist of all framing members and wall sections. Such shells include the structural inner walls, but not the non-bearing partitions. Usually the exterior surface of the outer walls is finished. But insulation and inner wall coverings are installed at the site. Door and window framing are factory-installed and window walls may include the sash already in place, as in Fig. 26-9. Ceilings, walls and floors are installed on the job.

Materials are delivered to the site after the foundation is poured.

Fig. 26-8 Precut frame sections ready for installation are delivered to job site after foundation is poured.

Fig. 26-9 On-site assembly of house shell from prefabricated components.

Panelized Prefabrication consists of building complete flat sections in such a way that they can be shipped and assembled on site. Basically, the procedure follows in this order: Each panel is first made with all framing parts fitted and strongly joined. Any interior wall surfacing materials are next bonded to the frame. Then the unit is inverted so that wiring, insulation and other facilities can be installed. Finally, the exterior surfacing is applied to finish the panel. The steps involved in building and erecting a small prefabricated house with these panels are shown in Figs. 26-10 through 26-21.

Types of Prefabrication

Fig. 26-10 Floor sections made from seamless 4 X 24-foot plywood sheets are pressure-glued to top and bottom edges of joists.

Fig. 26-11 After wall frames have been surfaced on one side, they are turned upside down for the next operation.

Fig. 26-12 Roof panels are insulated with 3-1/2" thick blankets; one worker installs them while the other secures them with fasteners.

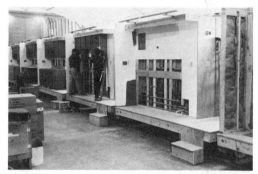

Fig. 26-13 Workers install wall unit as kitchen side of core rolls down production line with sink and cabinets already in place; before unit ships, plumbing will be given pressure test.

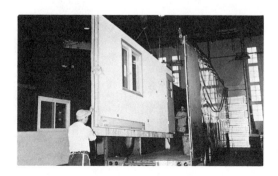

Fig. 26-14 Finished wall loads onto large open-top truck; completed panel includes door, window and heating unit; electric cable at left will connect to adjacent wall.

Fig. 26-15 Complete home packed neatly into several trucks rolls down the highway, with the shipment coordinated to arrive at the building site early in the work day.

Prefabrication

Fig. 26-16 Stressed-skin floor panels are lowered onto foundation; their finished underside (basement ceiling) is a feature not found in conventionally built homes.

Fig. 26-17 Wall section 24-feet long is easily installed by three workers; corners are joined with special connectors.

Fig. 26-18 Gable end wall is moved into position with windows whole and intact—evidence of efficient and careful loading and transport.

Fig. 26-19 Room layout in view before roof goes on shows double-plywood interior walls which add strength and durability to structure.

Fig. 26-20 Installed next, large ventilated roof panels cover house; large shingles will then be applied over these fully insulated panels.

Fig. 26-21 Clean, simple and attractive, completed house features 32-inch wide overhang for extra weather protection. (Figs. 26-10 through 26-21: Wausau Homes, Inc.)

Sectionalized Houses are factory-made in two or three sections complete with cabinetry, fixtures, wiring, plumbing, appliances and even carpeting. The sections are transported to the building site and erected like toy building blocks. As you can realize, they are ready for occupancy soon after they are trucked to the site, Fig. 26-22.

Fig. 26-22 Sectionalized house unit about to be loaded on a ferry for shipment to building site. *(American Plywood Assoc.)*

Prefabrication

GLOSSARY

prefabrication: the process of manufacturing standardized parts and sections beforehand at a factory so that house construction consists mainly of assembly.

truss: a preassembled combination of beams, bars and other members to form a rigid framework for supporting loads, especially over wide spans; based on the geometric rigidity of the triangle.

27

Wood Decks

Wood decks or roofless porches have become very popular in recent years. They extend the living area of a house outdoors at a modest cost. In effect, the deck becomes an outdoor "living-room", Fig. 27-1. Most decks are attached to the house for partial support. They may be surfaced with spaced boards or solidly with caulked planking or plywood.

Good drainage is important for decks. It keeps the ground firm and prevents the dampness which could promote decay in the wood posts and sills.

Fig. 27-1 Exposed low-level wood deck. *(Western Wood Products Assoc.)*

Wood Decks

Materials Specifications

You must give careful consideration to the type of lumber you use for the decking, as well as for the supporting members. For steps and decking, your lumber specifications should include decay resistance, good stiffness, strength and wear resistance. Be sure also that the lumber is nonsplintering and warp resistant. Heartwoods of cypress, white oak, locust, Douglas fir, western larch, redwood, cedar and southern pine are all well suited to this purpose. Any plywood used must be exterior grade if its edges are exposed to the weather.

The moisture content of wood during fabrication is very important. It should be maintained at the same degree as it will reach in use. Kiln or air-dried lumber is recommended.

Suitable Wood Preservatives

Treat posts and poles in direct contact with the soil so they comply with the American Wood Preservers Association (AWPA) standards. Indeed, any lumber and timber in contact with soil or water should be treated in the same way as the posts and poles. If paintability or appearance is a factor, creosote or similar treatments are not recommended. Instead, use non-leachable water-borne preservatives.

Wood not in contact with the ground may still require some preservative treatment if decay-resistant species are not available. Joints, connections and other critical areas need special attention. They should be treated with a penta solution and with a water repellant. Apply the solution by soaking, dipping, or flooding the wood so that the end grain is fully penetrated. Dipping the ends of all exterior framing members is strongly recommended. You may either pressure-treat or soak plywood decks the same way as lumber. Some plywood decks require special treatment, however. This should be checked with the plywood manufacturer.

Types of Construction for decking, joists and beams depend on both the type of wood used and on the grade and spacing of the various members. Figure 27-2 illustrates the arrangement of a typical deck. The posts support the beams which in turn support the joists; and the joists support the deck boards. If deck boards are thick enough and beams are spaced closely enough, you can eliminate the joists.

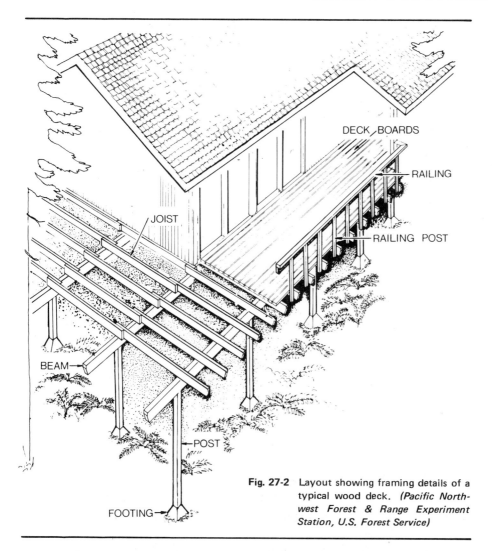

DECK BOARDS

RAILING

JOIST

RAILING POST

BEAM

POST

FOOTING

Fig. 27-2 Layout showing framing details of a typical wood deck. *(Pacific Northwest Forest & Range Experiment Station, U.S. Forest Service)*

Joists are not needed when beams are spaced 2 or 3 feet apart and 2 × 4 stock is used for deck boards. If the spans between beams are 5 feet or more, joists must be used between the beams; otherwise 2 × 3 or 2 × 4 boards on edge must be used for decking. When joists are not used, the deck boards must be nailed directly to the beams. Consult specification tables for allowable spans and sizes of beams, joists and deck boards. They are furnished by lumber manufacturers. Nor is post bracing ordinarily required if the deck is fastened to the house.

Wood Decks

Deck Footings: A masonry footing is used to support the posts and poles. Concrete footings below the surface are ordinarily used for treated posts or poles. Usually the footing should be laid 2'0" to 3'0" deep, and 4'0" in cold climates. The minimum size of the footing is 12" × 12" × 8" in normal soil. For entirely exposed posts, you can use a pedestal type footing, Fig. 27-3. Anchor the posts with bolts, angle iron or other steel straps and embed them in the footing when you pour the concrete. An excellent anchoring method is shown in Fig. 27-4. Galvanized or painted pipe with flanges is embedded in the concrete; then the flange on the top is bolted to the post bottom.

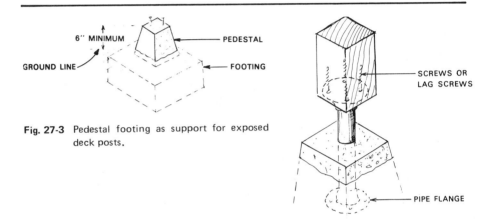

Fig. 27-3 Pedestal footing as support for exposed deck posts.

Fig. 27-4 right: Method for anchoring deck posts to pipe embedded in concrete footing.

Beam-to-Post Connections can be done in many ways. Two approved methods are shown in Fig. 27-5. In some construction, the top of the post may be exposed. In such cases, use flashing to protect the end of the post.

Beam-to-House Connection: When the deck is adjacent to the house, some means must be used to connect either the beams or the joists to the house frame. You can do this either with metal beam hangers or by resting the beam on a ledger bolted to the house frame. See Fig. 27-6. Joists that are perpendicular to the house can be connected to a ledge employing the same methods used for the beams.

Bracing: On free-standing decks not connected to a house, such as pool-side decks, some bracing may be needed to provide lateral support. Posts embedded in soil or concrete footings usually do not require additional bracing.

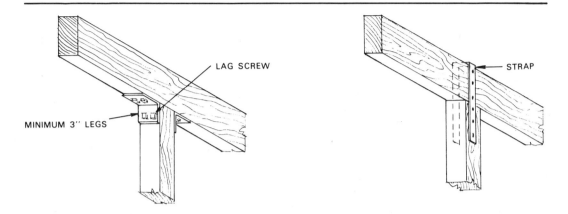

Fig. 27-5 Two industry-approved methods for anchoring deck posts to beams.

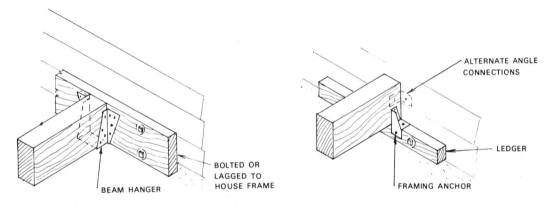

Fig. 27-6 Details of two methods for fastening deck members to house frame.

Bracing should be used on all sides of a free-standing deck. With single bracing, use 2 × 4's if the lengths are not over 8 feet. With lengths over 8 feet, use 2 × 6 braces. See Fig. 27-7 for single bracing. Because of its appearance, it is often called "W" bracing. When spans and post heights are great, 14 feet or more, the "X" brace may be used, in each bay or in every other bay. Braces are fastened with lag screws.

Partial Bracing may be utilized when post heights are small, say 5'0" to 7'0". Use plywood on both sides of the post and overlap it on the header slightly. Another type of brace is shown in Fig. 27-8. If the end grain is not protected on it, a vertical cut such as that indicated by the dotted line is recommended.

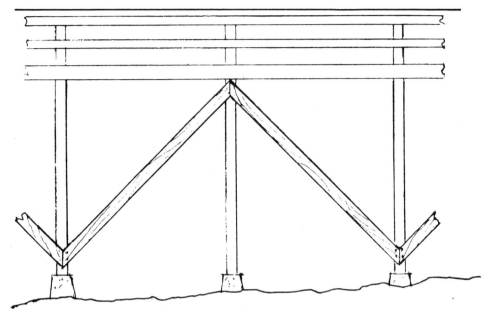

Fig. 27-7 "W" bracing for all sides of a free-standing deck.

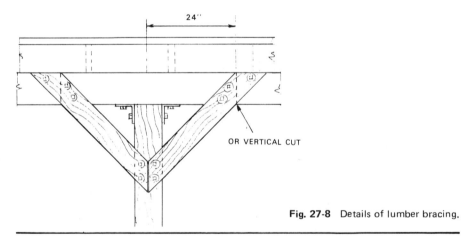

Fig. 27-8 Details of lumber bracing.

Deck Fastenings: Decks may be fastened with nails or screws. Screws are more costly but they provide greater withdrawal resistance. If you use nails, make sure that they are the annular or spiral groove type. For 2 X 3's or 2 X 4's laid flat, use two 12d nails per board. For 2 X 3's or 2 X 4's on edge, use one fastener per joist. For 2 X 3's use 5" nails; for 2 X 4's use 5" screws.

Suitable Wood Preservatives

Whether flat or vertical, spaced deck boards should be 1/8" to 1/4" apart. You can use nails themselves as an aid to spacing the boards, Fig. 27-9. Use 10d nails for 1/8" spacing. Be sure to place end joints over a joist or beam. If you use flat grain boards, place them with the bark side up. Space plywood panels a minimum of 1/16" apart between edge and end joints. To prevent moisture absorption, seal all plywood edges with an exterior primer. Do this before you install the panels. The sealing can be simplified by stacking the plywood so that you can coat the edges in series.

To provide drainage, slope decks at least one inch in every ten feet.

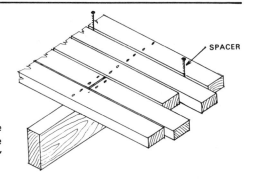

Fig. 27-9 Whether spaced deck boards are flat or vertical, they must be installed from 1/8" to 1/4" apart.

Railings and Stairs: Good design is important to both function and appearance. The top members should protect the end grain of the vertical members. The vertical members should be screwed or bolted to the edge of the deck and to the rail. The cap rail can be nailed to the rail as detailed in Fig. 27-10.

Stairs for decks are similar to those used in the house, except that exterior stairs must be designed so that edge grain is not exposed and moisture traps must be avoided. The tread supports consist of 2 × 4 cleats bolted to the stringers. By sloping the cleats back slightly, you make sure that rain water runs off the treads.

One type of stair railing is shown in Fig. 27-11. You may use 2 × 4's when spacing is 3'0" or less. For spacings from 3 to 6-feet, 2 × 6 or 3 × 4 posts should be used. Assembly is with lag screws or bolts.

Regardless of the design used, decks, stairs and railings should all be made with the safety of the user first in mind.

Finishes for wood decks include paint, stain and natural finishes.

Natural finishes with a preservative are highly recommended since they penetrate the wood, are easily renewed, and are made in many colors. Penetrating stains are similar to the natural finishes, but they contain less oil and more pigment.

Paint is used considerably outdoors. For lasting results, apply it to preservative-treated wood. Painted surfaces are subject to decay when not treated because the surface film cracks and lets moisture enter.

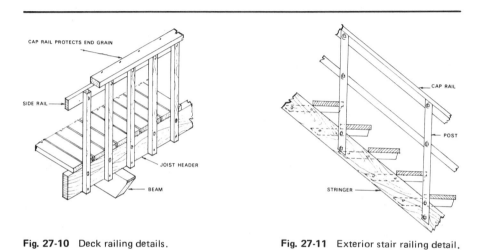

Fig. 27-10 Deck railing details. **Fig. 27-11** Exterior stair railing detail.

GLOSSARY

beam hanger: metal bracket for fastening beams; available in many sizes and shapes.
creosote: a wood preservative obtained by distilling coal tar.
penetrating stain: a stain which penetrates deeply into wood surfaces.

28

Estimating

The proper estimating of the materials needed is an important phase of carpentry, as it is for all construction. Ordering too much or too little can be costly. Underestimating can mean lost time waiting for deliveries. This in turn can hold up other trades who cannot operate until the carpenter has completed his work. Overestimating may not be as costly if the material can be used on another job, but the handling and storage of excess material can also take valuable time and effort.

The tables on pages 486 through 500 will assist you to estimate the materials needed for various phases of construction. There is a certain amount of unavoidable waste in all construction work; it must be taken into consideration also. Where practical, the tables (Figs. 28-2–28-16) list the percentage of waste to add to each final estimate. In addition to materials estimates, some tables list the approximate time needed for various operations. This labor-performance data also helps the carpenter or builder figure time costs in both new and remodeling work.

The diagrams and formulas in Fig. 28-1 below cover most of the geometric shapes encountered in house construction. For example, to estimate the square footage of siding for a building, you will probably have to figure the areas for rectangles or triangles or a combination of

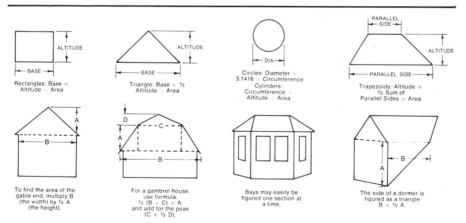

Rectangles: Base × Altitude = Area

Triangle: Base × ½ Altitude = Area

Circles: Diameter × 3.1416 = Circumference
Cylinders: Circumference × Altitude = Area

Trapezoids: Altitude × ½ Sum of Parallel Sides = Area

To find the area of the gable end, multiply B (the width) by ½ A (the height).

For a gambrel house, use formula, ½ (B + C) × A, and add for the peak (C × ½ D).

Bays may easily be figured one section at a time.

The side of a dormer is figured as a triangle B × ½ A.

Fig. 28-1   Method for finding the area of basic shapes encountered in house building.   *(Modern Materials Corp.)*

Estimating

both—whatever seems indicated by the design of the structure. No matter how broken up the face of a wall may be, you must figure each section separately as shown. A double-check on all calculations and estimates is advisable before you order materials.

EXCAVATION FACTORS

Depth	Cubic Yards per Square Foot
2"	.006
4"	.012
6"	.018
8"	.025
10"	.031
1' — 0"	.037
1' — 6"	.056
2' — 0"	.074
2' — 6"	.093
3' — 0"	.111
3' — 6"	.130
4' — 0"	.148
4' — 6"	.167
5' — 0"	1.85
5' — 6"	.204
6' — 0"	.222
6' — 6"	.241
7' — 0"	.259
7' — 6"	.278
8' — 0"	.296
8' — 6"	.314
9' — 0"	.332
9' — 6"	.350
10' — 0"	.369

Example: Assume an excavation 24 ft. x 30 ft. and 6 ft. deep. 24 x 30 = 720. In the table the 6 ft. depth has a factor of .222 (the number of cu. yd. in an excavation 1 ft. square and 6 ft. deep). 720 x .222 = 159.84 Cu. Yds.

TRENCH EXCAVATIONS
CU. YD. CONTENT PER 100 LINEAL FT.

Depth in Inches	Trench Width in Inches						
	12	18	24	30	36	42	48
6	1.9	2.8	3.7	4.6	5.6	6.6	7.4
12	3.7	5.6	7.4	9.3	11.1	13.0	14.8
18	5.6	8.3	11.1	13.9	16.7	19.4	22.3
24	7.4	11.1	14.8	18.5	22.2	26.0	29.6
30	9.3	13.8	18.5	23.2	27.8	32.4	37.0
36	11.1	16.6	22.2	27.8	33.3	38.9	44.5
42	13.0	19.4	25.9	32.4	38.9	45.4	52.0
48	14.8	22.2	29.6	37.0	44.5	52.0	59.2
54	16.7	25.0	33.3	41.6	50.0	58.4	66.7
60	18.6	27.8	37.0	46.3	55.5	64.9	74.1

STEEL SHAPES SUITABLE FOR LINTELS

The following chart shows the opening in a 4 in. masonry wall which may be spanned with various sized angles. 8 or 12 in. walls require two and three such lintels respectively.

Size of Angle	Maximum Opening
3-1/2 x 3-1/2 x 1/4 in.	8 ft. — 3 in.
4 x 4 x 1/4 in.	9 ft. — 3 in.
4 x 4 x 3/4 in.	12 ft. — 7 in.
5 x 3-1/2 x 5/16 in.	11 ft. — 1 in.
5 x 3-1/2 x 3/4 in.	14 ft. — 6 in.
6 x 3-1/2 x 3/8 in.	13 ft. — 4 in.
6 x 4 x 3/8 in.	13 ft. — 4 in.
6 x 3-1/2 x 3/4 in.	17 ft. — 2 in.
6 x 4 x 1 in.	17 ft. — 11 in.
7 x 4 x 3/8 in.	14 ft. — 8 in.
7 x 4 x 1 in.	19 ft. — 10 in.
8 x 4 x 7/16 in.	16 ft. — 9 in.
8 x 4 x 1 in.	21 ft. — 7 in.

MASONRY BLOCKS

SIZE	Wall Thickness	MATERIAL 100 SQUARE FEET OF WALL					LABOR
		CONCRETE BLOCK		LIGHTWEIGHT BLOCK			Blocks Per Hour
		No. of Units	Mortar Cubic Feet	No. of Units	Mortar Cubic Feet		
8 x 4 x 12	4"			146	4.0		24
8 x 4 x 16	4"			110	3.25		22
12 x 4 x 12	4"			100	3.25		30
8 x 6 x 16	6"			110	3.25		21
8 x 8 x 16	8"	110	3.25				18
8 x 10 x 16	10"	110	3.25				16
8 x 12 x 16	12"	110	3.25				13

NOTE: Mortar quantities based on 3/8" mortar joints, plus 25% waste. For 1/2" joints add 25%.

Fig. 28-2 Excavation factors, masonry block coverage and lintel shapes.

FOOTINGS	MATERIAL			LABOR	
SIZE	Cubic Feet Concrete Per Linear Foot	Cubic Feet Concrete Per 100 Lin. Feet	Cubic Yards Concrete Per 100 Lin. Feet	Excavation Hours per 100 Linear Feet	Placement Hours per Cubic Yard
6 x 12	0.50	50.00	1.9	3.8	2.3
8 x 12	0.67	66.67	2.5	5.0	2.3
8 x 16	0.89	88.89	3.3	6.4	2.3
8 x 18	1.00	100.00	3.7	7.2	2.3
10 x 12	0.83	83.33	3.1	6.1	2.0
10 x 16	1.11	111.11	4.1	8.1	2.0
10 x 18	1.25	125.00	4.6	9.1	2.0
12 x 12	1.00	100.00	3.7	7.2	2.0
12 x 16	1.33	133.33	4.9	9.8	2.0
12 x 20	1.67	166.67	6.1	12.1	1.8
12 x 24	2.00	200.00	7.4	15.8	1.8

NOTE: Excavation — Reduce hours by 1/4 for sand or loam.   Placement Labor based on ready-mixed concrete.
Soil — Increase hours by 1/4 for heavy clay soil.

WALLS	MATERIAL		FORMING			CONCRETE PLACEMENT
	Per 100 Square Feet Wall		Hours per 100 Square Feet			
WALL THICKNESS	Cubic Feet Required	Cubic Yards Required	Place		Remove	Hours Per Cubic Yard
			0'-4'	4'-8'		
4"	33.3	1.24	4.7	7.13	2.0	
6"	50.0	1.85	4.7	7.75	Varies as to Height	Average 3.25 Hours
8"	66.7	2.47	5.0	7.75		
10"	83.3	3.09	5.0	7.90		
12"	100.0	3.70	5.0	7.90	3.0	

SLABS	MATERIAL		LABOR	
	PER SQUARE FOOT			
THICKNESS	Cubic Feet of Concrete	Square Feet from One Cubic Yard	Forms and Screeds 100 Linear Feet	Placement 100 Sq. Ft. of Surface
2"	0.167	162		
3"	0.25	108	Average	Average
4"	0.333	81	30 Linear Feet	3.6
5"	0.417	65	Per Hour	Hours
6"	0.50	54		

NOTE: Placement includes finishing with topping. If topping omitted deduct 1.2 hours.

BRICK	Wall Thickness	MATERIAL				LABOR	
		BRICK		WALL TIES	MORTAR	100 Square Feet Wall	
MORTAR JOINT		100 Sq. Ft. Wall	Sq. Ft. per 1000 Brick	Per 100 Square Ft.	Cubic Ft. per 100 Sq. Ft.	MASON	LABORER
1/4	4"	698	143	100	4.48		
3/8	4"	655	153	93	6.56	Average	
1/2	4"	616	162	88	8.34	6-1/2	5 Hours
5/8	4"	581	172	83	10.52	Hours	
3/4	4"	549	182	78	12.60		

NOTE: Mortar includes 20% waste for all head and bed joints.

Fig. 28-3 Concrete requirements for footings, walls, slabs and brick mortar.

dimension (all species)

LIGHT FRAMING 2″ to 4″ Thick 2″ to 4″ Wide	CONSTRUCTION STANDARD UTILITY ECONOMY	This category for use where high strength values are **NOT** required; such as studs, plates, sills, cripples, blocking, etc.
STUDS 2″ to 4″ Thick 2″ to 4″ Wide	STUD ECONOMY STUD	An optional all-purpose stud grade limited to 10 feet and shorter. Characteristics affecting strength and stiffness values are limited so that the "Stud" grade is suitable for all stud uses, including load bearing walls.
STRUCTURAL LIGHT FRAMING 2″ to 4″ Thick 2″ to 4″ Wide	SELECT STRUCTURAL NO. 1 NO. 2 NO. 3 ECONOMY	These grades are designed to fit those engineering applications where higher bending strength ratios are needed in light framing sizes. Typical uses would be for trusses, concrete pier wall forms, etc.
APPEARANCE FRAMING 2″ to 4″ Thick 2″ and wider	APPEARANCE	This category for use where good appearance and high strength values are required. Intended primarily for exposed uses. Strength values are the same as those assigned to No. 1 Structural Light Framing and No. 1 Structural Joists and Planks.
STRUCTURAL JOISTS & PLANKS 2″ to 4″ Thick 6″ and wider	SELECT STRUCTURAL NO. 1 NO. 2 NO. 3 ECONOMY	These grades are designed especially to fit in engineering applications for lumber six inches and wider, such as joists, rafters and general framing use, etc.
BOARDS SHEATHING & FORM LUMBER	NO. 1 COMMON (IWP—COLONIAL) NO. 2 COMMON (IWP—STERLING) NO. 3 COMMON (IWP—STANDARD) NO. 4 COMMON (IWP—UTILITY) NO. 5 COMMON (IWP—INDUSTRIAL)	**ALTERNATE BOARD GRADES** SELECT MERCHANTABLE CONSTRUCTION STANDARD UTILITY ECONOMY

timbers

BEAMS & STRINGERS	SELECT STRUCTURAL NO. 1 NO. 2 (NO. 1 MINING) NO. 3 (NO. 2 MINING)	**POSTS & TIMBERS**	SELECT STRUCTURAL NO. 1 NO. 2 (NO. 1 MINING) NO. 3 (NO. 2 MINING)	

coverage estimator

The following estimator provides factors for determining the amount of material needed for the five basic types of wood paneling. Multiply square footage to be covered by factor (length x width x factor).

SHIPLAP	Nominal Size	WIDTH Dress	WIDTH Face	AREA FACTOR*	TONGUE AND GROOVE	Nominal Size	WIDTH Dress	WIDTH Face	AREA FACTOR*	S4S	Nominal Size	WIDTH Dress	WIDTH Face	AREA FACTOR*
	1 x 6	5½	5⅛	1.17		1 x 4	3⅜	3⅛	1.28		1 x 4	3½	3½	1.14
	1 x 8	7¼	6⅞	1.16		1 x 6	5⅜	5⅛	1.17		1 x 6	5½	5½	1.09
	1 x 10	9¼	8⅞	1.13		1 x 8	7⅛	6⅞	1.16		1 x 8	7¼	7¼	1.10
	1 x 12	11¼	10⅞	1.10		1 x 10	9⅛	8⅞	1.13		1 x 10	9¼	9¼	1.08
						1 x 12	11⅛	10⅞	1.10		1 x 12	11¼	11¼	1.07

*Allowance for trim and waste should be added.

Fig. 28-4 Framing materials: dimensions, grades and structural ratings. *(Georgia-Pacific)*

FLOOR JOIST	MATERIAL				NAILS	LABOR
	Board Feet Required for 100 Sq. Ft. of Surface Area				Per 1000 Bd. Ft.	Board Feet per Hr.
SIZE OF JOIST	12" O.C.	16" O.C.	20" O.C.	24" O.C.	Pounds	BMT
2 x 6	128	102	88	78	10	65
2 x 8	171	136	117	103	8	65
2 x 10	214	171	148	130	6	70
2 x 12	256	205	177	156	5	70

CEILING JOIST	MATERIAL				NAILS	LABOR
	Board Feet Required for 100 Sq. Ft. of Surface Area				Per 1000 Bd. Ft.	Board Feet per Hr.
SIZE OF JOIST	12" O.C.	16" O.C.	20" O.C.	24" O.C.	Pounds	Board Feet
2 x 4	78	59	48	42	19	60
2 x 6	115	88	72	63	13	65
2 x 8	153	117	96	84	9	65
2 x 10	194	147	121	104	7	70
2 x 12	230	176	144	126	6	70

TABLE OF SQUARE FEET PER SHEET OF CORRUGATED STRUCTURAL PLASTIC PANELS

Length of Sheet	26" Width	27-1/2" Width	33" Width	33-3/4" Width	35" Width	36" Width	40" Width	42" Width
* 6"	1.08	1.15	1.37	1.41	1.46	1.5	1.67	1.75
3'	6.50	6.87	8.25	8.44	8.75	9.0	10.00	10.50
4'	8.67	9.17	11.00	11.25	11.67	12.0	13.33	14.00
5'	10.83	11.46	13.75	14.06	14.58	15.0	16.67	17.50
6'	13.00	13.75	16.50	16.87	17.50	18.0	20.00	21.00
7'	15.17	16.04	19.25	19.69	20.42	21.0	23.33	24.50
8'	17.33	18.33	22.00	22.50	23.33	24.0	26.67	28.00
* 9'	19.50	20.62	24.75	25.31	26.25	27.0	30.00	31.50
10'	21.67	22.91	27.50	28.12	29.17	30.0	33.33	35.00
*11'	23.83	25.21	30.25	30.93	32.08	33.0	36.67	38.50
12'	26.00	27.50	33.00	33.75	35.00	36.0	40.00	42.00
*13'	28.17	29.79	35.75	36.56	37.92	39.0	43.33	45.50
14'	30.34	32.08	38.50	39.37	40.84	42.0	46.66	49.00

* Non-stock lengths.

TABLES OF SQUARE FEET PER SHEET OF FLAT PLASTIC PANELS

Length of Sheet	Width 16"	24"	32"	48"	Length of Sheet	Width 16"	24"	32"	48"
* 6"	.66	1.00	1.33	2.00	* 8'	10.66	16.00	21.33	32.00
* 1'	1.33	2.00	2.66	4.00	* 9'	12.00	18.00	24.00	36.00
* 2'	2.66	4.00	5.33	8.00	*10'	13.33	20.00	26.66	40.00
* 3'	4.00	6.00	8.00	12.00	*11'	14.66	22.00	29.33	44.00
4'	5.33	8.00	10.66	16.00	12'	16.00	24.00	32.00	48.00
5'	6.66	10.00	13.33	20.00	13'	17.33	26.00	34.66	52.00
6'	8.00	12.00	16.00	24.00	14'	18.66	28.00	37.33	56.00
7'	9.33	14.00	18.66	28.00					

* Non-stock lengths.

Following data applicable on most D&M lumber. Waste allowance shown includes width lost in dressing and lapping plus 5 percent waste in end-cutting and matching.

Measured Size, Inches	Finished Width, Inches	Add for Waste %	Quantity Lbr. Required, Multiply Area by	Feet of Lumber Required, 100 Sq. Ft. Surface
1 x 2	1-3/8	50	1.50	150
1 x 2-3/4	2	42-1/2	1.425	142-1/2
1 x 3	2-1/4	38-1/3	1.383	138
1 x 4	3-1/4	28	1.28	128
1 x 6	5-1/4	20	1.20	120
1 x 8	7-1/4	16	1.15	115
1-1/4 x 3	2-1/4	38-1/3	1.73	173
1-1/4 x 4	3-1/4	28	1.60	160
1-1/4 x 6	5-1/4	20	1.50	150
1-1/2 x 3	2-1/4	38-1/3	2.08	208
1-1/2 x 4	3-1/4	28	1.92	192
1-1/2 x 6	5-1/4	20	1.80	180
2 x 4	3-1/4	28	2.60	260
2 x 6	5-1/4	20	2.40	240
2 x 8	7-1/4	16	2.32	232
2 x 10	9-1/4	13	2.25	225
2 x 12	11-1/4	12	2.24	224
3 x 6	5-1/4	20	3.60	360
3 x 8	7-1/4	16	3.48	348
3 x 10	9-1/4	13	3.39	339
3 x 12	11-1/4	12	3.36	336

SQUARE-EDGED BOARDS

Following data is based on lumber surfaced 1 or 2 sides and 1 edge. The waste allowance shown includes width lost in dressing plus 5 percent waste in end-cutting. If laid diagonally, add 5 percent additional waste.

Measured Size, Inches	Finished Width, Inches	Add for Waste %	Quantity Lbr. Required, Multiply Area by	Feet of Lumber Required, 100 Sq. Ft. Surface
1 x 3	2-1/2	25	1.25	125
1 x 4 . . .	3-1/2	20	1.20	120
1 x 6 . . .	5-1/2	14	1.14	114
1 x 8 . . .	7-1/2	12	1.12	112
1 x 10 . . .	9-1/2	10	1.10	110
1 x 12 . . .	11-1/2	9-1/2	1.095	109-1/2
2 x 4 . . .	3-1/2	20	2.40	240
2 x 6 . . .	5-1/2	14	2.28	228
2 x 8 . . .	7-1/2	12	2.25	225
2 x 10 . . .	9-1/2	10	2.20	220
2 x 12 . . .	11-1/2	9-1/2	2.19	219
3 x 6 . . .	5-1/2	14	3.43	343
3 x 8 . . .	7-1/2	12	3.375	337-1/2
3 x 10 . . .	9-1/2	10	3.30	330
3 x 12 . . .	11-1/2	9-1/2	3.29	329

Fig. 28-5 Board feet quantities of floor and ceiling joists and board waste allowances.

BUILT-UP GIRDERS

Size of Girder	Bd. Ft. per Lin. Ft.	Nails per 1000 Bd. Ft.
4 x 6	2.15	53
4 x 8	2.85	40
4 x 10	3.58	32
4 x 12	4.28	26
6 x 6	3.21	43
6 x 8	4.28	32
6 x 10	5.35	26
6 x 12	6.42	22
8 x 8	5.71	30
8 x 10	7.13	24
8 x 12	8.56	20

SHEATHING AND SUBFLOORING
(HORIZONTAL APPLICATION)

Type	Size	Bd. Ft. per Sq. Ft. of Area	LBS. NAILS PER 1000 BD. FT. Spacing of Framing Members			
			12"	16"	20"	24"
T & G	1 x 4	1.32	66	52	44	36
	1 x 6	1.23	43	33	28	23
	1 x 8	1.19	32	24	21	17
	1 x 10	1.17	37	29	24	20
Shiplap	1 x 4	1.38	69	55	46	38
	1 x 6	1.26	44	34	29	24
	1 x 8	1.21	32	25	21	17
	1 x 10	1.18	37	29	25	20
S4S	1 x 4	1.19	60	47	40	33
	1 x 6	1.15	40	31	26	22
	1 x 8	1.15	30	23	20	17
	1 x 10	1.14	36	28	24	19

PARTITION STUDS
(Studs including top and bottom plates)

Size of Studs	Spacing on Centers	Bd. Ft. per Sq. Ft. of Area	Lbs. Nails per 1000 Bd. Ft.
2" x 3"	12"	.91	25
	16"	.83	
	24"	.76	
2" x 4"	12"	1.22	19
	16"	1.12	
	24"	1.02	
2" x 6"	16"	1.48	16
	24"	1.22	

PARTITION STUDS

Number of Feet of Lumber Required Per Sq. Ft. of Wood Stud Partition Using 2" x 4" Studs.

Studs spaced 16" on centers, with single top and bottom plates.

Length Partition in Feet	No. Studs Required	Ceiling Heights in Feet			
		8'-0"	9'-0"	10'-0"	12'-0"
2	3	1.25	1.167	1.13	1.13
3	3	0.833	.812	.80	.80
4	4	0.833	.812	.80	.80
5	5	0.833	.812	.80	.80
6	6	0.833	.812	.80	.80
7	6	0.833	.75	.75	.80
8	7	0.75	.75	.75	.70
9	8	0.75	.75	.75	.70
10	9	0.75	.75	.75	.70
11	9	0.75	.70	.70	.67
12	10	0.75	.70	.70	.67
13	11	0.75	.70	.70	.67
14	12	0.75	.70	.70	.67
15	12	0.70	.70	.70	.67
16	13	0.70	.70	.70	.67
17	14	0.70	.70	.70	.67
18	15	0.70	.70	.67	.67
19	15	0.70	.70	.67	.67
20	16	0.70	.70	.67	.67
For dbl. plate, add per sq. ft.		0.13	.11	.10	.083

For 2" x 8" studs, double above quantities.
For 2" x 6" studs, increase above quantities 50%.

EXTERIOR WALL STUDS
(Studs Including Corner Bracing.)

Size of Studs	Spacing on Centers	Bd. Ft. per Sq. Ft. of Area	Lbs. Nails per 1000 Bd. Ft.
2" x 3"	12"	.83	30
	16"	.78	
	20"	.74	
	24"	.71	
2" x 4"	12"	1.09	22
	16"	1.05	
	20"	.98	
	24"	.94	
2" x 6"	12"	1.66	15
	16"	1.51	
	20"	1.44	
	24"	1.38	

Example: Find the number of feet of lumber, b.m. required for a stud partition 18'-0" long and 9'-0" high. This partition would contain 18x9 = 162 sq. ft. The table gives 0.70 ft. of lumber, b.m. per sq. ft. of partition. Multiply 162 by 0.70 equals 113.4 ft. b.m.

Fig. 28-6 Girder, stud and subflooring quantites.

NUMBER OF WOOD JOISTS REQUIRED FOR ANY FLOOR AND SPACING

Length of Span	Spacing of Joists									
	12"	16"	20"	24"	30"	36"	42"	48"	54"	60"
6	7	6	5	4	3	3	3	3	2	2
7	8	6	5	5	4	4	3	3	3	2
8	9	7	6	5	4	4	3	3	3	3
9	10	8	6	6	5	4	4	3	3	3
10	11	9	7	6	5	4	4	4	3	3
11	12	9	8	7	5	5	4	4	3	3
12	13	10	8	7	6	5	4	4	4	3
13	14	11	9	8	6	5	5	4	4	4
14	15	12	9	8	7	6	5	5	4	4
15	16	12	10	9	7	6	5	5	4	4
16	17	13	11	9	7	6	6	5	5	4
17	18	14	11	10	8	7	6	5	5	4
18	19	15	12	10	8	7	6	6	5	4
19	20	15	12	11	9	7	6	6	5	5
20	21	16	13	11	9	8	7	6	5	5
21	22	17	14	12	9	8	7	6	6	5
22	23	18	14	12	10	8	7	7	6	5
23	24	18	15	13	10	9	8	7	6	6
24	25	19	15	13	11	9	8	7	6	6
25	26	20	16	14	11	9	8	7	7	6
26	27	21	17	14	11	10	8	8	7	6
27	28	21	17	15	12	10	9	8	7	6
28	29	22	18	15	12	10	9	8	7	7
29	30	23	18	16	13	11	9	8	7	7
30	31	24	19	16	13	11	10	9	8	7
31	32	24	20	17	13	11	10	9	8	7
32	33	25	20.	17	14	12	10	9	8	7
33	34	26	21	18	14	12	10	9	8	8
34	35	27	21	18	15	12	11	10	9	8
35	36	27	22	19	15	13	11	10	9	8
36	37	28	23	19	15	13	11	10	9	8
37	38	29	23	20	16	13	12	10	9	8
38	39	30	24	20	16	14	12	11	9	9
39	40	30	24	21	17	14	12	11	10	9
40	41	31	25	21	17	14	12	11	10	9

One joist has been added to each of the above quantities to take care of extra joist required at end of span.

Add for doubling joists under all partitions.

NUMBER OF FEET B.M. LUMBER REQUIRED PER 100 SQ. FT. OF SURFACE WHEN USED FOR STUDS, JOISTS, RAFTERS, WALL AND FLOOR FURRING STRIPS, ETC.

The following table does not include any allowance for waste in cutting, doubling joists under partitions or around stair wells, extra joists at end of each span, top or bottom plates, etc. These items vary with each job. Add as required.

Size of Lumber	12-Inch Centers	16-Inch Centers	20-Inch Centers	24-Inch Centers
1" x 2"	16-2/3	12-1/2	10	8-1/3
2" x 2"	33-1/3	25	20	16-2/3
2" x 4"	66-2/3	50	40	33-1/3
2" x 5"	83-1/3	62-1/2	50	41-2/3
2" x 6"	100	75	60	50
2" x 8"	133-1/3	100	80	66-2/3
2" x 10"	166-2/3	125	100	83-1/3
2" x 12"	200	150	120	100
2" x 14"	233-1/3	175	140	116-2/3
3" x 6"	150	112-1/2	90	75
3" x 8"	200	133-1/3	120	100
3" x 10"	250	187-1/2	150	125
3" x 12"	300	225	180	150
3" x 14"	350	262-1/2	210	175

FURRING

Size of Strips	Spacing on Centers	Board Feet per Square Feet of Area	Lbs. Nails per 1000 Board Feet
1 x 2	12"	.18	55
	16"	.14	
	20"	.11	
	24"	.10	
1 x 3	12"	.28	37
	16"	.21	
	20"	.17	
	24"	.14	
1 x 4	12"	.36	30
	16"	.28	
	20"	.22	
	24"	.20	

PLYWOOD THICKNESSES, SPANS AND NAILING RECOMMENDATIONS
(Plywood Continuous Over 2 or More Spans; Grain of Face Plys Across Supports)

PLYWOOD FLOOR CONSTRUCTION

Application	Recommended Thickness	Maximum Spac. of Supports (C. to C.)	Nail Size and Type	Nail Spacing	
				Panel Edges	Intermediate
Subflooring	1/2" (a)	16" (b)	6d Common (c)	6"	10"
	5/8" (a)	20"	8d Common (c)	6"	10"
	3/4" (a)	24"	8d Common (c)	6"	10"
	2.4.1	48"	8d Ring Shank (c)	6"	6"
Underlayment	3/8" (d)		6d Ring Shank or Cement Coated 8d Flathead	6"	8" Ea. Way
	5/8"				

(a) Provide blocking at panel edges for carpet, tile, linoleum or other non-structural flooring. No blocking required for 25/32" strip flooring.
(b) If strip flooring is perpendicular to supports 1/2" can be used on 24" span.
(c) If resilient flooring is to be applied without underlayment, set nails 1/16".
(d) FHA accepts 1/4" plywood.

If supports are not well seasoned, use ring-shank nails.

PLYWOOD ROOF SHEATHING

Recommended Thickness	Max. Spacing of Supports, (C. to C.)			Nail Size and Type	Nail Spacing	
	20 PSF	30 PSF	40 PSF		Panel Edge	Intermediate
5/16"	20" (b)	20"	20"	6d Common	6"	12"
3/8"	24" (b)	24"	24"	6d Common	6"	12"
1/2" (a)	32" (b)	32"	30"	6d Common	6"	12"
5/8" (a)	42" (b)	42"	39"	8d Common	6"	12"
3/4" (a)	48" (b)	47"	42"	8d Common	6"	12"

(a) Provide blocking or other means of suitable edge support when span exceeds 28" for 1/2"; 32" for 5/8" and 36" for 3/4".
(b) For special case of two span continuous beams, plywood spans can be increased 6-1/2% except as roof indicated by (b) in chart.

Fig. 28-7 Joist quantities, plywood thicknesses and nailing recommendations.

ESTIMATING FLOOR TILE

Square Feet	Number of Tiles			
	9" x 9"	12" x 12"	6" x 6"	9" x 18"
1	2	1	4	1
2	4	2	8	2
3	6	3	12	3
4	8	4	16	4
5	9	5	20	5
6	11	6	24	6
7	13	7	28	7
8	15	8	32	8
9	16	9	36	8
10	18	10	40	9
20	36	20	80	18
30	54	30	120	27
40	72	40	160	36
50	89	50	200	45
60	107	60	240	54
70	125	70	280	63
80	143	80	320	72
90	160	90	360	80
100	178	100	400	90
200	356	200	800	178
300	534	300	1200	267
400	712	400	1600	356
500	890	500	2000	445
Labor per 100 Sq. Ft.	2 Hours	1.3 Hours	3.3 Hours	1.3 Hours

FLOORING ADHESIVES

Type and Uses	Approximate Coverage in Square Feet
PRIMER — For treating on-or below-grade concrete subfloors before installing asphalt tile	250 to 350
ASPHALT CEMENT — For installing asphalt tile over primed concrete subfloors in direct contact with the ground	200
EMULSION ADHESIVE — Adhesive used for installing asphalt tile over lining felt	130 to 150
LINING PASTE — For cementing lining felt to wood subfloors	160
FLOOR AND WALL SIZE — Used to prime chalky or dusty suspended concrete sub-floors before installing resilient tiles other than asphalt	
WATERPROOF CEMENT — Recommended for installing linoleum tile, rubber and cork tile over any type of suspended subfloor in areas where surface moisture is a problem	130 to 150

TILE WASTE ALLOWANCES

1 to 50 sq. ft.	14%
50 to 100 sq. ft.	10%
100 to 200 sq. ft.	8%
200 to 300 sq. ft.	7%
300 to 1000 sq. ft.	5%
Over 1000 sq. ft.	3%

To find the number of tile needed for an area not shown on the chart, such as the number of 9" x 9" tile needed for 850 sq. ft., add the number of tile for 50 sq. ft. to the number of tile needed for 800 sq. ft. The result will then be 1513 to which must be added 5% for waste (see table). Total 1589.

WEIGHTS AND GAUGES: FLOORINGS

Material	Approximate Thickness, Inches	Finished Gauge, Inches	Average Net Wt. per Sq. Ft. in Lbs.	Roll Width in Feet	Material	Approximate Thickness, Inches	Finished Gauge, Inches	Average Net Wt. per Sq. Ft. in Lbs.	Roll Width in Feet
ASPHALT TILE	1/8	.125	1.16		LINOLEUM				
	3/16	.187	1.75		Battleship				
ASPHALT TILE	1/8	.125	1.17		Heavy Gauge	1/8	.125	.83	6
(Greaseproof)	3/16	.188	1.74		Embossed Inlaid				
CONDUCTIVE AS-	1/8	.125	.97		Standard Gauge	3/32	.0925	.60	6
PHALT TILE (Regular & Greaseproof types)	3/16	.187	1.45		Jaspe				
FELT — Lining Felt	1/25	.040	1.40	3	Heavy Gauge	1/8	.125	.83	6
INDUSTRIAL AS-	1/8	.125	.90		Standard Gauge	3/32	.0925	.65	6
PHALT TILE	3/16	.187	1.35		Marbleized				
RUBBER TILE	1/8	.125	1.24		Heavy Gauge	1/8	.125	.92	6
VINYL TILE	3/16	.187	1.86		Standard Gauge	3/32	.0925	.65	6
	3/32	.0925	.93		Light Gauge	1/16	.070	.46	6
					Plain				
					Heavy Gauge	1/8	.125	.83	6
					Standard Gauge	3/32	.0925	.60	6
					Straight Line Inlaid.				
					Standard Gauge	3/32	.0925	.62	6
					Light Gauge	1/16	.070	.46	6

*The weights and gauges in this table are manufacturing standards. Slight variations will occur, but for practical purposes, these figures are substantially correct.

Fig. 28-8 Estimating floor tiles, adhesives and waste allowances.

WOOD BLOCK FLOORING	MATERIAL			LABOR	
	BLOCK	ADHESIVE	NAILS	Per 100 Square Feet	
SIZE	Per 100 Sq. Ft.	Per 100 Sq. Ft.	Lbs. per 100 Sq. Ft.	Nailed	Adhesive
8 x 8	225	1 Gallon	4.0 Pounds	6.5 Hours	3.5 Hours
9 x 9	178	1 Gallon	3.5 Pounds	5.0 Hours	2.2 Hours
12 x 12	100	1 Gallon	2.8 Pounds	3.5 Hours	1.8 Hours

RESILIENT FLOORING	LABOR						MATERIAL
	Hours per 100 Square Feet						MASTIC
TYPE	4 x 4	6 x 6	9 x 9	6 x 12	12 x 12	9 x 18	Per 100 Square Feet
Rubber Tile	5.0	3.3	2.2	2.2	1.8	1.3	0.75 Gallon
Asphalt Tile		3.3	2.0	2.0	1.3	1.3	0.75 Gallon
Linotile		7.0	6.4		5.7		1.50 Gallon
Plastic Tile			2.4				1.35 Gallon
Cork Tile		6.0	3.5		2.5		1.50 Gallon

STRIP FLOORING	MATERIAL		NAILS	LABOR		
	Board Feet per 100 Square Feet	1000 Board Feet will lay Square Feet	Per 100 Square Feet	Hours per 100 Square Feet		
SIZE				Laying	Sanding	Finishing
25/32 x 1-1/2	155.0	645.0	3.7 Pounds	3.7 Hours		
25/32 x 2	142.5	701.8	3.0 Pounds	3.4 Hours		
25/32 x 2-1/4	138.3	723.0	3.0 Pounds	3.0 Hours	Average	Average
25/32 x 3-1/4	129.0	775.2	2.3 Pounds	2.6 Hours	1.3	2.6
3/8 x 1-1/2	138.3	723.0	3.7 Pounds	3.7 Hours	Hours	Hours
3/8 x 2	130.0	769.2	3.0 Pounds	3.4 Hours		
1/2 x 1-1/2	138.3	723.0	3.7 Pounds	3.7 Hours		
1/2 x 2	130.0	769.2	3.0 Pounds	3.4 Hours		

NAIL SCHEDULE

Tongued and Grooved Flooring Must Always Be Blind-Nailed, Square-Edge Flooring Face-Nailed

Size Flooring	Type and Size of Nails	Spacing	Size Flooring	Type and Size of Nails	Spacing
(Tongued & Grooved) 25/32 x 3-1/4	7d or 8d screw type, cut steel nails or 2" barbed fasteners*	10-12 in. apart	Following flooring must be laid on wood sub-floor		
(Tongued & Grooved) 25/32 x 2-1/4	Same as above	Same as above	(Tongued & Grooved) 3/8 x 2, 3/8 x 1-1/2	4d bright casing, wire, cut, screw nail or 1-1/4" barbed fasteners*	6-8 in. apart
(Tongued & Grooved) 25/32 x 1-1/2	Same as above	Same as above	(Square-Edge) 5/16 x 2, 5/16 x 1-1/2	1-in. 15 gauge fully barbed flooring brad, preferably cement coated.	2 nails every 7 in.
(Tongued & Grooved) 1/2 x 2, 1/2 x 1-1/2	5d screw, cut or wire nail. Or, 1-1/2" barbed fasteners*	8-10 in. apart	*If steel wire flooring nails are used they should be 8d, preferably cement coated. Newly developed machine-driven barbed fasteners, used as recommended by the manufacturer, are acceptable.		

Fig. 28-9 Flooring quantities and nailing schedule.

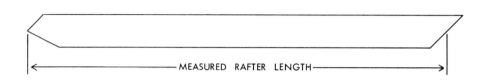

MEASURED RAFTER LENGTH

RAFTERS		RAFTER LENGTHS			
BUILDING WIDTH	RISE	3"	4"	5"	6"
10		5' - 2"	5' - 3"	5' - 5"	5' - 7"
12		6' - 2"	6' - 4"	6' - 6"	6' - 8"
14		7' - 3"	7' - 5"	7' - 7"	7' - 10"
16		8' - 3"	8' - 5"	8' - 8"	9' - 0"
18		9' - 3"	9' - 6"	9' - 9"	10' - 1"
20		10' - 4"	10' - 7"	10' - 10"	11' - 2"
22		11' - 4"	11' - 7"	11' - 11"	12' - 4"
24		12' - 4"	12' - 8"	13' - 0"	13' - 5"
26		13' - 5"	13' - 8"	14' - 1"	14' - 6"
28		14' - 5"	14' - 9"	15' - 2"	15' - 8"
30		15' - 6"	15' - 10"	16' - 3"	16' - 9"
32		16' - 6"	16' - 10"	17' - 4"	17' - 11"

NOTE: Tables accurate only to the nearest inch.

RAFTERS	BOARD FEET REQUIRED 100 Square Feet Surface Area			NAILS	LABOR
	12" O.C.	16" O.C.	24" O.C.	Per 1000 Board Feet	Board Feet Per Hour
2 x 4	89	71	53	17	
2 x 6	129	102	75	12	See
2 x 8	171	134	112	9	Table
2 x 10	212	197	121	7	Below
2 x 12	252	197	143	6	

NOTE: Includes common, hip and valley rafters, ridge boards and collar beams.

RAFTERS	LABOR					
	Common	Hip	Jack	Valley	Ridge	Collars
Board Feet Per Hour	35	35	25	35	35	65

Fig. 28-10 Rafter lengths, quantities and estimated installation times.

LUMBER SCALE
BOARD FEET PER TIMBER

Length of Timber	8	10	12	14	16	18	20	22	24
1 x 2	1-1/3	1-2/3	2	2-1/3	2-2/3	3	3-1/3	3-2/3	4
1 x 3	2	2-1/2	3	3-1/2	4	4-1/2	5	5-1/2	6
1 x 4	2-2/3	3-1/3	4	4-2/3	5-1/3	6	6-2/3	7-1/3	8
1 x 6	4	5	6	7	8	9	10	11	12
1 x 8	5-1/3	6-2/3	8	9-1/3	10-2/3	12	13-1/3	14-2/3	16
1 x 10	6-2/3	8-1/3	10	11-2/3	13-1/3	15	16-2/3	18-1/3	20
1 x 12	8	10	12	14	16	18	20	22	24
2 x 4	5-1/3	6-2/3	8	9-1/3	10-2/3	12	13-1/3	14-2/3	16
2 x 6	8	10	12	14	16	18	20	22	24
2 x 8	10-2/3	13-1/3	16	18-2/3	21-1/3	24	26-2/3	29-1/3	32
2 x 10	13-1/3	16-2/3	20	23-1/3	26-2/3	30	33-1/3	36-2/3	40
2 x 12	16	20	24	28	32	36	40	44	48
2 x 14	18-2/3	23-1/3	28	32-2/3	37-1/3	42	46-2/3	51-1/3	56
2 x 16	21-1/3	26-2/3	32	37-1/2	42-2/3	48	53-1/3	58-2/3	64
3 x 6	12	15	18	21	24	27	30	33	36
3 x 8	16	20	24	28	32	36	40	44	48
3 x 10	20	25	30	35	40	45	50	55	60
3 x 12	24	30	36	42	48	54	60	66	72
3 x 14	28	35	42	49	56	63	70	77	84
3 x 16	32	40	48	56	64	72	80	88	96
4 x 4	10-2/3	13-1/3	16	18-2/3	21-1/3	24	26-2/3	29-1/3	32
4 x 6	16	20	24	28	32	36	40	44	48
4 x 8	21-1/3	26-2/3	32	37-1/3	42-2/3	48	53-1/3	58-2/3	64
4 x 10	26-2/3	33-1/3	40	46-2/3	53-1/3	60	66-2/3	73-1/3	80
4 x 12	32	40	48	56	64	72	80	88	96
4 x 14	37-1/3	46-2/3	56	65-1/3	74-2/3	84	93-1/3	102-2/3	112
6 x 6	24	30	36	42	48	54	60	66	72
6 x 8	32	40	48	56	64	72	80	88	96
6 x 10	40	50	60	70	80	90	100	110	120
6 x 12	48	60	72	84	96	108	120	132	144

Fig. 28-11 Board-feet measure for various timbers.

CALCULATING ROOM AND WALL DIMENSIONS

Determine the perimeter. This is merely the total of the widths of each wall in the room. Use the below conversion table to figure the number of panels needed.

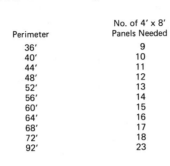

Perimeter	No. of 4' x 8' Panels Needed
36'	9
40'	10
44'	11
48'	12
52'	13
56'	14
60'	15
64'	16
68'	17
72'	18
92'	23

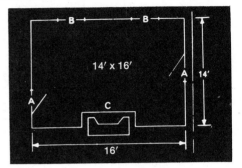

For example, if your room walls measured 14' + 14' + 16' + 16', this would equal 60' or 15 panels required. To allow for areas such as windows, doors, fireplaces, etc., use these deductions listed below:

Deductions

Door	1/2 panel (A)
Window	1/4 panel (B)
Fireplace	1/2 panel (C)

Thus, the actual number of panels for this room would be 13 pieces (15 pieces minus 2 total deductions). If the perimeter of the room falls in between the figures in the above table, use the next highest number to determine panels required. These figures are for rooms with 8' ceiling heights or less.

WALL AREAS

Ceiling Height	6	8	10	12	14	16	18	20	22	24	26	28	30
7' - 6"	45	60	75	90	105	120	135	150	165	180	195	210	225
8' - 0"	48	64	80	96	112	128	144	160	176	192	208	224	240
8' - 6"	51	68	85	102	119	136	153	170	187	204	221	238	255
9' - 0"	54	72	90	108	126	144	162	180	198	216	234	252	270
9' - 6"	57	76	95	114	133	152	171	190	209	228	247	266	285
10' - 0"	60	80	100	120	140	160	180	200	220	240	260	280	300

NOTE: Square Feet of Area per Wall Length.

WALL BOARDS	MATERIAL			ADHESIVE or NAILS	LABOR
MATERIAL	SIZE	Fastened by		Per 100 Sq. Ft.	Hours per 100 Sq. Ft.
Gypsumboard	48" x 96"	Nailing to Studs		5 Pounds	2.2
Plank T&G Board	8" to 12" x 96"	Nailing to Studs		2 Pounds	2.4
Tempered Tileboard	48" x 48"	Nailing to Studs		1 Pound	6
Tempered Tileboard	48" x 48"	Adhesive to Walls		1.5 Gallons	8
Plywood Panels	48" x 96"	Nailing to Studs or Wall		1.25 Pounds	1.8
Rock Lath	16" x 48"	Nailing to Studs		5 Pounds	0.8
Perforated Hardboard	48" x 96"	Nailing to Studs		4 Pounds	1.6

Fig. 28-12 Wall areas and panel requirements.

CALCULATING PAINT QUANTITIES FOR ROOMS

ROOM AREAS

SQUARE FEET — 4 WALLS and CEILINGS

Lineal Ft. Per Wall		6	8	10	12	14	16	18	20	22	24	26	28	30
6	C	36	48	60	72	84	96	108	120	132	144	156	168	180
	W	192	224	256	288	320	352	384	416	448	480	512	544	576
8	C	48	64	80	96	112	128	144	160	176	182	198	224	240
	W	224	256	288	320	352	384	416	448	480	512	544	576	608
10	C	60	80	100	120	140	160	180	200	220	240	260	280	300
	W	256	288	320	352	384	416	448	480	512	544	576	608	640
12	C	72	96	120	144	168	192	216	240	264	288	312	336	360
	W	288	320	352	384	416	448	480	512	544	576	608	640	672
14	C	84	112	140	168	196	224	252	280	308	336	364	392	420
	W	320	352	384	416	.448	480	512	544	576	608	640	672	704
16	C	96	128	160	192	224	256	288	320	352	384	416	448	480
	W	352	384	416	448	480	512	544	576	608	640	672	704	736
18	C	108	144	180	216	252	288	324	360	396	432	468	504	540
	W	384	416	448	480	512	544	576	608	640	672	704	736	768
20	C	120	160	200	240	280	320	360	400	440	480	520	560	600
	W	416	448	480	512	544	576	608	640	672	704	736	768	800
22	C	132	176	220	264	308	352	396	440	484	528	572	616	660
	W	448	480	512	544	576	608	640	672	704	736	768	800	832
24	C	144	182	240	288	336	384	432	480	528	576	624	672	720
	W	480	512	544	576	608	640	672	704	736	768	800	832	864
26	C	156	198	260	312	364	416	468	520	572	624	676	728	780
	W	512	544	576	608	640	672	704	736	768	800	832	864	896
28	C	168	224	280	336	392	448	504	560	616	672	728	784	840
	W	544	576	608	640	672	704	736	768	800	832	864	896	928
30	C	180	240	300	360	420	480	540	600	660	720	780	840	900
	W	576	608	640	672	704	736	768	800	832	864	896	928	960

NOTE: Based on wall height of 8' — 0". C = Ceiling area. W = Wall Area = 4 Walls.

APPROXIMATE PAINT REQUIREMENTS FOR INTERIORS AND EXTERIORS

On interior work, for rough, sand-finished walls or unpainted gypsumboard, add 50% to quantities; for each door or window deduct 1/2 pint of materials for walls. For trim, add 1/8 to 1/5 of the amount required for the body. For exterior blinds 1/2 gallon will cover 12 to 14 blinds, one coat.

Distance Around the Room	Ceiling Height 8 Feet	Ceiling Height 8-1/2 Feet	Ceiling Height 9 Feet	Ceiling Height 9-1/2 Feet	Paint for Ceiling	Finish for Floors	For Each Door or Window	
30 Feet	5/8 Gallon	5/8 Gallon	3/4 Gallon	3/4 Gallon	1 Pint	1 Pint		
35 Feet	3/4 Gallon	3/4 Gallon	3/4 Gallon	7/8 Gallon	1 Quart	1 Pint		
40 Feet	7/8 Gallon	7/8 Gallon	7/8 Gallon	1 Gallon	1 Quart	1 Quart	Each Window and Frame Requires 1/4 Pint	Each Door and Frame Requires 1/2 Pint
45 Feet	7/8 Gallon	1 Gallon	1 Gallon	1-1/8 Gallons	3 Pints	1 Quart		
50 Feet	1 Gallon	1-1/8 Gallons	1-1/8 Gallons	1-1/4 Gallons	3 Pints	1 Quart		
55 Feet	1-1/8 Gallons	1-1/8 Gallons	1-1/4 Gallons	1-1/4 Gallons	2 Quarts	3 Pints		
60 Feet	1-1/4 Gallons	1-1/4 Gallons	1-3/8 Gallons	1-3/8 Gallons	2 Quarts	3 Pints		
70 Feet	1-3/8 Gallons	1-1/2 Gallons	1-1/2 Gallons	1-5/8 Gallons	3 Quarts	2 Quarts		
80 Feet	1-1/2 Gallons	1-5/8 Gallons	1-3/4 Gallons	1-7/8 Gallons	1 Gallon	5 Pints		

Distance Around the House	Average Height 12 Feet	Average Height 15 Feet	Average Height 18 Feet	Average Height 21 Feet	Average Height 24 Feet
60 Feet	1 Gallon	1-1/4 Gallons	1-1/2 Gallons	1-3/4 Gallons	2 Gallons
76 Feet	1-1/4 Gallons	1-1/2 Gallons	2 Gallons	2-1/4 Gallons	2-1/2 Gallons
92 Feet	1-1/2 Gallons	2 Gallons	2-1/2 Gallons	2-3/4 Gallons	3 Gallons
108 Feet	1-3/4 Gallons	2-1/4 Gallons	2-3/4 Gallons	3-1/4 Gallons	3-3/4 Gallons
124 Feet	2 Gallons	2-1/2 Gallons	3-1/4 Gallons	3-3/4 Gallons	4-1/4 Gallons
140 Feet	2-1/2 Gallons	3 Gallons	3-1/2 Gallons	4 Gallons	4-1/2 Gallons
156 Feet	2-3/4 Gallons	3-1/4 Gallons	4 Gallons	4-1/2 Gallons	5-1/4 Gallons
172 Feet	3 Gallons	3-3/4 Gallons	4-1/2 Gallons	5 Gallons	5-3/4 Gallons

Fig. 28-13 Room areas and paint quantities.

OBTAINING ROOF AREA FROM PLAN AREA

Rise	Factor	Rise	Factor
3"	1.031	8"	1.202
3-1/2"	1.042	8-1/2"	1.225
4"	1.054	9"	1.250
4-1/2"	1.068	9-1/2"	1.275
5"	1.083	10"	1.302
5-1/2"	1.100	10-1/2"	1.329
6"	1.118	11"	1.357
6-1/2"	1.137	11-1/2"	1.385
7"	1.158	12"	1.414
7-1/2"	1.179		

When a roof has to be figured from a plan only, and the roof pitch is known, the roof area may be fairly accurately computed from the table to the left. The horizontal or plan area (including overhangs) should be multiplied by the factor shown in the table opposite the rise, which is given in inches per horizontal foot. The result will be the roof area.

PRODUCT SELECTION DATA

PRODUCT	Saturated Felt	Smooth Roll	Mineral Surfaced Roll	Pattern Edge Roll	19" Selvage Double Coverage	3 Tab Square Butt Strip Shingle	2 and 3 Tab Hex Strip	Individual Lock Down	Individual Staple Down	Giant Individual American	Giant Individual Dutch Lap
Approximate Shipping Weight per Square	15 lb. 30 lb.	65 lb. 55 lb. 45 lb.	90 lb. 90 lb. 90 lb.	105 lb. 105 lb.	140 to 144 lb.	210 lb. 262 lb.	167 lb.	135 lb.	135 lb.	325 lb.	162 lb.
Packages Per Square	1/4 1/2	1 1 1	1.0	1 1	2	2 or 3	2	2	2	4	2
Length	144' 72'	36' 36' 36'		42' 48'	36'	36" 36"	36"	16"	16"	16"	16"
Width	36" 36"	36" 36" 36"		36" 32"	36"	12" 12"	11-1/3"	16"	16"	12"	12"
Units Per Square			1.0 1.075 1.15			80 100	86	80	80	226	113

WOOD SHINGLES	MATERIAL			NAILS		LABOR
	Per 100 Square Feet of Surface			per 100 Square Feet		Hours
LAID TO WEATHER	Shingles per 100 Sq. Feet	Waste	Shingles per 100 Sq. Ft. with Waste	3d Nails	4d Nails	per 100 Square Feet
4"	900	10%	990	3-3/4 Pounds	6-1/2 Pounds	3-3/4
5"	720	10%	792	3 Pounds	5-1/4 Pounds	3
6"	600	10%	660	2-1/2 Pounds	4-1/4 Pounds	2-1/2

NOTE: Nails based on using 2 nails per shingle. Increase time factor 25% for hip roofs.

BEVEL SIDING	MATERIAL			NAILS	LABOR
	Siding for 100 Square Foot Wall			Per	Board Feet
SIZE	Exposed to Weather	Add for Lap	BM per 100 Sq. Ft.	100 Square Feet	Per Hour
1/2 x 4	2-3/4	46%	151	1-1/2 Pounds	30
1/2 x 5	3-3/4	33%	138	1-1/2 Pounds	40
1/2 x 6	4-3/4	26%	131	1 Pound	45
1/2 x 8	6-3/4	18%	123	3/4 Pound	50
5/8 x 8	6-3/4	18%	123	3/4 Pound	50
3/4 x 8	6-3/4	18%	123	3/4 Pound	50
5/8 x 10	8-3/4	14%	119	1/2 Pound	55
3/4 x 10	8-3/4	14%	119	1/2 Pound	55
3/4 x 12	10-3/4	12%	117	1/2 Pound	55

NOTE: Quantities include 5% for endcutting and waste. Deduct for all openings over ten square feet.

DROP SIDING	MATERIAL			NAILS	LABOR
	Siding for 100 Square Foot Wall			Per	Board Feet
SIZE	Exposed to Weather	Add for Lap	BM per 100 Sq. Ft.	100 Square Feet	Per Hour
1 x 6	5-1/4	14%	119	2-1/2 Pounds	50
1 x 8	7-1/4	10%	115	2 Pounds	55

NOTE: Quantities include 5% for encutting and waste. Deduct for all openings over ten square feet.

Fig. 28-14 Roofing and siding quantities.

INSULATION (BATTS)

| SIZE | Square Feet | MATERIAL | | STAPLES | LABOR |
		Number of Batts Per 100 Square Feet	Number of Square Feet for 100 Sq. Ft. Wall	per 100 Square Feet	Square Feet per Hour
15 x 24	2.5	40	95	160	80
15 x 48	5.	20	95	160	85
19 x 24	3.7	32	96	160	90
19 x 48	6.33	16	95	160	95
23 x 24	3.84	26	95	160	90
23 x 48	7.67	13	100	160	100

NOTES: Glass, Mineral or Rock Wool batts with paper back roll insulation or strip insulation.
Studding or joist excluded.
Batts stapled @ 6" O.C.

INSULATION (LOOSE FILL)

FILL THICKNESS	MATERIAL					LABOR
	Number of Square Feet Covered per Cubic Foot					Square Feet per Hour
	Density					
	6 Pounds	7 Pounds	8 Pounds	9 Pounds	10 Pounds	
1 "	21.2	18.0	15.9	14.1	13.0	110
2 "	10.6	9.1	8.0	7.1	6.4	100
3 "	7.1	6.1	5.3	4.7	4.2	90
4 "	5.3	4.6	4.0	3.5	3.2	85

MINIMUM PITCH REQUIREMENTS
FOR ASPHALT ROOFING PRODUCTS

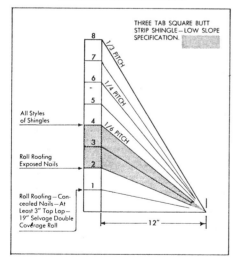

NAIL REQUIREMENTS FOR
ASPHALT ROOFING PRODUCTS

| Type of Roofing | Shin- gles per Sq. | Nails per Shin- gle | Length of Nail * | Nails per Square | Pounds per Square (approximate) | | LABOR |
					12 ga. by 7/16" head	11 ga. by 7/16" head	Hours per Square
Roll Roofing on new deck			1"	252**	.73	1.12	1
Roll Roofing over old roof'g			1-3/4"	252**	1.13	1.78	1-1/4
19" Selvage over old shing.			1-3/4"	181	.83	1.07	1
3 Tab Sq. Butt on new deck	80	4	1-1/4"	336	1.22	1.44	1-1/2
3 Tab Sq. Butt reroofing	80	4	1-3/4"	504	2.38	3.01	1-5/6
Hex Strip on new deck	86	4	1-1/4"	361	1.28	1.68	1-1/2
Hex Strip reroofing	86	4	1-3/4"	361	1.65	2.03	2
Giant Amer.	226	2	1-1/4"	479	1.79	2.27	2-1/2
Giant Dutch Lap	113	2	1-1/4"	236	1.07	1.39	1-1/2
Individ. Hex	82	2	1-3/4"	172	.79	1.03	1-1/2

(*) Length of nail should always be sufficient to penetrate at least 3/4" into sound wood. Nails should show little, if any, below underside of deck.
(**) This is the number of nails required when spaced 2" apart.

Fig. 28-15 Insulation quantities and nailing requirements for roofing.

Item	Size or Kind	Est. Labor Performance
Wall plates	2 x 6 — 2 x 8	65 b.f. per hour
	2 x 10 — 2 x 12	70 b.f. per hour
Basement beam (girder)	2 x 8	30 b.f. per hour
	2 x 10	40 b.f. per hour
Basement posts		75 b.f. per hour
Box sills		35 b.f. per hour
Floor joists	2 x 6 — 2 x 8	65 b.f. per hour
	2 x 10 — 2 x 12	70 b.f. per hour
Headers, tail joists and trimmers	2 x 6 — 2 x 8	65 b.f. per hour
	2 x 10 — 2 x 12	70 b.f. per hour
Bridging		5 sets per hour
Sub-flooring (Boards)	Straight	75 b.f. per hour
	Diagonal	65 b.f. per hour
(Plywood)	4' x 8' Sheets	10-12 hours per 1000 sq. ft.
Building paper	(Included in flooring)	
Finish flooring (Softwood)	25/32 x 3-1/4	50 b.f. per hour
(Hardwood)	25/32 x 2-1/4	30 b.f. per hour
Ceiling joists	2 x 6 — 2 x 8	65 b.f. per hour
	2 x 10 — 2 x 12	70 b.f. per hour
Ceiling backing	2 x 6 — 2 x 8	65 b.f. per hour
	2 x 10 — 2 x 12	70 b.f. per hour
Attic floor		75 b.f. per hour
Outside wall plates and shoe	2 x 4	40 b.f. per hour
	2 x 6	50 b.f. per hour
Outside studs	2 x 4	40 b.f. per hour
	2 x 6	50 b.f. per hour
Headers for wall openings	2 x 4	40 b.f. per hour
	2 x 6	50 b.f. per hour
Gable-end studs		50 b.f. per hour
Fire-stopping		50 b.f. per hour
Corner braces		50 b.f. per hour
Partition plates and shoe		50 b.f. per hour
Partition studs		50 b.f. per hour
Wall backing		50 b.f. per hour
Grounds		85 lin. ft. per hour
Knee wall plates	2 x 4	40 b.f. per hour
	2 x 6	40 b.f. per hour
Knee wall studs	2 x 4	40 b.f. per hour
	2 x 6	50 b.f. per hour
Sheathing (Boards)	1 x 6 diag.	65 b.f. per hour
	1 x 8 diag.	70 b.f. per hour
	1 x 10 diag.	75 b.f. per hour
(Plywood)	4' x 8' Sheets	10-12 hours per 1000 sq. ft.
Sheathing paper	(Included in sheathing)	
Siding (Boards)	1/2 x 6 bevel	40 b.f. per hour
	1/2 x 8 bevel	50 b.f. per hour
	3/4 x 10 bevel	60 b.f. per hour
(Plywood)	4' x 8' Sheets	10-12 hours per 1000 sq. ft.
Corner boards		25 b.f. per hour
Common Rafters		35 b.f. per hour
Hip rafters		35 b.f. per hour
Jack rafters		35 b.f. per hour
Valley rafters		35 b.f. per hour
Ridge pole		35 b.f. per hour
Collar beams (rafter ties)		65 b.f. per hour
Roof sheathing (Boards)	1 x 6 S4S	65 b.f. per hour
	1 x 6 center match	55 b.f. per hour
	1 x 8 shiplap	60 b.f. per hour
	1 x 10 shiplap	75 b.f. per hour
(Plywood)	4' x 8' Sheets	10-12 hours per 1000 sq. ft.

Item	Size or Kind	Est. Labor Performance
Roofing felt	(Included in roof covering)	
Roof covering (Asphalt — 4-in-1 shingle)		2.5 hrs. per sq.
Cornice	2 member	20 lin. ft. per hour
	3 member	12 lin. ft. per hour
	4 member	10 lin. ft. per hour
Exterior door frames	(Included in exterior doors)	
Window and casement units		2.8 hours each
Stationary sash frames and sash		2.8 hours each
Louvers		1.0 hours each
Exterior doors, complete		4.5 hours each
Interior door jambs	(Included in interior doors)	
Interior door trim	(Included in window and casement units)	
Interior window and casement trim	(Included in window and casement units)	
Base (Softwood — add 25% for hardwood)	1 member	20 lin. ft. per hour
	2 member	15 lin. ft. per hour
Picture moulding		40 lin. ft. per hour
Closet shelves, hook strip and pole		Depends on size, type and design
Kitchen cupboards		Depends on size, type and design
Medicine cabinets		Depends on size, type and design
Wood mantels		Depends on size, type and design
China cabinets		Depends on size, type and design
Book cases		Depends on size, type and design
Wardrobes		Depends on size, type and design
Linen cabinets and other cabinet work		Depends on size, type and design
Attic stairways (box)		10 hours each
Basement stairs		7 hours each
Storm sash		1 hour each
Combination doors		1 hour each
Window screens		1 hour each
WALL FINISHES		
Wood paneling 4' high — 7' high		2.2 hrs. per 100 sq. ft.
8' high		2.5 hrs. per 100 sq. ft.
Gypsumboard with joint system		2.5 hrs. per 100 sq. ft.
wood grained		2. hrs. per 100 sq. ft.
Insulating plank		2.5 hrs. per 100 sq. ft.
Insulating wall board		2. hrs per 100 sq. ft.
Standard hardboard		2. hrs. per 100 sq. ft.
Asbestos wall board		2. hrs. per 100 sq. ft.
1/8" perf. hardboard		2. hrs. per 100 sq. ft.
1/4" plywood		2. hrs. per 100 sq. ft.
CEILING FINISHES		
Gypsumboard with joint system		3. hrs. per 100 sq. ft.
Ceiling tile, 12 x 12 — 16 x 32		2.5 hrs. per 100 sq. ft.
INSULATION		
Loose Wool, 4" thick		.70 hrs. per 100 sq. ft.
Wool blanket or batt		1.30 hrs. per 100 sq. ft.

Fig. 28-16 Estimated labor performance for constructing various parts of a house.

REFERENCE MATTER

APPENDICES

AD	air-dried	clr.	clear
aggr.	aggregate	CM	center matched
A.I.A.	American Institute of Architects	cm.	centimeter
		com.	common
air cond.	air conditioned	conc.	concrete
a.l.	all lengths	conc. blk.	concrete block
Al.	aluminum	csg.	casing
alt.	alternate	C. to C.	center to center
approx.	approximate	cu. ft.	cubic feet
arch.	architect, architecture, architectural	cu. yd.	cubic yard
		d	penny (nail size)
asb.	asbestos	D & M	dressed and matched
A.S.A.	American Standards Assoc.	D.S.	drop siding
		D & SM	dressed and standard
A.S.T.M.	American Society for Testing Materials	D 2S & SM	dressed two sides and standard matched
atm.	atmosphere	dbl.	double
av.	average	deg.	degree
av.l.	average length	dia.	diameter
av.w.	average width	diag.	diagonal
bd.	board	dim.	dimension
bd. ft.	board foot	dsg.	double strength glass
bdl.	bundle	dup.	duplicate
bds.	boards	dz.	dozen
bev.	beveled	E	edge
bldg.	building	EG	edge grain
bldr.	builder	exc.	excavate
blk.	block	F	Fahrenheit
bm.	beam	FAS	firsts and seconds (lumber grades)
B.M.	board measure		
bsmt.	basement	FBM	foot board measure
btr.	better	FG	flat grain
Btu	British thermal unit	F.H.A.	Federal Housing Authority
C	Centigrade, or one hundred		
		fdn.	foundation
cwt	hundredweight	flg.	flooring
CL	center line	ftg.	footing
clg.	ceiling	ft.	foot

Appendix 1 Commonly-used abbreviations in building construction.

503

gl.	glass	rfg.	roofing
hdwd.	hardwood	RH	right hand
hp.	horsepower	r.l.	random length
hrtwd.	heartwood	r.w.	random width
in.	inch, inches	rm.	room
KD	kiln-dried, or knocked down	rnd.	round
		R O	rough opening
lb.	pound	rpm	revolutions per minute
lbr.	lumber		
lg.	large	rps	revolutions per second
lgth.	length		
LH	left hand	rt.	right
lin.	lineal, linear	sdg.	siding
M	thousand	sel.	select
MC	moisture content	SF	surface foot
max.	maximum	SM	surface measure
min.	minimum	sq.	square
mldg.	molding	S & E	surfaced one side and one edge
M.R.	mill run		
N.G.	no good	S1S	surfaced one side
no.	north, number	S2S	surfaced two sides
oz.	ounce, ounces	S4S	surfaced four sides
O.C.	on center	S2S & CM	surfaced two sides and center matched
O.G.	ogee		
pc.	piece	std.	standard
pn.	partition	str.	structure, structural
psf	pounds per square foot	stwy.	stairway
		T & G	tongue and groove
psi	pounds per square inch	typ.	typical
		wt.	weight
rd.	round	wd.	wood
rdm.	random		

NAILING SPECIFICATIONS FOR BUILDING WOOD HOUSE

Joining	Nailing method	Nails Number	Size	Placement
Header to joist	End-nail	3	16d	
Joist to sill or girder	Toenail	2 3	10d or 8d	
Header and stringer joist to sill	Toenail		10d	16 in. on center
Bridging to joist	Toenail each end	2	8d	
Ledger strip to beam, 2 in. thick		3	16d	At each joist
Subfloor, boards:				
1 by 6 in. and smaller		2	8d	To each joist
1 by 8 in.		3	8d	To each joist
Subfloor, plywood:				
At edges			8d	6 in. on center
At intermediate joists			8d	8 in. on center
Subfloor (2 by 6 in., T&G) to joist or girder	Blind-nail (casing) and face-nail	2	16d	
Soleplate to stud, horizontal assembly	End-nail	2	16d	At each stud
Top plate to stud	End-nail	2	16d	
Stud to soleplate	Toenail	4	8d	
Soleplate to joist or blocking	Face-nail		16d	16 in. on center
Doubled studs	Face-nail, stagger		10d	16 in. on center
End stud of intersecting wall to exterior wall stud	Face-nail		16d	16 in. on center
Upper top plate to lower top plate	Face-nail		16d	16 in. on center
Upper top plate, laps and intersections	Face-nail	2	16d	
Continuous header, two pieces, each edge			12d	12 in. on center
Ceiling joist to top wall plates	Toenail	3	8d	
Ceiling joist laps at partition	Face-nail	4	16d	
Rafter to top plate	Toenail	2	8d	
Rafter to ceiling joist	Face-nail	5	10d	
Rafter to valley or hip rafter	Toenail	3	10d	
Ridge board to rafter	End-nail	3	10d	
Rafter to rafter through ridge board	Toenail Edge-nail	4 1	8d 10d	
Collar beam to rafter:				
2 in. member	Face-nail	2	12d	
1 in. member	Face-nail	3	8d	
1-in. diagonal let-in brace to each stud and plate (4 nails at top)		2	8d	
Built-up corner studs:				
Studs to blocking	Face-nail	2	10d	Each side
Intersecting stud to corner studs	Face-nail		16d	12 in. on center
Built-up girders and beams, three or more members	Face-nail		20d	32 in. on center, each side
Wall sheathing:				
1 by 8 in. or less, horizontal	Face-nail	2	8d	At each stud
1 by 6 in. or greater, diagonal	Face-nail	3	8d	At each stud
Wall sheathing, vertically applied plywood:				
⅜ in. and less thick	Face-nail		6d	6 in. edge
½ in. and over thick	Face-nail		8d	12 in. intermediate
Wall sheathing, vertically applied fiberboard:				
½ in. thick	Face-nail			1½ in. roofing nail) 3 in. edge and
²⁵⁄₃₂ in. thick	Face-nail			1¾ in. roofing nail) 6 in. intermediate
Roof sheathing, boards, 4-, 6-, 8-in. width	Face-nail	2	8d	At each rafter
Roof sheathing, plywood:				
⅜ in. and less thick	Face-nail		6d }	6 in. edge and 12 in. intermediate
½ in. and over thick	Face-nail		8d }	

Appendix 2 Nailing schedule for wood frame house. *(Forest Products Lab.)*

SYMBOLS FOR MATERIALS		EXTERIOR & INTERIOR WALLS	
	CONCRETE		EXTERIOR BRICK – CONCRETE BLOCK
	CUT STONE		EXTERIOR STONE – BRICK
	CLAY TILE		INTERIOR GLAZED FACE TILE
	GLAZED TILE		INTERIOR GYP. WALLBOARD
	FACE BRICK	**FLOOR SECTIONS**	
	PLASTER		WOOD ON WOOD
	STEEL		TERRAZZO ON CONCRETE
	FINISH LUMBER	**PLUMBING SYMBOLS**	
	TERRAZZO	———— COLD WATER	FLOOR DRAIN
	MARBLE	– – – – HOT WATER	SHOWER DRAIN
	INSULATION	–o o—o o– ICE WATER	HW HOT WATER TANK
	EARTH	–x—x—x– GAS LINE	WM WASHING MACHINE
	ROUGH LUMBER	PIPE CHASE	WATER CLOSET
	GYPSUM WALLBOARD	HOSE BIB	URINAL
STRUCTURAL STEEL SECTIONS		HOSE RACK	LAVATORY

STRUCTURAL STEEL SECTIONS		HEATING & VENTILATING	ELECTRICAL
PLATE	CHANNEL (C)	UNIT VENTILATOR	CEILING OUTLET
		CONVECTOR RADIATOR	DROP CORD
ANGLE (L)	STANDARD BEAM (I)	SUPPLY DUCT	WALL BRACKET
		RETURN DUCT	WALL SWITCH (1)
TEE (T)	WIDE FLANGE (WF)	STEAM PIPE	WALL SWITCH (2)
		RETURN	WALL PLUG
		EXHAUST	TELEPHONE
		DRIP LINE	

Appendix 3 Architectural symbols for walls, floors, structural steel, plumbing, electrical, heating and ventilating units.

BUILDING MATERIAL WEIGHTS

ASBESTOS — 110-120 lbs. per cu. ft.
BRICK (Common) — 2-1/2"x4"x8-1/4", 5.4 lbs. each; 2.7 tons — M.
BRICK (Fire) (Standard) — 9"x4-1/2"x2-1/2", 7.0 lbs. each; 3.5 tons — M.
BRICK (Hard) — 2-1/4"x4-1/4"x8-1/2", 6.48 lbs. each; 3.24 tons — M.
BRICK (Paving) — 2-1/4"x4"x8-1/2", 6.75 lbs. each; 3.37 tons — M.
BRICK (Paving Block) — 3-1/4"x4"x8-1/2", 8.75 lbs. each; 4.37 tons — M.
BRICK (Soft) — 2-1/4"x4"x8-1/4", 4.32 lbs. each; 2.6 tons — M.
CEMENT — Bag — 94 lbs. each; bbl. weighs 376 lbs.
CLAY (Dry) — 63-95 lbs. — cu. ft.; 1700-2295 lbs. — cu. yard.
CLAY (Fire) — 130 lbs. — cu. ft.; 3510 lbs. — cu. yard.
CLAY (Wet) — 120-140 lbs. — cu. ft.; 2970-3200 lbs. — cu. yard.
CONCRETE — 138 lbs. — cu. ft., 3726 lbs. cu. yard.
CONCRETE: Cinder concrete — 112 lbs. per cu. ft.
    Gravel and limestone concrete — 150 lbs. per cu. ft.
    Trap-rock concrete — 155 lbs. per cu. ft.
CRUSHED STONE — 100 lbs. cu. ft.; 2700 lbs. — cu. yard.
GRAVEL — 95 lbs. — cu. ft.; 2565 lbs. — cu. yard.
HYDRATED LIME — Abt. 40 lbs. per cu. ft.
LIME — 75 lbs. — bu.; 320 lbs. — bbl. large; 220 lbs. small.
MORTAR — 103 lbs. per cu. ft.
PLASTER OF PARIS — 98 lbs. per cu. ft.
REINFORCED CONCRETE — 150 lbs. per cu. ft.
SAND (Dry) — 97-117 lbs. — cu. ft.; 2619-3159 lbs. — cu. yard.
SAND (Wet) — 120-140 lbs. — cu. ft.; 3240-3780 lbs. — cu. yard.
SHINGLES — Bundles 24"long, 20"wide, 10" high — weighs 50 lbs.
    Approx. 250 per bl.
SLAG — 1755-1890 lbs. per cu. yd.; 65-70 lbs. per cu. ft.
SLAG CONCRETE — 135 lbs. per cu. ft.
STONE RIPRAP — 65 lbs. — cu. ft.; 1775 lbs. — cu. yard.

WEIGHT OF VARIOUS MATERIALS IN POUNDS PER SQUARE FOOT FOR USE IN DETERMINING DEAD LOADS ON FLOORS AND ROOFS

Block, creosoted wood, 3"	15.00
Boards, fiber insulating, 1"	1.50
Boards, fiber insulating, 3/4"	1.10
Boards, fiber insulating, 1/2"	0.80
Ceiling, wood, 3/4"	2.50
Ceiling, wood, 5/8"	1.80
Ceiling, wood, 1/2"	1.40
Ceiling, wood, 3/8"	1.10
Copper, sheet	1.00
Iron, corrugated	1.00 to 4.00
Iron, flat seam, galvanized	1.00 to 3.00
Lead, sheet	4.00 to 8.00
Plaster, wood lath, 3/4" grounds	5.00
Plaster, fiber board lath, 1" grounds	5.00
Plaster, gypsumboard lath, 7/8" grounds	6.00
Plaster, metal lath, 3/4" grounds	6.00
Plywood, 1/4"	0.70
Plywood, 5/16"	1.00
Plywood, 3/8"	1.10
Roofing, roll, light	0.50
Roofing, heavy	1.00
Roofing, tar and gravel	6.00
Sheathing, wood, 3/4"	2.50
Sheathing, wood, 1-5/8"	5.40
Shingles, asphalt	2.00 to 3.00
Shingles, wood	2.50
Slate	10.00
Tile, plain	9.00 to 12.00
Tin, painted	1.00
Zinc, sheet	1.00 to 2.00

ACREAGE AND AREAS
SQUARE TRACTS OF LAND

Acres	One Side Square Tract	Area
1/10	66.0 lin. ft.	4,356 sq. ft.
1/8	73.8 lin. ft.	5,445 sq. ft.
1/6	85.2 lin. ft.	7,260 sq. ft.
1/4	104.4 lin. ft.	10,890 sq. ft.
1/3	120.5 lin. ft.	14,520 sq. ft.
1/2	147.6 lin. ft.	21,780 sq. ft.
3/4	180.8 lin. ft.	32,670 sq. ft.
1	208.7 lin. ft.	43,560 sq. ft.
1-1/2	255.6 lin. ft.	65,340 sq. ft.
2	295.2 lin. ft.	87,120 sq. ft.
2-1/2	330.0 lin. ft.	108,900 sq. ft.
3	361.5 lin. ft.	130,680 sq. ft.
5	466.7 lin. ft.	217,800 sq. ft.

LONG MEASURE

12 inches . . . . 1 foot
3 feet . . . . 1 yard
5-1/2 yards . . . . 1 rod
40 rods . . . . 1 furlong
8 furlongs . . . . 1 sta. mile
3 miles . . . . 1 league

SQUARE MEASURE

1 sq. centimeter . . . 0.1550 sq. in.
1 sq. decimeter . . 0.1076 square feet
1 sq. meter . . . . . . 1.196 sq. yd.
1 acre . . . . 3.954 sq. rods
1 hectare . . . . 2.47 acres
1 sq. kilometer . . . 0.386 sq. mile
1 sq. inch . . . 6.452 sq. centimeters
1 sq. foot . . 9.2903 sq. decimeters
1 sq. yard . . 0.8361 square meter
1 square rod . . . . . . 0.259 acre
1 acre . . . . 0.4047 hectare
1 sq. mile . . . 2.59 sq. kilometers
144 sq. inches . . . . . 1 sq. foot
9 square feet . . . . 1 square yard
30-1/4 sq. yds. . . . . 1 square rod
40 sq. rods . . . . 1 rood
4 roods . . . . 1 acre
640 acres . . . . 1 square mile

SURVEYOR'S MEASURE

7.92 inches . . . . 1 link
25 links . . . . 1 rod
4 rods . . . . 1 chain
10 sq. chains or 160 sq. rods . 1 acre
640 acres . . . . 1 square mile
36 sq. mi. or 6 mi. square 1 township

CUBIC MEASURE

1.728 cubic inches . . . 1 cubic foot
128 cubic feet . . . . 1 cord wood
27 cubic feet . . . . 1 cubic yard
40 cubic feet . . . . . 1 ton shpg.
2,150.42 cu. in. 1 standard bushel
268.8 cu. in. 1 standard gallon dry
231 cu. in. 1 standard gallon liquid
1 cubic foot about 4/5 of a bushel
1 Perch . . . . . . A mass 16-1/2 ft.
  long, 1 ft. high and 1-1/2 ft.
  wide, containing 24-2/3 cu.ft.

APPROXIMATE METRIC EQUIVALENT

1 decimeter . . . . 4 inches
1 meter . . . . 1.1 yards
1 kilometer . . . . 5/8 of mile
1 hectare . . . . 2-1/2 acres
1 stere, or cu. meter . 1/4 of a cord
1 liter 1.06 qt. liquid or 0.9 qt. dry
1 hektoliter . . . . 2.8 bushels
1 kilogram . . . . 2.2 pounds
1 metric ton . . . . . 2,200 pounds

METRIC EQUIVALENTS — LINEAR MEASURE

1 centimeter . . . . . . 0.3937 in.
1 decimeter 3.937 in. or 0.328 ft.
1 meter . . 39.37 in. or 1.0936 yards
1 dekameter . . . . 1.9884 rods
1 kilometer . . . . . . 0.62137 mile
1 inch . . . . . . 2.54 centimeters
1 foot . . . . . . 3.048 decimeters
1 yard . . . . 0.9144 meter
1 rod . . . . . . 0.5028 dekameter
1 mile . . . . . . 1.6093 kilometers

DEAD LOADS
APPROX. WEIGHTS PER SQUARE FOOT

ROOFS	Pounds	ROOFS	Pounds
Asphalt felt & gravel (3-5 ply built-up)	5-6-1/2	Sheathing boards (1" WP, Spr., Hmlk.)	2-1/2-3
Asphalt, felt & slag (3-5 ply built-up)	4-1/2– 5-1/2	Shingles	
		Asbestos	3-6
Composition 3-ply (ready roofing) . .	1	Asphalt, slate-covered	2
Concrete, cinder (per inch thickness) . .	9	Wood	2-3
		Slate	
Concrete, nailing (per inch thickness)	8	(3/16" –3/8" thick	7-14
Corrugated aluminum (.024" thick) . .	1/2	Tile, clay	10-15
Corrugated asbestos (1/4"–3/8" thick)	3-4-1/2	CEILINGS	
Corrugated iron-steel (20-18 gauge) . .	2-3	Gypsum lath & plaster (3/8" plus 1/2" thick)	5-1/2
Gypsum slab (per inch thickness) . .	8	Lath & 3/4" plaster.	8
		Suspended metal lath and plaster	10
		Gypsumboard (1/2" thick)	2

Appendix 4 Measures and weights used in construction estimating.

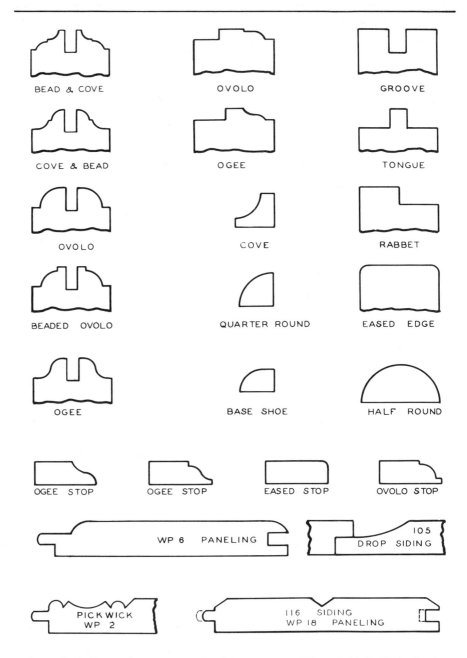

Appendix 5 Names of common woodworking patterns. *(Wisconsin Knife Works, Inc.)*

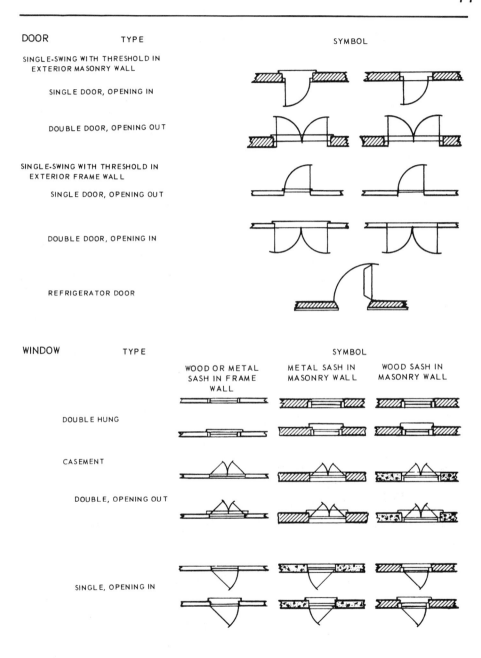

DOOR   TYPE   SYMBOL

SINGLE-SWING WITH THRESHOLD IN
EXTERIOR MASONRY WALL

    SINGLE DOOR, OPENING IN

    DOUBLE DOOR, OPENING OUT

SINGLE-SWING WITH THRESHOLD IN
EXTERIOR FRAME WALL

    SINGLE DOOR, OPENING OUT

    DOUBLE DOOR, OPENING IN

    REFRIGERATOR DOOR

WINDOW   TYPE   SYMBOL

WOOD OR METAL   METAL SASH IN   WOOD SASH IN
SASH IN FRAME   MASONRY WALL   MASONRY WALL
WALL

    DOUBLE HUNG

    CASEMENT

    DOUBLE, OPENING OUT

    SINGLE, OPENING IN

Appendix 6 Architectural symbols for doors and windows.

PAINTING MATERIALS FOR 1000 SQ. FT. OF SURFACE

REPAINTING (FIGURES ARE TOTALS FOR TWO COATS)

Kind of Surface	Gallons Paint	White Lead	Linseed Oil	Lead Mixing or Reducing Oil	Turpentine	Liquid Drier
Exterior Wood	3	45 Pounds	1-1/8 Gallons	.....	3 Pints	3/4 Pint
Wood Shingles	3-3/4	55 Pounds	1-3/8 Gallons	.....	1/2 Gallon	1 Pint
Interior Wood, Plaster and Wallboard	2-1/2	40 Pounds	.....	1-1/4 Gallons	.....	.....
Exterior Stucco, Concrete, Stone, and Brick	5	70 Pounds	3 Quarts	2 Gallons	.....	.....

NEW WORK (FIGURES ARE TOTALS FOR THREE COATS)

Kind of Surface	Gallons Paint	White Lead	Linseed Oil	Lead Mixing or Reducing Oil	Turpentine	Liquid Drier
Exterior Wood	4-3/4	62 Pounds	1-7/8 Gallons	.....	3 Quarts	1-1/4 Pints
Wood Shingles	8-5/8	105 Pounds	3-5/8 Gallons	.....	1-5/8 Gallons	2-1/4 Pints
Interior Wood	4	56 Pounds	2 Quarts	1-1/4 Gallons	3 Pints	1/4 Pint
Interior Plaster and Gypsumboard	3-3/4	57 Pounds	.....	2 Gallons	.....	.....
Exterior Stucco, Concrete and Stone	10	135 Pounds	2-1/4 Gallons	3-1/2 Gallons	.....	.....
Exterior Brick	10	120 Pounds	3-1/2 Gallons	2 Gallons	5/8 Gallons	1-1/4 Pints

WALL PAPER REQUIREMENTS

SINGLE ROLLS OF PAPER						SINGLE ROLLS OF PAPER					
Size of Room	Different Heights of Ceiling			Yards of Border	Rolls for Ceiling	Size of Room	Different Heights of Ceiling			Yards of Border	Rolls for Ceiling
	8 Ft.	9 Ft.	10 Ft.				8 Ft.	9 Ft.	10 Ft.		
4 x 8	6	7	8	9	2	16 x 18	17	19	21	25	10
4 x 10	7	8	9	11	2	16 x 20	18	20	22	26	10
4 x 12	8	9	10	12	2	16 x 22	19	21	23	28	11
6 x 10	8	9	10	12	2	16 x 24	20	22	25	29	12
6 x 12	9	10	11	13	3	16 x 26	21	23	26	31	13
8 x 12	10	11	13	15	4	17 x 22	19	22	24	23	12
8 x 14	11	12	14	16	4	17 x 25	21	23	26	31	13
10 x 14	12	14	15	18	5	17 x 28	22	25	28	32	15
10 x 16	13	15	16	19	6	17 x 32	24	27	30	35	17
12 x 16	14	16	17	20	7	17 x 35	26	29	32	37	18
12 x 18	15	17	19	22	8	18 x 22	20	22	25	29	12
14 x 18	16	18	20	23	8	18 x 25	21	24	27	31	14
14 x 22	18	20	22	26	10	18 x 28	23	26	28	33	16
15 x 16	15	17	19	23	8	20 x 26	23	26	28	33	17
15 x 18	16	18	20	24	9	20 x 28	24	27	30	34	18
15 x 20	17	20	22	25	10	20 x 34	27	30	33	39	21
15 x 23	19	21	23	28	11						

In this chart the standard roll of wall paper, 8 yds. long and 18" wide, was used in computing the estimates.
Deduct one roll of side wall paper for estimated requirements for every two doors or windows of ordinary dimensions, or for each 50 square feet of opening.

COVERING CAPACITIES*

Material	Square Feet per Gal.
Bleaching Solutions	250-300
Lacquer	200-300
Lacquer Sealer	250-300
Paste Wood Filler	40-50 (per lb.)
Liquid Filler	250-400
Water Stain	350-400
Oil Stain	300-350
Pigment Oil Stain	350-400
Non-Grain Raising Stain	275-325
Paint	650-750
Spirit Stain	250-300
Shellac	300-350
Rubbing Varnish	450-500
Flat Varnish	300-350
Paste Wax	125-175
Liquid Wax	600-700

DRYING TIMES*

Material	Touch	Recoat	Rub
Lacquer	1-10 min.	1-1/2 - 3 hrs.	16-24 hrs.
Lacquer Sealer	1-10 min.	30-45 min.	1 hr. (sand)
Paste Wood Filler....	.....	24-48 hrs.	.....
Paste Wood Filler (Q.D.)	.....	3-4 hrs.	.....
Water Stain	1 hr.	12 hrs.	.....
Oil Stain	1 hr.	24 hrs.	.....
Spirit Stain........	Zero	10 min.	.....
Shading Stain	Zero	Zero	.....
Non-Grain Raising Stain	15 min.	3 hrs.	.....
NGR Stain (Quick-Dry)	2 min.	15 min.	.....
Pigment Oil Stain ...	1 hr.	12 hrs.	.....
Pigment Oil Stain (Q.D.)	1 hr.	3 hrs.	.....
Shellac	15 min.	2 hrs.	12-18 hrs.
Shellac (Wash Coat) ..	2 min.	30 min.	.....
Varnish	1-1/2 hrs.	18-24 hrs.	24-48 hrs.
Varnish (Q.D. Synthet.)	1/2 hr.	4 hrs.	12-24 hrs.

Appendix 7 Paint and wallpaper coverage and drying times.

"U" VALUES OF SIDING, DECK & ROOFING COMBINATIONS

MATERIAL COMBINATIONS
(INCLUDING ROOFING)

APPROXIMATE
TOTAL "U" VALUE

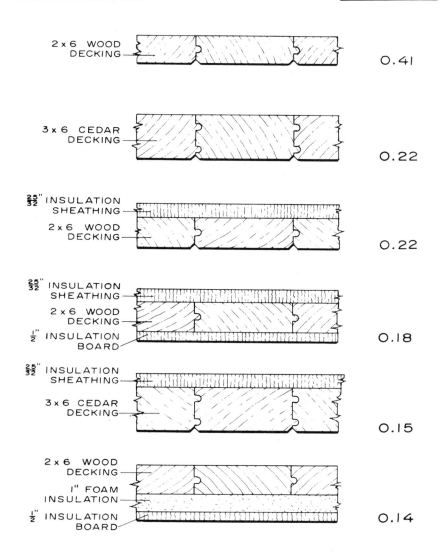

2 x 6 WOOD DECKING — 0.41

3 x 6 CEDAR DECKING — 0.22

$\frac{25}{32}$" INSULATION SHEATHING
2 x 6 WOOD DECKING — 0.22

$\frac{25}{32}$" INSULATION SHEATHING
2 x 6 WOOD DECKING
$\frac{1}{2}$" INSULATION BOARD — 0.18

$\frac{25}{32}$" INSULATION SHEATHING
3 x 6 CEDAR DECKING — 0.15

2 x 6 WOOD DECKING
1" FOAM INSULATION
$\frac{1}{2}$" INSULATION BOARD — 0.14

Appendix 8 Insulating ("U") values of siding, deck and roofing material combinations.

Appendices

MENSURATION

Area of a square = length x breadth or height.

Area of a rectangle = length x breadth or height.

Area of a triangle = base x 1/2 altitude.

Area of parallelogram = base x altitude.

Area of trapezoid = altitude x 1/2 the sum of parallel sides.

Area of trapezium = divide into two triangles, total their areas.

Circumference of circle = diameter x 3.1416.

Circumference of circle = radius x 6.283185.

Diameter of circle = circumference x .3183.

Diameter of circle = square root of area x 1.12838.

Radius of a circle = circumference x .0159155.

Area of a circle = half diameter x half circumference.

Area of a circle = square of diameter x .7854.

Area of a circle = square of circumference x .07958.

Area of a sector of circle = length of arc x 1/2 radius.

Area of a segment of circle = area of sector of equal radius — area of a triangle, when the segment is less, and plus area of triangle, when segment is greater than the semi-circle.

Area of circular ring = sum of the diameter of the two circles x difference of the diameter of the two circles and that product x .7854.

Side of square that shall equal area of circle = diameter x .8862.

Side of square that shall equal area of circle = circumference x .2821.

Diameter of circle that shall contain area of a given square = side of square x 1.1284.

Side of inscribed equilateral triangle = diameter x .86

Side of inscribed square = diameter x .7071.

Side of inscribed square = circumference x .225.

Area of ellipse = product of the two diameters x .7854.

Area of a parabola = base x 2/3 of altitude.

Area of a regular polygon = sum of its sides x perpendicular from its center to one of its sides divided by 2.

Surface of cylinder or prism = area of both ends plus length and x circumference.

Surface of sphere = diameter x circumference.

Solidity of sphere = surface x 1/6 diameter.

Solidity of sphere = cube of diameter x .5236.

Solidity of sphere = cube of radius x 4.1888.

Solidity of sphere = cube of circumference x .016887.

Diameter of sphere = cube root of solidity x 1.2407.

Diameter of sphere = square root of surface x .56419.

Circumference of sphere = square root of surface x 1.772454.

Circumference of sphere = cube root of solidity x 3.8978.

Contents of segment of sphere = (height squared plus three times the square of radius of base) x (height x .5236).

Contents of a sphere = diameter x .5236.

Side of inscribed cube of sphere = radius x 1.1547.

Side of inscribed cube of sphere = square root of diameter.

Surface of pyramid or cone = circumference of base x 1/2 of the slant height plus area of base.

Contents of pyramid or cone = area of base x 1/3 altitude.

Contents of frustum of pyramid or cone = sum of circumference at both ends x 1/2 slant height plus area of both ends.

Contents of frustum of pyramid or cone = multiply areas of two ends together and extract square root. Add to this root the two areas and x 1/3 altitude.

Contents of a wedge = area of base x 1/2 altitude.

LIVE LOAD ALLOWANCES

Minimum Uniformly Distributed Live Loads

REQUIRED LIVE LOADS. The live loads to be assumed in the design of buildings and other structure should be the greatest loads that probably will be produced by the intended use or occupancy, but in no case less than the minimum uniformly distributed unit loads shown in the chart below. These loads are not intended to supersede local building codes, however they may be brought to the attention of local building authorities for approval and as a guide where requirements are not specified.

LOADS NOT SPECIFIED. For occupancies or uses not listed in the chart, the live loads should be determined in a manner satisfactory to the building official.

Occupancy or Use		Live Load Lb. per Sq. Ft.
APARTMENTS (see Residential)		
ASSEMBLY HALLS and other places of assembly:		
Fixed seats		60
Movable seats		100
BALCONY (exterior)		100
BOWLING ALLEYS, poolrooms, etc.		75
CORRIDORS:		
First floor		100
Other floors, same as occupancy served		
DANCE HALLS, dining rooms and restaurants		100
DWELLINGS (see Residential)		
GARAGES (Passenger cars)		100
Floors shall be designed to carry 150 percent of the maximum wheel load anywhere on the floor.		
GRANDSTANDS (see Reviewing stands)		
GYMNASIUMS, main floors and balconies		100
HOSPITALS:		
Operating rooms		60
Private rooms		40
Wards		40
HOTELS (see Residential)		
LIBRARIES:		
Reading rooms		60
Stack rooms		150
MANUFACTURING		125
MARQUEES		75
OFFICE BUILDINGS:		
Offices		80
Lobbies		100
RESIDENTIAL:		
Multifamily houses	Private apartments	40
	Public rooms	100
	Corridors	60
Dwellings	First Floor	40
	Second floor and habitable attics	30
	Uninhabitable attics	20
Hotels	Guest rooms	40
	Public rooms	100
	Corridors serving public rooms	100
	Public corridors	60
	Private corridors	40
REVIEWING STANDS and BLEACHERS		100
SCHOOLS:		
Classrooms		40
Corridors		100
SIDEWALKS, vehicular driveways, and yards, subject to trucking		250
SKATING RINKS		100
STAIRS, fire escapes, and exitways		100
STORAGE warehouse, light		125
STORAGE warehouse, heavy		250
STORES:		
Retail:	First-floor, rooms	100
	Upper floors	75
Wholesale		125
THEATERS:		
Aisles, corridors, and lobbies		100
Orchestra floors		60
Balconies		60
Stage floors		150
YARDS and TERRACES, pedestrians		100

Appendix 9 Rules of mensuration and live load allowances.

REFERENCES

Badzinski, Stanley. *Carpentry in Residential Construction*. Englewood Cliffs, N.J.: Prentice-Hall, 1972.

Bayliss, Robert. *Carpentry & Joinery*. 3 vols. London: Hutchinson Technical Foundation, 1967.

Burbank, Nelson, and Pfister, Herbert. *House Construction Details.* 6th ed. New York: Simmons-Boardman Co., 1968.

Central Mortgage and Financing Corporation. *Canadian Wood-Frame House Construction*. Ottawa, N.d.

Durbahn, Walter Edward. *Fundamentals of Carpentry*. 2 vols., 4th ed., revised by Elmer W. Sundberg. Chicago: American Technical Society, 1967-1969.

Feirer, John Louis. *Cabinetmaking and Millwork*. Peoria, Ill.: C.A. Bennett Co., 1970.

Hjorth, Herman. *Basic Woodworking Processes*. Revised by Ewell Fowler. Milwaukee: Bruce, 1961.

National Forest Products Association. *Manual for House Framing* (book no. 1); *Plank & Beam Framing* (book no. 4). Washington, D.C.: NFPA, 1948.

Schmidt, John; Olin, Harold B.; and Lewis, Walter. *Construction: Principles, Materials, & Methods*. Danville, Ill.: Interstate Printers & Publishers, n.d.

Townsend, Gilbert. *The Steel Square*. Published with Ross & Macdonald, Architects (Montreal). Chicago: American Technical Society, 1939.

References

U.S. Army, Technical Manuals. *Carpentry and Building Construction* (TM 5-460 [O.P.]), revised as *Carpenter* (TM 5-551 B). Washington, D.C.: Government Printing Office, 1971.

U.S. Department of Agriculture, Forest Products Laboratory Handbooks. *Wood* (no.72, rev. ed., 1974); *Wood-Frame House Construction* (no.73, rev. ed., 1970; new ed. in press); *Low Cost Wood Houses* (no.364, 1969); *Construction Guide for Wood Decks* (no. 432, 1972). Washington, D.C.: Government Printing Office.

U.S. Navy, Bureau of Naval Personnel. *Basic Construction Techniques for Small Houses & Small Buildings* (originally *Builder* 3-2, 6th rev. ed.; republished unabridged). New York: Dover Publications, 1972.

Wagner, Willis, H. *Modern Carpentry*. South Holland, Ill.: Goodheart-Willcox Co., 1973.

Wass, Alonzo. *Methods and Materials of Residential Construction*. Reston, Va.: Reston Publishing, 1973.

Wilson, John Douglas. *Practical House Carpentry*. 2nd ed. New York: McGraw-Hill, 1973.

Wilson, John Douglas and Werner, [A.?] . *Simplified Roof Framing*. 2nd ed. New York: McGraw-Hill, Webster Division, 1948.

INDEX

Index

Index

Index

Index

Index